Journeys Beyond the Standard Model

Journeys Beyond the Standard Model

Pierre Ramond

THE ADVANCED BOOK PROGRAM

PERSEUS BOOKS
Cambridge, Massachusetts

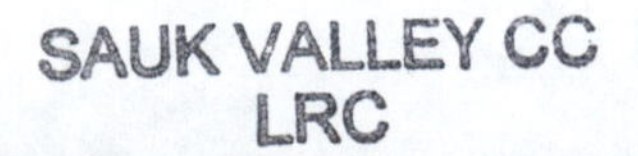

Many of the designations used by manufacturers and sellers to distinguish their products are claimed as trademarks. Where those designations appear in this book and Perseus Books was aware of a trademark claim, the designations have been printed in initial capital letters.

ISBN: 0-7382-0116-2

Library of Congress Catalog Card Number: 99-066460

Perseus Books is a member of the Perseus Books Group.

Cover design by Lynne Reed

1 2 3 4 5 6 7 8 9 10—03 02 01 00 99

Find us on the World Wide Web at
http://www.perseusbooks.com

Frontiers in Physics

David Pines, Editor

Volumes of the Series published from 1961 to 1973 are not officially numbered. The parenthetical numbers shown are designed to aid librarians and bibliographers to check the completeness of their holdings.

Titles published in this series prior to 1987 appear under either the W. A. Benjamin or the Benjamin/Cummings imprint; titles published since 1986 appear under the Addison-Wesley imprint.

1.	N. Bloembergen	Nuclear Magnetic Relaxation: A Reprint Volume, 1961
2.	G. F. Chew	S-Matrix Theory of Strong Interactions: A Lecture Note and Reprint Volume, 1961
3.	R. P. Feynman	Quantum Electrodynamics: A Lecture Note and Reprint Volume
4.	R. P. Feynman	The Theory of Fundamental Processes: A Lecture Note Volume, 1961
5.	L. Van Hove, N. M. Hugenholtz L. P. Howland	Problem in Quantum Theory of Many-Particle Systems: A Lecture Note and Reprint Volume, 1961
6.	D. Pines	The Many-Body Problem: A Lecture Note and Reprint Volume, 1961
7.	H. Frauenfelder	The Mössbauer Effect: A Review—With a Collection of Reprints, 1962
8.	L. P. Kadanoff G. Baym	Quantum Statistical Mechanics: Green's Function Methods in Equilibrium and Nonequilibrium Problems, 1962
9.	G. E. Pake	Paramagnetic Resonance: An Introductory Monograph, 1962 [cr. (42)—2nd edition]G. E. Pake
10.	P. W. Anderson	Concepts in Solids: Lectures on the Theory of Solids, 1963
11.	S. C. Frautschi	Regge Poles and S-Matrix Theory, 1963
12.	R. Hofstadter	Electron Scattering and Nuclear and Nucleon Structure: A Collection of Reprints with an Introduction, 1963
13.	A. M. Lane	Nuclear Theory: Pairing Force Correlations to Collective Motion, 1964
14.	R. Omnès M. Froissart	Mandelstam Theory and Regge Poles: An Introduction for Experimentalists, 1963
15.	E. J. Squires	Complex Angular Momenta and Particle Physics: A Lecture Note and Reprint Volume, 1963
16.	H. L. Frisch J. L. Lebowitz	The Equilibrium Theory of Classical Fluids: A Lecture Note and Reprint Volume, 1964

17.	M. Gell-Mann Y. Ne'eman	The Eightfold Way (A Review—With a Collection of Reprints), 1964
18.	M. Jacob G. F. Chew	Strong-Interaction Physics: A Lecture Note Volume, 1964
19.	P. Nozières	Theory of Interacting Fermi Systems, 1964
20.	J. R. Schrieffer	Theory of Superconductivity, 1964 (revised 3rd print ing, 1983)
21.	N. Bloembergen	Nonlinear Optics: A Lecture Note and Reprint Volume, 1965
22.	R. Brout	Phase Transitions, 1965
23.	I. M. Khalatnikov	An Introduction to the Theory of Superfluidity, 1965
24.	P. G. deGennes	Superconductivity of Metals and Alloys, 1966
25.	W. A. Harrison	Pseudopotentials in the Theory of Metals, 1966
26.	V. Barger D. Cline	Phenomenological Theories of High Energy Scattering: An Experimental Evaluation, 1967
27.	P. Choquàrd	The Anharmonic Crystal, 1967
28.	T. Loucks	Augmented Plane Wave Method: A Guide to Performing Electronic Structure Calculations—A Lecture Note and Reprint Volume, 1967
29.	Y. Ne'eman	Algebraic Theory of Particle Physics: Hadron Dynamics In Terms of Unitary Spin Current, 1967
30.	S. L. Adler R. F. Dashen	Current Algebras and Applications to Particle Physics, 1968
31.	A. B. Migdal	Nuclear Theory: The Quasiparticle Method, 1968
32.	J. J. J. Kokkedee	The Quark Model, 1969
33.	A. B. Migdal	Approximation Methods in Quantum Mechanics, 1969
34.	R. Z. Sagdeev	Nonlinear Plasma Theory, 1969
35.	J. Schwinger	Quantum Kinematics and Dynamics, 1970
36.	R. P. Feynman	Statistical Mechanics: A Set of Lectures, 1972
37.	R. P. Feynman	Photon-Hadron Interactions, 1972
38.	E. R. Caianiello	Combinatorics and Renormalization in Quantum Field Theory, 1973
39.	G. B. Field H. Arp J. N. Bahcall	The Redshift Controversy, 1973
40.	D. Horn F. Zachariasen	Hadron Physics at Very High Energies, 1973
41.	S. Ichimaru	Basic Principles of Plasma Physics: A Statistical Approach, 1973 (2nd printing, with revisions, 1980)
42.	G. E. Pake T. L. Estle	The Physical Principles of Electron Paramagnetic Resonance, 2nd Edition, completely revised, enlarged, and reset, 1973 [cf. (9)—1st edition

Volumes published from 1974 onward are being numbered as an integral part of the bibliography.

43.	R. C. Davidson	Theory of Nonneutral Plasmas, 1974

44. S. Doniach, E. H. Sondheimer — Green's Functions for Solid State Physicists, 1974
45. P. H. Frampton — Dual Resonance Models, 1974
46. S. K. Ma — Modern Theory of Critical Phenomena, 1976
47. D. Forster — Hydrodynamic Fluctuation, Broken Symmetry, and Correlation Functions, 1975
48. A. B. Migdal — Qualitative Methods in Quantum Theory, 1977
49. S. W. Lovesey — Condensed Matter Physics: Dynamic Correlations, 1980
50. L. D. Faddeev, A. A. Slavnov — Gauge Fields: Introduction to Quantum Theory, 1980
51. P. Ramond — Field Theory: A Modern Primer, 1981 [cf. 74—2nd ed.]
52. R. A. Broglia, A. Winther — Heavy Ion Reactions: Lecture Notes Vol. I, Elastic and Inelastic Reactions, 1981
53. R. A. Broglia, A. Winther — Heavy Ion Reactions: Lecture Notes Vol. II, 1990
54. H. Georgi — Lie Algebras in Particle Physics, 2nd edition, 1999
55. P. W. Anderson — Basic Notions of Condensed Matter Physics, 1983
56. C. Quigg — Gauge Theories of the Strong, Weak, and Electromagnetic Interactions, 1983
57. S. I. Pekar — Crystal Optics and Additional Light Waves, 1983
58. S. J. Gates, M. T. Grisaru, M. Rocek, W. Siegel — Superspace *or* One Thousand and One Lessons in Supersymmetry, 1983
59. R. N. Cahn — Semi-Simple Lie Algebras and Their Representations, 1984
60. G. G. Ross — Grand Unified Theories, 1984
61. S. W. Lovesey — Condensed Matter Physics: Dynamic Correlations, 2nd Edition, 1986
62. P. H. Frampton — Gauge Field Theories, 1986
63. J. I. Katz — High Energy Astrophysics, 1987
64. T. J. Ferbel — Experimental Techniques in High Energy Physics, 1987
65. T. Appelquist, A. Chodos, P. G. O. Freund — Modern Kaluza-Klein Theories, 1987
66. G. Parisi — Statistical Field Theory, 1988
67. R. C. Richardson, E. N. Smith — Techniques in Low-Temperature Condensed Matter Physics, 1988
68. J. W. Negele, H. Orland — Quantum Many-Particle Systems, 1987
69. E. W. Kolb, M. S. Turner — The Early Universe, 1990
70. E. W. Kolb, M. S. Turner — The Early Universe: Reprints, 1988
71. V. Barger, R. J. N. Phillips — Collider Physics, 1987
72. T. Tajima — Computational Plasma Physics, 1989

73.	W. Kruer	The Physics of Laser Plasma Interactions, 1988
74.	P. Ramond	Field Theory: A Modern Primer, 2nd edition, 1989 [cf. 51—1st edition]
75.	B. F. Hatfield	Quantum Field Theory of Point Particles and Strings, 1989
76.	P. Sokolsky	Introduction to Ultrahigh Energy Cosmic Ray Physics, 1989
77.	R. Field	Applications of Perturbative QCD, 1989
80.	J. F. Gunion H. E. Haber G. Kane S. Dawson	The Higgs Hunter's Guide, 1990
81.	R. C. Davidson	Physics of Nonneutral Plasmas, 1990
82.	E. Fradkin	Field Theories of Condensed Matter Systems, 1991
83.	L. D. Faddeev A. A. Slavnov	Gauge Fields, 1990
84.	R. Broglia A. Winther	Heavy Ion Reactions, Parts I and II, 1990
85.	N. Goldenfeld	Lectures on Phase Transitions and the Renormalization Group, 1992
86.	R. D. Hazeltine J. D. Meiss	Plasma Confinement, 1992
87.	S. Ichimaru	Statistical Plasma Physics, Volume I: Basic Principles, 1992
88.	S. Ichimaru	Statistical Plasma Physics, Volume II: Condensed Plasmas, 1994
89.	G. Grüner	Density Waves in Solids, 1994
90.	S. Safran	Statistical Thermodynamics of Surfaces, Interfaces, and Membranes, 1994
91.	B. d'Espagnat	Veiled Reality: An Analysis of Present Day Quantum Mechanical Concepts, 1994
92.	J. Bahcall R. Davis, Jr. P. Parker A. Smirnov R. Ulrich	Solar Neutrinos: The First Thirty Years
93.	R. Feynman F. Morinigo W. Wagner	Feynman Lectures on Gravitation
94.	M. Peskin D. Schroeder	An Introduction to Quantum Field Theory
95.	R. Feynman	Feynman Lectures on Computation
96.	M. Brack R. Bhaduri	Semiclassical Physics
97.	D. Cline	Weak Neutral Currents
98.	T. Tajima K. Shibata	Plasma Astrophysics
99.	J. Rammer	Quantum Transport Theory
100.	R. Hazeltine F. Waelbroeck	The Framework of Plasma Physics
101.	P. Ramond	Journeys Beyond the Standard Model

Editor's Foreword

The problem of communicating in a coherent fashion recent developments in the most exciting and active fields of physics continues to be with us. The enormous growth in the number of physicists has tended to make the familiar channels of communication considerably less effective. It has become increasingly difficult for experts in a given field to keep up with the current literature; the novice can only be confused. What is needed is both a consistent account of a field and the presentation of a definite "point of view" concerning it. Formal monographs cannot meet such a need in a rapidly developing field, while the review article seems to have fallen into disfavor. Indeed, it would seem that the people who are most actively engaged in developing a given field are the people least likely to write at length about it.

Frontiers in Physics was conceived in 1961 in an effort to improve the situation in several ways. Leading physicists frequently give a series of lectures, a graduate seminar, or a graduate course in their special fields of interest. Such lectures serve to summarize the present status of a rapidly developing field and may well constitute the only coherent account available at the time. One of the principal purposes of the *Frontiers in Physics* series is to make notes on such lectures available to the wider physics community.

As *Frontiers in Physics* has evolved, a second category of book, the informal text/monograph, an intermediate step between lecture notes and formal texts or monographs, has played an increasingly important role in the series. In an informal text or monograph, an author has reworked his or her lecture notes to the point that the manuscript represents a coherent summation of a newly developed field, complete with references and problems, suitable for either classroom teaching or individual study.

Journeys Beyond the Standard Model is just such a volume. Pierre Ramond, one of the world's leading high energy theorists, has provided the reader with an unusually lucid introduction to the Standard Model which describes quite accurately present-day particle physics experiments. He then examines three ways of going beyond the Standard Model to understand the experiments of tomorrow. His emphasis, throughout the text, on a pedagogical presentation makes his book readily accessible to graduate students and experienced researchers alike. It gives me great pleasure to welcome him back to *Frontiers in Physics*.

Aspen, Colorado
July, 1999
David Pines
pines@santafe.edu

PREFACE

Over the last half of the twentieth century, through heroic efforts of experimental and theoretical physicists, the investigations of fundamental interactions and constituents of Nature have produced a remarkably concise picture, called the Standard Model of Particle Physics. Its appreciation requires intimacy with of the data, and a working knowledge of the methods of quantum field theory. In this book, the Standard Model is presented, not along historical lines, but through its Lagrangian, using the language of quantum field theory. As such it does not do justice to the generations of theoretical and experimental physicists who have brought the model to its present glory. This book has been written over a ten-year period, starting with a course I gave at the University of Florida in 1989. In the belief that the Standard Model was but a stage in particle physicists' travails towards physics at shorter distances, this book started as a description of possible physical theories beyond the Standard Model, but I quickly realized that a concise introduction was needed. The result is a book that is mostly on the standard model, with only the three extensions which I find most interesting. This choice of topics reflects only my own interest at the time of writing. I have added an extensive bibliography, which contains references to the many books and review articles on the subject, and apologize to the many physicists whose work has not been cited. I have used the Particle Data Group numbers when quoting experimental results, unless otherwise indicated.

I wish to thank the many students who have labored under various versions of the manuscript. I reserve very special thanks to Dr. N. Irges, and to Professors D. Kennedy S. Martin, J. Rosner, and P. Sikivie, for their many constructive suggestions and critical reading of the manuscript.

Pierre Ramond

Gainesville, Summer 1999

To Lillian

Contents

CHAPTER 1

INTRODUCTION

As the study of the standard model requires many tools and techniques, it is useful to gather in one place salient facts about relativistic invariance, the description of fields of different spin, and the construction of gauge theories, spontaneous breaking, and group theory. This introductory chapter offers a short review of these basic ingredients. It also serves as a brief description of the notations and conventions we will follow throughout the book. The reader unfamiliar with these concepts is encouraged to consult standard texts on Advanced Quantum Mechanics and Field Theory.

1.1 LORENTZ INVARIANCE

We set $\hbar = c = 1$. Our metric is $g_{\mu\nu} = \mathrm{diag}(1, -1, -1, -1)$. The space-time coordinates are given by

$$x^\mu = (x^0, x^i) = (t, \vec{x}) \; ; \qquad x_\mu = g_{\mu\nu} x^\nu = (t, -\vec{x}) \; , \tag{1.1}$$

$$\partial_\mu = (\frac{\partial}{\partial t}, \vec{\nabla}) \; ; \qquad \partial^\mu = (\frac{\partial}{\partial t}, -\vec{\nabla}) \; . \tag{1.2}$$

The Lorentz-invariant inner product of two vectors is

$$x^2 = g_{\mu\nu} x^\mu x^\nu = x_\mu x^\mu = t^2 - \vec{x} \cdot \vec{x}. \tag{1.3}$$

We also have

$$\partial_\mu x^\rho = \delta_\mu^\rho \ , \tag{1.4}$$

where δ_μ^ρ is the diagonal Kronecker delta.

In four space-time dimensions, the algebra of the Lorentz group is isomorphic to that of $SU_2 \times SU_2$ (up to factors of i), the first generated by $\vec{J} + i\vec{K}$, the second by $\vec{J} - i\vec{K}$; $\vec{J}$ are the generators of angular momentum and $\vec{K}$ are the boosts. These two SU_2 are thus connected by complex conjugation ($i \to -i$) and/or parity ($\vec{K} \to -\vec{K}, \vec{J} \to \vec{J}$), and are therefore left invariant by the combined operation of CP. In accordance with this algebraic structure, spinor fields appear in two varieties, left-handed spinors, transforming under the first SU_2 as spin $\frac{1}{2}$ representations, and right-handed spinors transforming only under the second SU_2. They are represented by two-component complex spinor fields, called Weyl spinors,

$$\psi_L(x) \sim (\mathbf{2}, \mathbf{1}) \ , \qquad \psi_R(x) \sim (\mathbf{1}, \mathbf{2}) \ . \tag{1.5}$$

The spinor fields must be taken to be anticommuting Grassman variables, in accordance with the Pauli exclusion principle. Their Lorentz transformation properties can be written in terms of the Pauli spin matrices

$$\sigma_1 = \begin{pmatrix} 0 & 1 \\ 1 & 0 \end{pmatrix} \ , \ \sigma_2 = \begin{pmatrix} 0 & -i \\ i & 0 \end{pmatrix} \ , \ \sigma_3 = \begin{pmatrix} 1 & 0 \\ 0 & -1 \end{pmatrix} \ , \tag{1.6}$$

which satisfy

$$\sigma_i \sigma_j = \delta_{ij} + i \ \epsilon_{ijk} \sigma_\kappa \ . \tag{1.7}$$

It is useful to express Fierz transformations in Weyl language. Let ζ and η be complex two-component Weyl spinors, each transforming as (**2**,**1**) of the Lorentz group. The first Fierz decomposition is

$$\zeta \eta^t \sigma_2 = -\frac{1}{2} \sigma_2 \eta^t \sigma_2 \zeta - \frac{1}{2} \sigma^i \eta^t \sigma_2 \sigma^i \zeta \ , \tag{1.8}$$

where t means transpose, corresponding to

$$(\mathbf{2}, \mathbf{1}) \otimes (\mathbf{2}, \mathbf{1}) = (\mathbf{1}, \mathbf{1}) \oplus (\mathbf{3}, \mathbf{1}) \ . \tag{1.9}$$

We will find it useful to use the notation

$$\widehat{\eta} \equiv \eta^t \sigma_2 \ . \tag{1.10}$$

Since the combinations $\sigma_2\zeta^*$ and $\sigma_2\eta^*$ transform according to the **(1,2)** representation, we have another Fierz relation

$$\zeta\eta^\dagger = -\frac{1}{2}\eta^\dagger\zeta - \frac{1}{2}\sigma^i\eta^\dagger\sigma^i\zeta \ , \tag{1.11}$$

associated with

$$(\mathbf{2},\mathbf{1}) \otimes (\mathbf{1},\mathbf{2}) = (\mathbf{2},\mathbf{2}). \tag{1.12}$$

The right hand side of this equation does indeed correspond to the Lorentz vector representation, since it can be written succintly as

$$\zeta\eta^\dagger = -\frac{1}{2}\overline{\sigma}^\mu\eta^\dagger\sigma_\mu\zeta \ , \tag{1.13}$$

in terms of the matrices

$$\sigma^\mu = (\sigma^0 = 1, \sigma^i) \ ; \qquad \overline{\sigma}^\mu = (\sigma^0 = 1, -\sigma^i) \ . \tag{1.14}$$

The Pauli matrices property

$$\sigma_2\sigma^i\sigma_2 = -(\sigma^i)^t = -\sigma^{i*} \ ,$$

translates in the covariant language as

$$\sigma_2\overline{\sigma}^\mu\sigma_2 = (\sigma^\mu)^t \ . \tag{1.15}$$

The Lorentz group acts on the spinor fields as

$$\psi_{L,R} \to \Lambda_{L,R}\psi_{L,R} \equiv e^{\frac{i}{2}\vec{\sigma}\cdot(\vec{\omega}\mp i\vec{\nu})}\psi_{L,R} \ , \tag{1.16}$$

where $\vec{\omega}$ and $\vec{\nu}$ are the real rotation and boost angles, respectively. This corresponds to the representation where

$$\vec{J} = \frac{\vec{\sigma}}{2} \ ; \qquad \vec{K} = -i\frac{\vec{\sigma}}{2} \ . \tag{1.17}$$

Note that the two-component Weyl spinors transform non-unitarily, as expected: unitary representations of non-compact groups, such as the Lorentz group, are infinite-dimensional.

This allows us to construct by simple conjugation left-handed spinors out of right-handed antispinors, and vice-versa. One checks that

$$\overline{\psi}_L \equiv \sigma_2\psi_R^* \sim (\mathbf{2},\mathbf{1}) \ , \qquad \overline{\psi}_R \equiv \sigma_2\psi_L^* \sim (\mathbf{1},\mathbf{2}) \ . \tag{1.18}$$

Under charge conjugation the fields behave as

$$C\ :\ \psi_L \to \sigma_2\psi_R^*\ ,\qquad \psi_R \to -\sigma_2\psi_L^*\ , \tag{1.19}$$

and under parity

$$P\ :\ \psi_L \to \psi_R\ ,\qquad \psi_R \to \psi_L\ . \tag{1.20}$$

This purely left-handed notation for the fermions which we use in most of this book is especially suited to describe the parity violating weak interactions. For instance the neutrinos appear only as left-handed fields, while the antineutrinos are purely right-handed. On the other hand, fermions which interact in a parity invariant way as in QED and QCD, have both left- and right-handed parts. In that case, it is far more convenient to use the Dirac four-component notation. The fields ψ_L and ψ_R are then assembled in a four-component Dirac spinor

$$\Psi = \begin{pmatrix} \psi_L \\ \psi_R \end{pmatrix}\ , \tag{1.21}$$

on which the operation of parity is well-defined. This is the Weyl representation, in which the anticommuting Dirac matrices are (in 2×2 block form)

$$\gamma_0 = \begin{pmatrix} 0 & 1 \\ 1 & 0 \end{pmatrix}\ , \qquad \gamma_i = \begin{pmatrix} 0 & -\sigma_i \\ \sigma_i & 0 \end{pmatrix}\ , \qquad \gamma_5 = \begin{pmatrix} 1 & 0 \\ 0 & -1 \end{pmatrix}\ .$$

Since one can generate right-handed fields starting from left-handed ones, it is enough to consider only polynomials made out of left-handed fields. For instance, the well-known group-theoretic product rule of the SU_2 representation

$$\mathbf{2} \otimes \mathbf{2} = \mathbf{1} \oplus \mathbf{3}\ , \tag{1.22}$$

tells us that we can build a Lorentz scalar bilinear in the left-handed fields

$$\text{spin } 0:\quad \widehat{\eta}_L^a \eta_L^b\ ,\quad \text{symmetric under } a \leftrightarrow b\ , \tag{1.23}$$

where a and b label the fields, or else a space-time tensor bilinear

$$\text{tensor}:\quad \widehat{\eta}_L^a \sigma^i \eta_L^b\ ,\quad \text{antisymmetric under } a \leftrightarrow b\ . \tag{1.24}$$

On the other hand, one needs both left and right handed fields to generate the currents

$$i(\eta_L^\dagger \eta_L,\ \eta_L^\dagger \vec{\sigma} \eta_L) \equiv i\ \eta_L^\dagger \sigma^\mu \eta_L\ . \tag{1.25}$$

One can check that their four components transform as $(\mathbf{2},\mathbf{2})$, *i.e.* like a four-vector. The invariant actions built out of η_L and η_R are given by

$$\mathcal{L}_L = \frac{1}{2}\eta_L^\dagger \sigma^\mu \overleftrightarrow{\partial}_\mu \eta_L \ , \qquad \mathcal{L}_R = \frac{1}{2}\eta_R^\dagger \overline{\sigma}^\mu \overleftrightarrow{\partial}_\mu \eta_R \ , \tag{1.26}$$

representing the kinetic terms of the fermion fields. For charged fermions which have both left- and right-handed components, the kinetic term is just the parity-invariant sum of the two, leading to the massless Dirac equation. Since the action has units of $\hbar$, it is dimensionless in our natural system. It follows that the Lagrangian has units of $(\text{length})^{-4}$, which is the same as $(\text{mass})^4$. Since the derivative operator has the dimension of inverse length, we deduce that the fermion fields have engineering dimension of $(\text{length})^{-3/2}$. Two Lorentz invariant spinor bilinears can appear in the Lagrangian. Using only one left-handed field, we can form only one complex spin-zero combination, $\widehat{\eta}_L \eta_L$. In order to appear in the Lagrangian, it must be added to its complex conjugate. It has engineering dimension of $(\text{length})^{-3}$, and can enter the Lagrangian density with a prefactor with the dimension of mass. This type of mass is called the Majorana mass. It is only relevant for neutral fermions.

For charged fermions, which have independent left- and right-handed parts, the diagonal Majorana mass term is not allowed because of charge conservation. However, the off-diagonal Majorana mass is nothing but the Dirac mass term. To see this, consider the combination

$$\widehat{\eta}_L^1 \eta_L^2 = \psi_R^\dagger \psi_L \ , \tag{1.27}$$

where we have made the identifications $\eta_L^1 = \sigma_2 \psi_R^*$, $\eta_L^2 = \psi_L$. We can now form two real combinations

$$i\overline{\Psi}\Psi = i(\psi_R^\dagger \psi_L + \psi_L^\dagger \psi_R) \ , \qquad \overline{\Psi}\gamma_5 \Psi = \psi_R^\dagger \psi_L - \psi_L^\dagger \psi_R \ ,$$

corresponding to scalar (Dirac mass) and pseudoscalar combinations, respectively.

Our notation is simply related to that introduced long ago by Van der Waerden. A left-handed spinor, transforming as the $(\mathbf{2},\mathbf{1})$, is assigned an two-valued lower undotted index α, while the right-handed $(\mathbf{1},\mathbf{2})$ spinor is assigned a two-valued dotted index $\dot{\alpha}$. Each index can be raised by the antisymmetric $\epsilon^{\alpha\beta}$, or $\epsilon^{\dot{\alpha}\dot{\beta}}$. Thus the two-spinor singlet

$$\text{singlet} \ : \qquad \psi_\alpha \zeta_\beta \epsilon^{\alpha\beta} = \psi_\alpha \zeta^\alpha \ . \tag{1.28}$$

In this notation, the Lorentz vectors have both dotted and undotted indices.

PROBLEMS

A. Show that the combination $\overline{\psi}_L \equiv \sigma_2 \psi_L^*$ transforms as a right handed field, given that ψ_L is left-handed.

B. Given the transformation properties on the fields, show that the current combinations $i\eta_L^\dagger \sigma_\mu \eta_L$ does indeed transform as a four-vector.

C. Show that the kinetic term for a neutrino field is CP invariant. What happens to the Majorana mass under CP?

1.2 GAUGE FIELDS

Gauge fields and their interactions are introduced through the covariant derivatives

$$\mathcal{D}_\mu = \partial_\mu + i\mathbf{A}_\mu \ , \tag{1.29}$$

where $\mathbf{A}_\mu$ is a matrix in the Lie algebra $\mathcal{G}$,

$$\mathbf{A}_\mu(x) = \sum_{A=1}^{N} A_\mu^A(x)\mathbf{T}^A \ , \tag{1.30}$$

and $\mathbf{T}^A$ are the matrices which generate $\mathcal{G}$; they are hermitian and satisfy the commutation relations

$$[\mathbf{T}^A, \mathbf{T}^A] = if^{ABC}\mathbf{T}^C \ . \tag{1.31}$$

In the fundamental representation of the algebra, they are normalized as

$$\mathrm{Tr}\,(\mathbf{T}^A\mathbf{T}^B) = \frac{1}{2}\delta^{AB} \ . \tag{1.32}$$

Under the gauge transformations generated by the unitary matrices

$$\mathbf{U}(x) = e^{i\omega^A(x)\mathbf{T}^A} \ , \tag{1.33}$$

the covariant derivatives transform (as their name indicates) covariantly

$$\mathcal{D}_\mu \to \mathcal{D}'_\mu = \mathbf{U}\mathcal{D}_\mu\mathbf{U}^\dagger \ , \tag{1.34}$$

leading to the transformations on the gauge fields

$$\mathbf{A}_\mu \to \mathbf{A}'_\mu = \mathbf{U}\mathbf{A}_\mu\mathbf{U}^\dagger - i\mathbf{U}\partial_\mu\mathbf{U}^\dagger \ . \tag{1.35}$$

For infinitesimal gauge transformations, these correspond to

$$\delta A_\mu^A = -\partial_\mu \omega^A - f^{ABC} \omega^B A_\mu^C \; . \tag{1.36}$$

The covariant Yang-Mills field strengths are built from the covariant derivatives

$$\mathbf{F}_{\mu\nu} = -i[\mathcal{D}_\mu, \mathcal{D}_\nu], \tag{1.37}$$

$$= \partial_\mu \mathbf{A}_\nu - \partial_\nu \mathbf{A}_\mu + i[\mathbf{A}_\mu, \mathbf{A}_\nu] \; , \tag{1.38}$$

or in component form,

$$\mathbf{F}_{\mu\nu} = F_{\mu\nu}^A \mathbf{T}^A \; , \tag{1.39}$$

$$F_{\mu\nu}^A = \partial_\mu A_\nu^A - \partial_\nu A_\mu^A - f^{ABC} A_\mu^B A_\nu^C \; . \tag{1.40}$$

The field strengths transform covariantly,

$$\mathbf{F}_{\mu\nu} \to \mathbf{U}\mathbf{F}_{\mu\nu}\mathbf{U}^\dagger, \tag{1.41}$$

and satisfy the integrability conditions (Bianchi identities)

$$\epsilon^{\alpha\beta\gamma\delta} \mathcal{D}_\beta \mathbf{F}_{\gamma\delta} = 0 \; . \tag{1.42}$$

Finally, the Yang-Mills action is given by

$$S_{YM} = -\frac{1}{4g^2} \int d^4x \mathrm{Tr}(\mathbf{F}_{\mu\nu}\mathbf{F}^{\mu\nu}) \; ; \tag{1.43}$$

it is invariant under both Lorentz and gauge transformations. The dimensionless Yang-Mills coupling constant g appears in the denominator, but it can be absorbed by redefining the potentials $\mathbf{A}_\mu \to g\mathbf{A}_\mu$, in which case it reappears in the interaction terms. One may build another quadratic invariant,

$$\int d^4x \mathrm{Tr}(\mathbf{F}_{\mu\nu}\mathbf{F}_{\rho\sigma})\epsilon^{\mu\nu\rho\sigma}. \tag{1.44}$$

It is easy to show that it does not affect the equations of motion because the integrand is a total derivative. It also differs from the Yang-Mills action in its properties under discrete symmetries. In quantum theory, where there is much more than the classical equations of motion, terms of this type do play an important role.

The gauge fields couple to matter (fermions, scalars) only through the covariant derivative, thus insuring gauge invariance. Consider a complex boson field $\Phi(x)$, with components $\varphi_a(x)$, $a = 1, 2, \cdots, n$, which transforms according to the n-dimensional representation of a gauge group G. Its coupling to the gauge fields is achieved by simply replacing the normal derivative by the covariant derivative

$$\partial_\mu \varphi_a \to (\mathcal{D}_\mu \varphi)_a = \partial_\mu \varphi_a + i A^A_\mu \mathbf{T}^A_{ab} \varphi_b \; . \tag{1.45}$$

Here the $\mathbf{T}^A$ are the $n \times n$ matrices which represent the Lie algebra in the representation of the scalar field. The invariant Lagrangian is just the absolute square

$$\mathcal{L} = (\mathcal{D}_\mu \Phi)^\dagger \mathcal{D}^\mu \Phi \; . \tag{1.46}$$

The coupling to fermions proceeds in the same way, by changing the normal derivative into the covariant derivative. This can be done equally well for Weyl and Dirac fermions. This results in the Weyl-Dirac Lagrangian

$$\mathcal{L}_{WD} = \mathbf{f}^\dagger_L \sigma^\mu \mathcal{D}_\mu \mathbf{f}_L, \tag{1.47}$$

where the left-handed fermion fields $\mathbf{f}$ transform as some representation of the gauge algebra,

$$\mathbf{f}_L \to \mathbf{f}'_L = \mathbf{U}\mathbf{f}_L \; . \tag{1.48}$$

One can form a similar Lagrangian for right-handed fields,

$$\mathcal{L}_{WD} = \mathbf{f}^\dagger_R \bar{\sigma}^\mu \mathcal{D}_\mu \mathbf{f}_R \; , \tag{1.49}$$

where the right-handed fields need not transform in the same way as the left-handed fields. When both helicities transform the same, the theory is invariant under parity, and it is said to be *vector-like*; otherwise they are said to be *chiral*. Chiral gauge theories run the risk of developing perturbative anomalies which spoil their renormalizability. Most groups have no such anomalies, except $U(1)$, $SU(n)$, for $n > 2$. The standard model is a chiral theory, which contains a $U(1)$ gauge group. Although it could in principle be anomalous, it escapes through a remarkable cancellation between quarks and leptons.

PROBLEMS

A. Show that with the gauge transformation Eq. (1.35), the covariant derivative does transform as advertised in the text.

B. Show that the quadratic invariant given by Eq. (1.44) is a divergence of a four-vector. Find its expression in terms of the gauge potentials.

1.3 SPONTANEOUS SYMMETRY BREAKING

First invented by Heisenberg to describe ferromagnetism, spontaneous symmetry breaking (SSB) is a ubiquitous phenomenon. It appears in superconductivity, superfluidity, is the mechanism by which the strong interactions break chiral invariance, and also describes the breaking of electroweak symmetry, to name but a few of its applications.

In quantum mechanics, symmetries play a central role. The dynamics of a quantum system is determined by the Hamiltonian $\mathcal{H}$ whose eigenvalues describe its allowed energies; to each energy eigenvalue correspond one or more states which contain the detailed information about the system at that energy, and satisfy

$$\mathcal{H}|\Psi_a\rangle = E_a|\Psi_a\rangle \ . \tag{1.50}$$

If the Hamiltonian is invariant under some symmetry operation represented by the operator $\mathcal{S}$, then

$$\mathcal{H} \rightarrow \mathcal{H}' = \mathcal{S}\mathcal{H}\mathcal{S}^{-1} = \mathcal{H} \ ; \tag{1.51}$$

if the symmetry is continuous and analytic, we have

$$[\mathcal{H}, T] = 0 \ , \tag{1.52}$$

where T is any of the generators of the symmetry. In that case, the state functions $\mathcal{S}|\Psi_a\rangle$ belong to the same eigenvalue as $|\Psi_a\rangle$. How does this analysis apply to the ground state of the system?

The ground state of a physical system is special in the sense that it is unique. Suppose it were not, and a student finds that the Hamiltonian he is considering has two states of lowest energy. She may envisage two possibilities. In the first, she finds that he made a mistake, and closer scrutiny indicates that there is a non-vanishing transition amplitude between the two "vacua". This means that the starting Hamiltonian was not properly analyzed, and the effect of this transition must be included. This effect can be subtle, as in tunnelling. Rediagonalization breaks the degeneracy: the linear combination with a lower energy is the unique ground state. It could be that no mistake has been made, and there is no physical transition from one "vacuum" state to the other. Starting with each state, she may build a tower of excited states, with no physical transitions between the two towers. These two sets of states span different Hilbert spaces, and both describe a consistent physical system. In the absence of any criterion to distinguish them, they must yield the same physics. A trivial example is the Bohr atom with electron spin included in its kets, but not in the Hamiltonian, resulting in two non-communicating equivalent Hilbert spaces. Of course, when immersed in a magnetic field, the two Hilbert spaces mesh into one, and one unique ground state emerges after diagonalization of the new Hamiltonian.

Since the ground state $|\Omega\rangle$ is unique, what then is $S|\Omega\rangle$ when S is a symmetry of the Hamiltonian? The first possibility is that the ground state is invariant under all symmetries, in which case $S|\Omega\rangle = |\Omega\rangle$, up to a phase, that is $T|\Omega\rangle = 0$ for infinitesimal generators. This is true for many physical systemson which the symmetry is then said to be *linearly realized.* The symmetry is manifest in the degeneracy of the excited states, and can be even be deduced from the multiplicity of the degenerate states.

A prototypical example is the degeneracy structure of the states of the Bohr Hydrogen atom. It has a unique ground state, on which the symmetry operations are trivial. The first excited state contains four states; one s-wave, and three p-wave states. The rotation symmetry of the Hamiltonian is identified by recognizing the degenerate states as representations of the rotation group. Since there are two different representations of angular momentum at the first excited level, it hints of a symmetry more general than the rotation symmetry. There must be three extra symmetry operations, each rotating the s-wave singlet into one component of the p-wave triplet. This symmetry with six generators (three rotations and three "others") is the hidden, but well-known, Runge-Lenz symmetry of the Hydrogen atom. It is identified with the group $SO(4)$, and the four states of the first excited level act as the components of its four dimensional representation. The symmetry of the Hamiltonian has been linearly realized on its eigenstates.

However, another possibility may arise. It may happen that under the action of the symmetry operation on the ground state, a different state is produced, so that the symmetry generators do not vanish on the ground state, even though the Hamiltonian is symmetric. If the symmetry is indeed present in the Hamiltonian, but not on its ground state, the symmetry is said to be *spontaneously* broken. The vacuum happens not to be an eigenstate of the symmetry operations, whose action on the vacuum generates other states of zero energy. If the symmetry is discrete, there is a finite number of such vacuum states; when the symmetry operation is continuous and compact, the action of the symmetry yields a closed family of equivalent ground states parametrized by the parameters of the symmetry. All of these vacuum states are equivalent, each belonging to a different Hilbert space. Since there can be only one state of lowest energy, the resulting physics of this system must not depend on the angles that parametrize these states.

Since to each possible "ground state" there corresponds an equivalent physical theory, *any* one of them can serve as *the* ground state. This chosen vacuum state is no longer invariant under the action of the symmetry group; the symmetry is spontaneously broken, an unfortunate description since the symmetry is not really broken, just expressed differently. The symmetry is *non-lineraly realized* on the states; it is still encoded in the system, but in a more subtle way. In fact, there is a limit in which one can recover the symmetry in its linear realization. The degeneracy of highly excited states, with energies much larger than the energy scale associated with the breaking of the symmetry, will reflect the original symmetry of the system.

The celebrated example of spontaneous symmetry breaking is the Heisenberg ferromagnet, described by a lattice of spins in interaction with one another. These spins can only take two values; they can be either "up" or "down". The Hamiltonian is invariant under the "up-down" symmetry. In the ground state, the spins are either all aligned or anti-aligned, depending on the sign of the coupling between the spins. If the anti-aligned configuration is favored, there is no preferred "up" or "down" direction, and the vacuum preserves the "up-down" symmetry. On the other hand, there are two possible ground state configurations with all the spins aligned, "up-up" and "down-down". For either choice, the "up-down" symmetry has been broken, since there is a preferred direction in both configurations. The Heisenberg Hamiltonian is given by

$$\mathcal{H} = -J \sum_{<ij>} \sigma_i \cdot \sigma_j \ , \tag{1.53}$$

where $\sigma_i = \pm$, and the sum is over nearest neighbors. It is invariant under a change of sign of the two spins. For $J > 0$, the Hamiltonian is at minimum when all the spins are aligned, breaking the "up-down" invariance. In this example, the broken symmetry is discrete, but it is not hard to generalize the model to continuous symmetry by replacing the elementary spins by normalized three-dimensional spin vectors at each site, $\vec{S}_i$.

We are primarily interested in describing spontaneous breaking in the context of field theory. To do so, we start with a simple system with an Abelian symmetry, a complex scalar field $\phi(x)$, with Lagrangian

$$\mathcal{L} = \partial_\mu \phi^* \partial^\mu \phi - V(\phi^* \phi) , \tag{1.54}$$

where the potential is chosen to be of the form

$$V(\phi^* \phi) = \lambda(\phi^* \phi - \frac{v^2}{2})^2 ; \tag{1.55}$$

This Lagrangian is clearly invariant under the $U(1)$ transformation

$$\phi \rightarrow e^{i\alpha} \phi , \tag{1.56}$$

where α is a constant phase. Using Noether methods, we obtain a conserved current

$$j_\mu = i \phi^* \overleftrightarrow{\partial}_\mu \phi , \qquad \partial_\mu j^\mu = 0 . \tag{1.57}$$

The field configurations which yields a minimum energy density are constant in space time; otherwise the kinetic energy term would give a positive contribution. Its actual values are determined by minimizing the potential,

$$\phi^* \phi = \frac{v^2}{2} . \tag{1.58}$$

There are therefore an infinite number of possible vacuum field configurations, all with zero energy,

$$\phi_0 = \frac{1}{\sqrt{2}} v e^{i\theta} , \tag{1.59}$$

each labelled by an angle θ. Since the symmetry is compact, they form a closed circle of states. We note that while the continuous phase symmetry is broken by these configurations, the vacuum condition leaves one discrete symmetry,

$$\Phi_0 \rightarrow -\Phi_0 , \tag{1.60}$$

so that the angle θ is defined *modulo* π. We expand the field away from this minimum configuration

$$\phi(x) = \frac{1}{\sqrt{2}} e^{i\frac{\xi(x)}{v}} (v + \rho(x)) \ , \tag{1.61}$$

a parametrization introduced by Kibble. The phase θ has been absorbed in the definition of the field $\xi(x)$. Rewriting the Lagrangian in terms of these new fields, we obtain

$$\mathcal{L} = \frac{1}{2}\partial_\mu \rho \partial^\mu \rho + \frac{1}{2}\partial_\mu \xi \partial^\mu \xi (1 + \frac{\rho}{v})^2 - \frac{1}{2}m^2 \rho^2 - \frac{\lambda}{4}\rho^4 - v\lambda\rho^3 \ . \tag{1.62}$$

We recognize kinetic terms for two fields, $\rho(x)$ and $\xi(x)$. The first has a mass $m = v\sqrt{2\lambda}$, the second is massless. The minimum of the energy density now corresponds to zero values for both fields. The Lagrangian is invariant under the constant shift

$$\xi(x) \to \xi(x) + \theta \ , \tag{1.63}$$

with the ρ field unchanged. This shift is in one-to-one correspondance with the original $U(1)$ phase symmetry. Its linear realization on ϕ is replaced by a non-linear realization (by a shift) on the massless ξ field, which represents the Nambu-Goldstone boson. The invariance of the Lagrangian under a constant shift indicates that all vacua differing by their value of θ are indeed physically equivalent.

The Nambu-Goldstone boson couples to itself and to the other field in a very special way, restrited by the shift invariance. To see it, we rewrite (after integration by parts) the coupling of the Nambu-Goldstone boson in the more transparent form

$$\mathcal{L}_{int} = -\frac{1}{v}\xi(x)\partial_\mu j^\mu(x) \ , \tag{1.64}$$

since the current is just

$$j_\mu = -v(1 + \frac{\rho(x)}{v})^2 \partial_\mu \xi(x) \ . \tag{1.65}$$

Under a constant shift of the Nambu-Goldstone field, the Lagrangian density picks up a total divergence, which does not affect the equations of motion. This is a general feature: Nambu-Goldstone bosons couple to the divergence of the current that was broken, with strength proportional to the inverse of the scale of breaking, v. It is the dynamical variable associated with the angle that parametrizes the continuous circle of minima.

It is instructive to verify that these features generalize when we add fermions. This is done through the Yukawa coupling

$$\mathcal{L}_{Yu} = y\chi_L^\dagger(x)\chi_R(x)\phi(x) + \text{c.c;} \tag{1.66}$$

invariance is retained as long as the fermions transform as

$$\chi_L \to e^{-i\alpha}\chi_L \ ; \qquad \chi_R \to \chi_R \ . \tag{1.67}$$

It is a bit more involved to show that the NG boson still couples to the divergence of the current, and we leave it as a problem.

In generalizing this discussion to non-Abelian symmetries, we find the mathematics to be slightly more involved, but the conclusions to be the same as in the Abelian case. To each broken continuous symmetry corresponds a massless Nambu-Goldstone boson, which couples to the divergence of the broken current. One new feature is that generally not all the continuous symmetry is spontaneously broken, and it is sometimes tricky to identify the unbroken symmetry.

To study this case, we consider the example of a real field $\Phi(x)$, with components $\varphi_a(x)$, $a = 1, 2, \ldots, N$, with interactions described by the Lagrangian

$$\mathcal{L} = \frac{1}{2}\partial_\mu\varphi_a\partial^\mu\varphi^a - V(\varphi_a\varphi^a) \ , \tag{1.68}$$

where the sum over a is implicit. It is clearly invariant under the transformations

$$\Phi \to \Phi' = \mathbf{R}\Phi(x) \ , \tag{1.69}$$

where $\mathbf{R}$ is a constant rotation matrix

$$\mathbf{R} = e^{i\vec{\theta}\cdot\vec{\mathbf{T}}} \ , \qquad \mathbf{R}\mathbf{R}^t = 1, \tag{1.70}$$

written in terms of the $N(N-1)/2$ antisymmetric real matrices $\mathbf{T}^A$ that generate the Lie algebra of $SO(N)$. To each of these generators corresponds a conserved current

$$J_\mu^A = i\Phi^t\mathbf{T}^A\overleftrightarrow{\partial}_\mu\Phi, \qquad \partial^\mu J_\mu^A = 0 \ . \tag{1.71}$$

The potential is chosen to be

$$V = \lambda(\Phi^t\Phi - \frac{v^2}{2})^2 \ , \tag{1.72}$$

so as to produce an infinite number of minimum field configurations written in the form

$$\Phi_0 = \frac{1}{\sqrt{2}} e^{i\vec{\theta}\cdot\vec{\mathbf{K}}} \begin{pmatrix} 0 \\ \vdots \\ 0 \\ v \end{pmatrix} . \tag{1.73}$$

The vacuum expectation value v has been aligned along the Nth component of the vector; this entails no loss of generality, since it can always be aligned this way by a suitable rotation. The vacuum value of Φ clearly singles out one direction in the N dimensional internal space. It breaks the $SO(N)$ symmetry, but all rotations in the $(N-1)$-dimensional plane perpendicular to that direction are left invariant. The generators of the $(N-1)$ broken symmetries are denoted by $\vec{\mathbf{K}}$. The vacuum configuration depends on the $(N-1)$ angles which parametrize these broken rotations. The generators of $SO(N)$ can be decomposed in two classes

$$\vec{\mathbf{T}} = \vec{\mathbf{K}} \oplus \vec{\mathbf{H}} , \tag{1.74}$$

where the generators $\vec{\mathbf{H}}$ generate the Lie algebra of the unbroken subalgebra, $SO(N-1)$,

$$[H_a, H_b] = i f_{abc} H_c \ , \quad a, b, c, = 1 \cdots, \frac{(N-1)(N-2)}{2}, \tag{1.75}$$

and the $\mathbf{K}_i$, $i = 1, \cdots, N-1$ no longer form a Lie algebra, and transform as a $(N-1)$ vector under the unbroken symmetry, $SO(N-1)$. They span a space called the *coset* space. Their commutators yield generators of the unbroken subalgebra. Symbolically we may write

$$[H, H] \subset H \ ; \qquad [H, K] \subset K \ ; \qquad [K, K] \subset H \ . \tag{1.76}$$

Note that the generators of the unbroken subalgebra vanish on the vacuum. The coset generators do not vanish on the vacuum, although their commutators do.

We expand $\Phi(x)$ away from its vacuum configuration *à la* Kibble,

$$\Phi(x) = \frac{1}{\sqrt{2}} e^{i\frac{\vec{\theta}}{v}\cdot\vec{\mathbf{K}}} \begin{pmatrix} 0 \\ \vdots \\ 0 \\ v + \rho(x) \end{pmatrix} , \tag{1.77}$$

$$\equiv \frac{1}{\sqrt{2}} \mathbf{U}(x)(v + \rho(x))\mathbf{n}_0 \ , \tag{1.78}$$

where $\mathbf{n}_0$ is a unit vector pointing along the Nth direction. As before the potential is simply written in terms of the ρ field alone,

$$V = \frac{1}{2}m^2\rho^2 + \frac{\lambda}{4}\rho^4 + v\lambda\rho^3. \tag{1.79}$$

The kinetic part of the Lagrangian yields

$$\frac{1}{2}\partial_\mu\rho\partial^\mu\rho + \frac{1}{2}\partial_\mu\xi^i\partial^\mu\xi^j(1+\frac{\rho}{v})^2\mathbf{n}_0^t\frac{\partial}{\partial\xi^i}\mathbf{U}^\dagger\frac{\partial}{\partial\xi^j}\mathbf{U}\mathbf{n}_0 \ , \tag{1.80}$$

which again exhibits the canonical kinetic term for the ρ field. The indices i, j run over the $(N-1)$ dimensions of the coset. The kinetic piece for the ξ^i fields can be simplified considerably. For a general group element

$$\mathbf{U} = e^{i\vec{\theta}\cdot\vec{\mathbf{T}}} \ , \tag{1.81}$$

it can be shown that

$$\frac{\partial}{\partial\theta^A}\mathbf{U} = i\mathbf{U}X_{AB}\mathbf{T}^B \ , \tag{1.82}$$

where the indices A, B run over the whole algebra, where the matrix $\mathbf{X}$ is given by

$$X_{AB} = \delta_{AB} + \frac{1}{2!}(\vec{\theta}\cdot\vec{F})_{AB} + \frac{1}{3!}((\vec{\theta}\cdot\vec{F})^2)_{AB} + \cdots \ , \tag{1.83}$$

where the $\mathbf{F}^A$ are the antisymmetric matrices that represent the algebra in its adjoint representation; they are simply related to the structure functions

$$(F^A)_{BC} = -if^A{}_{BC} \ . \tag{1.84}$$

It follows that

$$\frac{\partial}{\partial\theta^A}\mathbf{U}^\dagger = -i\mathbf{T}^B X^{-1}_{BA}\mathbf{U}^\dagger \ , \tag{1.85}$$

Applying these to the case in hand, we obtain

$$\begin{aligned}\mathbf{n}_0^t\frac{\partial}{\partial\xi^i}\mathbf{U}^\dagger\frac{\partial}{\partial\xi^j}\mathbf{U}\mathbf{n}_0 =& X^{-1}_{li}X_{jk}\mathbf{n}_0^t\mathbf{K}^l\mathbf{K}^k\mathbf{n}_0 \ , \\ =& X^{-1}_{Ci}X_{jD}\delta^{CD} = \delta^{ij} \ ,\end{aligned}$$

since the unbroken group generators vanish on the unit vector $\mathbf{n}_0$. Thus the full Lagrangian looks much the same as in the Abelian case,

$$\frac{1}{2}\partial_\mu\rho\partial^\mu\rho + \frac{1}{2}(1+\frac{\rho}{v})^2\partial_\mu\xi^i\partial^\mu\xi_i - \frac{1}{2}m^2\rho^2 - \frac{\lambda}{4}\rho^4 - v\lambda\rho^3. \tag{1.86}$$

The only difference is that there are now as many Nambu-Goldstone bosons as there are broken symmetries, and we have invariance under $(N-1)$ constant shifts

$$\xi_i(x) \to \xi_i(x) + \theta_i \ . \tag{1.87}$$

These invariances are made obvious made manifest by writing the couplings of the Nambu-Goldstone bosons in terms of the divergence of the broken currents,

$$\mathcal{L}_{int} = \frac{1}{v}\xi^i(x)\partial^\mu J^i_\mu(x) = -\frac{1}{v}J^i_\mu(x)\partial^\mu\xi^i(x) \ , \tag{1.88}$$

where we have integrated by parts, and dropped the surface terms. In this case, we see that the unbroken symmetry $SO(N-1)$ is realized linearly on the Nambu-Goldstone bosons which transform as a $(N-1)$-dimensional vector,

$$\Xi(x) \to e^{i\vec{\theta}\cdot\vec{\mathbf{K}}}\Xi(x) \ , \tag{1.89}$$

where Ξ is the vector with components $\xi(x)$. The number of degrees of freedom has not changed; we started with N fields, and we end up with $(N-1)$ massless Nambu-Goldstone bosons and one massive Higgs boson.

This form of the interaction between the Nambu-Goldstone bosons and the curents is quite suggestive, when compared to the interaction of a gauge field and the same current.

Suppose we gauge the starting symmetry, in this case $SO(N)$. We expect as many gauge potentials as there are generators of $SO(N)$. They interact with the currents J^A_μ via the interaction term

$$\mathcal{L}_{int} = gA^A_\mu J^{A\mu} \ . \tag{1.90}$$

By comparing Eqs. (1.88) and (1.90) for the broken directions, we note that the interaction term of the Nambu-Goldstone bosons can be cancelled exactly by performing the gauge transformation

$$\delta A^i_\mu = -\frac{1}{g}\partial_\mu\xi^i + f^{ijk}A^j_\mu\xi^k \ . \tag{1.91}$$

Hence the Nambu-Goldstone bosons are to be viewed as gauge artifacts. They no longer correspond to physical particles, since they can be absorbed by redefining the gauge fields. Each gauge field that corresponds to a broken symmetry acquires an extra degree of freedom, and becomes massive, to keep the description relativistic.

In order to see exactly what happens, we start from the Lagrangian density (1.68), with the normal derivative replaced by the covariant derivative

$$\partial_\mu \Phi \to (\partial_\mu + ig\mathbf{A}_\mu)\Phi \ , \tag{1.92}$$

with

$$\mathbf{A}_\mu = \mathbf{T}^A A^A_\mu \ ,$$

where the $N \times N$ matrices $\mathbf{T}^A$ generate the Lie algebra of $SO(N)$. We now simply retrace the steps starting with the vacuum configuration (1.73). The only difference is that the covariant derivative on the vacuum configuration leaves behind a term linear in those gauge potentials associated with the broken generators

$$\mathcal{D}_\mu \Phi = \left(\partial_\mu + igA^i_\mu(x)\mathbf{T}^i\right) \frac{v + \rho(x)}{\sqrt{2}} \hat{n}_0 \ . \tag{1.93}$$

When squared in the Lagrangian, these yield a mass term for these gauge fields which correspond to the broken directions. This is the Higgs mechanism: the gauge potential associated to each broken generator becomes massive, after "eating" the Nambu-Goldstone boson, which then serves as its third degree of freedom. The gauge ppotentials along unbroken directions stay massless. The advantage of giving a mass to the gauge bosons in this particular way is that it does not affect the ultraviolet properties of the theory. This is equivalent to the solid state case where spontaneous symmetry breaking, being a property of the vacuum is more apparent on states near the vacuum and become less relevant for the highly excited states. Thus a spontaneously broken gauge theory has the same renormalizability properties as an unbroken one.

PROBLEMS

A. When fermions are added in the Abelian model with the Yukawa coupling of Eq. (1.66) , find the new expression for the current and show that the Nambu-Goldstone boson still couples to the divergence of the broken current.

B. Show that the vacuum configuration Φ_0 is left invariant by $SO(N-1)$ rotations.

C. Verify the form of the $\mathbf{X}$ matrix given by Eq. (1.83), and show that the kinetic term for the Nambu-Goldstone bosons is indeed invariant under constant shifts.

D. 1-) Show that, in the Non-Abelian case,.the couplings of the Nambu-Goldstone bosons can be expressed solely in terms of the divergences of the currents associated with the broken generators.
2-)Add to the theory a set of N left- and right-handed fermions which transform as the vector representation of $SO(N)$. Write an $SO(N)$ invariant Yukawa coupling of these fermions to the Higgs. Show that the Nambu-Goldstone bosons still couple to the divergences of the broken currents.

1.4 GROUP THEORY

This section summarizes the essentials group-theoretic facts the reader will need to follow some of the material in this book. Much of it follows the presentation that can be found in R. Slansky, *Phys. Reports* **79**, 1(1981). Continuous groups of physical interest are generated by Lie algebras. The semisimple Lie algebras, classified by Cartan a century ago, come in four infinite families, A_n, B_n, C_n, D_n, $n = 1, 2...$, and in five exceptional algebras, G_2, F_4, E_6, E_7, E_8. The subscript denotes the the *rank* of the algebra, the number of its commuting generators. Here follows a brief description:

• The unitary series, A_n ,$n = 1, 2, \cdots$, is called SU_{n+1} by physicists. This Lie algebra is generated by the $n^2 - 1$ independent $n \times n$ traceless hermitian matrices. They generate transformations that leaves the real quadratic form

$$z_1^* z_1 + z_2^* z_2 + \cdots + z_n^* z_n \ , \tag{1.94}$$

invariant, where the z_i are complex numbers. Acting on the complex n-dimensional vector, these transformations are represented by unitary matrices. They are expressed as exponentials of i times hermitian matrices

$$U(\theta_A) = e^{i\theta_A \mathbf{T}^A} \ , \tag{1.95}$$

where θ_A are real parameters, $A = 1, 2, \ldots, n^2 - 1$, and

$$U^\dagger = U^{-1} \quad \to \quad \mathbf{T}^{A\dagger} = \mathbf{T}^A \ . \tag{1.96}$$

• The infinite orthogonal series which generates transformations that leave the quadratic form

$$x_1^2 + x_2^2 + \cdots + x_n^2 \ , \tag{1.97}$$

invariant, where the x_i are real numbers. These transformations are represented by orthogonal *real* matrices which are exponentials of antisymmetric hermitian $n \times n$ matrices. There are $n(n-1)/2$ such matrices.

$$U(\omega_A) = e^{i\omega_A \mathbf{T}^A} \ , \tag{1.98}$$

where the ω_A are the $n(n-1)/2$ real rotation angles, and

$$U^T = U^{-1} \ \to \ \mathbf{T}^{A\dagger} = \mathbf{T}^A \ . \tag{1.99}$$

These transformations are called SO_n by physicists. However they have very different structure depending on whether n is even or odd. Accordingly, mathematicians split them up in two infinite series, B_n and D_n, corresponding to SO_{2n+1} and SO_{2n}, with $n(2n+1)$ and $n(2n-1)$ generators respectively. While identical in their tensor representations, their spinor representations are very different. Each B_n has one real fundamental spinor irreducible representation (irrep). D_n have two fundamental spinor irreps; for n even they are real, and for n odd they are conjugate of one another.

• The symplectic series C_n, which we denote by Sp_{2n} generate transformations that leave invariant the *real antisymmetric* quadratic form

$$\sum_{i,j}^{n} x_i J_{ij} y_j \ , \tag{1.100}$$

with the antisymmetric matrix $\mathbf{J}$ given by

$$\mathbf{J} = \begin{pmatrix} 0 & I \\ -I & 0 \end{pmatrix} \ , \tag{1.101}$$

where I is the unit matrix in n dimensions. Symplectic transformations are represented by symmetric $2n \times 2n$ real matrices with $n(2n+1)$ parameters.

• There are five exceptional Lie algebras, G_2, F_4, E_6, E_7, E_8, with dimensions 14, 26, 78, 133, and 248, respectively. It is not as easy to characterize the groups generated by these algebras as they leave invariant not only quadratic but also higher-order polynomials.

All these Lie algebras have common features as they consist of $d_{\mathbf{a}}$ hermitian matrices $\mathbf{T}^A$, $A = 1, 2..., d_{\mathbf{a}}$, which satisfy an algebra of the form

$$[\mathbf{T}^A, \mathbf{T}^B] = if^{ABC}\mathbf{T}^C \ , \tag{1.102}$$

implying that these matrices must be traceless. The real antisymmetric f^{ABC} are called the *structure functions* of the algebra. Since matrices naturally satisfy the Jacobi identity

$$[\mathbf{T}^A, [\mathbf{T}^B, \mathbf{T}^C]] + [\mathbf{T}^B, [\mathbf{T}^C, \mathbf{T}^A]] + [\mathbf{T}^C, [\mathbf{T}^A, \mathbf{T}^B]] = 0 \ , \tag{1.103}$$

it follows that the f^{ABC} obey the relations

$$f^{ADE}f^{BCD} + f^{BDE}f^{CAD} + f^{CDE}f^{ABD} = 0 \ . \tag{1.104}$$

We can therefore define a $d_{\mathbf{a}} \times d_{\mathbf{a}}$ matrix

$$(T^C)_{AB} \equiv -if_{ABC} \ , \tag{1.105}$$

which satisfies the same Lie algebra. These matrices form a real representation of the Lie Algebra in a $d_{\mathbf{a}}$-dimensional vector space called the *adjoint* representation; its dimension is equal to the number of generators. Each Lie algebra has its own unique adjoint representation.

Lie algebras can also be represented in different finite-dimensional Hilbert space, in infinitely many ways. The properties of these representations are the starting point for physical applications. Each representation is labelled by a number of invariant combinations of the generators, called Casimir operators; a Lie algebra of rank l contains l Casimir operators. The $\mathbf{T}^A_{\mathbf{r}}$ matrices which represent the algebra in the $\mathbf{r}$ representation satisfy a normalization condition

$$\mathrm{Tr}(\mathbf{T_r}^A \mathbf{T_r}^B) = C_{\mathbf{r}}\delta^{AB} \ , \tag{1.106}$$

where the coefficient $C_{\mathbf{r}}$ is called the Dynkin index of the representation. By multiplying by δ^{AB} and summing over A, B, we find

$$d_{\mathbf{r}}C^{[2]}(\mathbf{r}) = C_{\mathbf{r}}d_{\mathbf{a}} \ , \tag{1.107}$$

where $d_{\mathbf{r}}$ is the dimension of the representation, $d_{\mathbf{a}}$ is the number of generators of the Lie algebra, and $C^{[2]}(\mathbf{r})$ is the quadratic Casimir operator of the representation $\mathbf{r}$, defined through

$$\sum_A^{d_\mathbf{a}} \mathbf{T}^A \mathbf{T}^A = C^{[2]}(\mathbf{r}) \mathbf{I} \; , \tag{1.108}$$

where $\mathbf{I}$ is the $d_\mathbf{r} \times d_\mathbf{r}$ unit matrix. The Dynkin index is the first of the group-theoretic numbers associated with each representation. We single it out because of its importance in physical applications.

Representations can be multiplied together to yield other representations. An elementary example is that of vector multiplication. A real vector in three dimensions transforms as the **3** representation of the orthogonal group $SO(3)$. We learn in high school that, given two vectors, we can either form another vector by taking their antisymmetric (cross) product, or form a rotation invariant by taking their symmetric dot product, or else form a five-component symmetric traceless second rank tensor (quadrupole). In group-theoretic terms, this would simply read

$$\mathbf{3} \otimes \mathbf{3} = \mathbf{3}_A \oplus (\mathbf{5} \oplus \mathbf{1})_S \; . \tag{1.109}$$

Each algebra has a fundamental representation out of which one can generate all of its representations by repeatedly taking their (Kronecker) product.

There is a useful graphical way to represent group-theoretic numbers associated with any representation. We represent the $\mathbf{r}$ representation by a solid line, and the adjoint representation by a wavy line. The matrices in the representation $\mathbf{r}$ can then be written as a vertex between two solid lines and a wavy line:

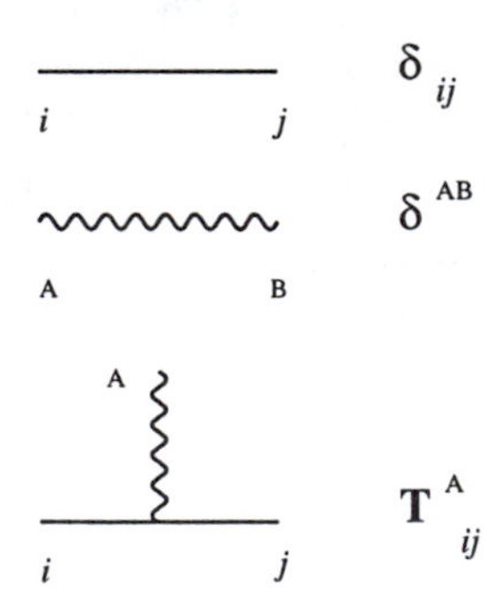

The Dynkin index is represented by the vacuum polarization

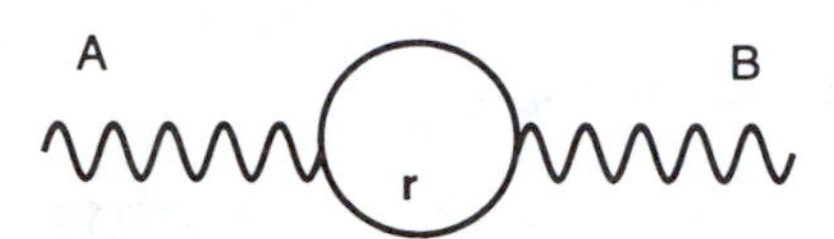

This graphical representation provides an easy way to deduce the Dynkin index of other representations. The Dynkin indices of two representation $\mathbf{r}$ and $\mathbf{s}$ of dimensions $d_{\mathbf{r}}$ and $d_{\mathbf{s}}$ are related to that of the compound $\mathbf{r} \otimes \mathbf{s}$ reducible representation through

$$C_{\mathbf{r}\otimes\mathbf{s}} = d_{\mathbf{r}} C_{\mathbf{s}} + d_{\mathbf{s}} C_{\mathbf{r}} \ , \tag{1.110}$$

which can be understood graphically

s s
r x s r r

To derive this, we have used two facts: a closed line without any external line is equal to the dimension of the representation, and a closed loop with one adjoint external line is zero, since the matrices are traceless. We can go further and compute the Dynkin index of the symmetric or antisymmetric products of representations, using our graphical representation.

Consider the group SU_n. It is represented in a complex n-dimensional vector space by one-half times the Gell-Mann matrices, $\boldsymbol{\lambda}^A$. Thus

$$\mathrm{Tr}\left(\frac{\boldsymbol{\lambda}^A}{2}\frac{\boldsymbol{\lambda}^B}{2}\right) = \frac{1}{2}\delta^{AB} \ , \tag{1.111}$$

so that the Dynkin index of the fundamental representation is $C_{\mathbf{n}} = \frac{1}{2}$, independent of n. One can check that the matrices $-(\boldsymbol{\lambda}^A)^t/2$ represent the algebra in the conjugate representation, $\overline{\mathbf{n}}$, so that $C_{\mathbf{n}} = C_{\overline{\mathbf{n}}}$. It is now easy to deduce the Dynkin index of the adjoint representation $\mathbf{n^2 - 1}$ which appears in the product

$$\mathbf{n} \otimes \overline{\mathbf{n}} = (\mathbf{n^2 - 1}) \oplus \mathbf{1} \ . \tag{1.112}$$

The composition law yields

$$C_{\mathbf{n^2-1}} = C_{\mathbf{n}\otimes\bar{\mathbf{n}}} = n\frac{1}{2} + n\frac{1}{2} = n \ . \tag{1.113}$$

One can proceed to build the indices of other representations using this technique. This graphical construction can be extended to diagrams with more than two wavy lines. Consider the case of three wavy lines. This corresponds to the symmetric product of three adjoint representations; algebraically

$$\mathrm{Tr}\,(\mathbf{T}_\mathbf{r}^A \mathbf{T}_\mathbf{r}^B \mathbf{T}_\mathbf{r}^C) = \mathcal{A}_\mathbf{r} d^{ABC} \ . \tag{1.114}$$

For most Lie algebras, it is zero, because the symmetric product of three adjoint representations does not contain an invariant. It is only for the unitary series $SU(n)$ with $n \geq 3$ that it does not vanish (as well as for the Abelian algebra $U(1)$). It is a number, called the anomaly number associated with every representation. Our graphical rules give us an easy way to derive its composition law

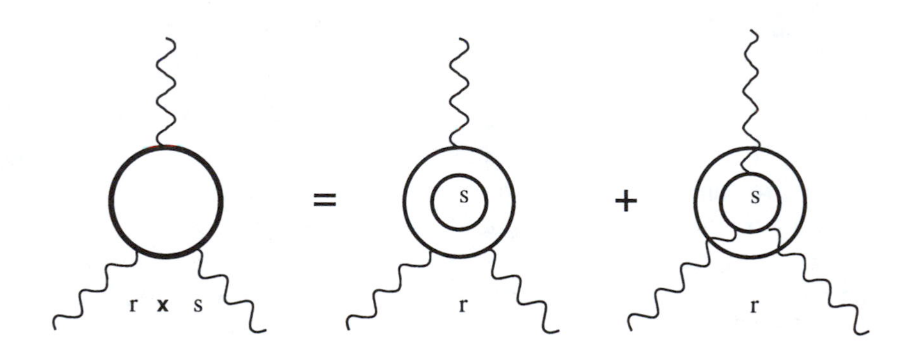

One uses the fact that all three wavy lines must connect to the same internal line, to find the composition law for the symmetric and antisymmetric products of representations (see problem). Applying this to $SU(3)$,

$$\mathbf{3} \otimes \mathbf{3} = \bar{\mathbf{3}}_A \oplus \mathbf{6}_S \ , \tag{1.115}$$

it is easy to show that the anomaly of the triplet is minus that of the antitriplet. We can proceed to deduce that the anomaly of the real adjoint representation is zero. In fact these results generalize: the anomaly of any representation is equal to minus that of its conjugate, and thus it is zero for real representations.

One may apply this construction to diagrams with many external adjoint lines. We leave this as an exercise. The nice feature is that one gets composition rules for higher indices that are independent of the algebra.

While we have proceeded by giving salient examples, it is clear that a more systematic approach to the study of the representations of Lie algebras is warranted. While many techniques can be found in the literature, they are often patterned after a particular algebra,. Below we present the method of Dynkin, as the most general and simplest way to study the representations of Lie algebras.

Dynkinese

The essence of this approach is to summarize the relevant properties of all Lie Algebras, using only two-dimensional diagrams, called Dynkin diagrams.

We present only the necessary mathematical facts, leaving proofs to more learned texts. Any Lie algebra of rank l contains l commuting generators, H_i, $i = 1, \ldots, l$,

$$[H_i, H_j] = 0 \ , \qquad H_i^\dagger = H_i \ . \tag{1.116}$$

The remaining $(d_{\mathbf{a}} - l)$ generators of the algebra split in two conjugate groups. These are the $E_{\boldsymbol{\alpha}}$, labelled by a $l-$dimensional *root* vector,

$$\boldsymbol{\alpha} = (\alpha_1, \alpha_2, \ldots, \alpha_l) \ , \tag{1.117}$$

which satisfy

$$[H_i, E_{\boldsymbol{\alpha}}] = \alpha_i E_{\boldsymbol{\alpha}} \ . \tag{1.118}$$

The remaining are their conjugates

$$E_{\boldsymbol{\alpha}}^\dagger = E_{-\boldsymbol{\alpha}} \ , \tag{1.119}$$

so that $-\boldsymbol{\alpha}$ is also a root. We have the further commutation relations

$$[E_{\boldsymbol{\alpha}}, E_{-\boldsymbol{\alpha}}] = \alpha^i H_i \ , \tag{1.120}$$

where the α^i are related to the components of the root vector by the metric g_{ij},

$$\alpha^i \equiv g^{ij}\alpha_j \ . \qquad g^{ij}g_{jk} = \delta^i_k \tag{1.121}$$

Finally, we have

$$[E_{\boldsymbol{\alpha}}, E_{\boldsymbol{\beta}}] = \begin{cases} 0 & \text{if } \boldsymbol{\alpha} + \boldsymbol{\beta} \text{ is not a root }, \\ N_{\boldsymbol{\alpha\beta}} E_{\boldsymbol{\alpha}+\boldsymbol{\beta}} & \text{if } \boldsymbol{\alpha} + \boldsymbol{\beta} \text{ is a root }. \end{cases} \tag{1.122}$$

Thus a Lie algebra is characterized by its root vectors, the metric function, and the coefficients $N_{\boldsymbol{\alpha\beta}}$.

To see that this abstract notation is actually very physical, consider $SU(2) = A_1$, with its generators represented by the Pauli spin matrices. The rank is one, and we identify

$$H_1 = \mathbf{T}^3 = \frac{1}{2}\begin{pmatrix} 1 & 0 \\ 0 & -1 \end{pmatrix} ,$$

as well as

$$E_\alpha = \mathbf{T}^1 + i\mathbf{T}^2 = \begin{pmatrix} 0 & 1 \\ 0 & 0 \end{pmatrix} , \qquad E_{-\alpha} = \mathbf{T}^1 - i\mathbf{T}^2 = \begin{pmatrix} 0 & 0 \\ 1 & 0 \end{pmatrix} .$$

The commutation relations are

$$[H_1, E_{\pm\alpha}] = \pm E_{\pm\alpha} , \tag{1.123}$$

so that there are two one-dimensional roots, one is positive, $\alpha = 1$, and the other negative, $\alpha = -1$. Physicists recognize E_α as the ladder operator which raises the magnetic quantum number by one unit. Since

$$[E_\alpha, E_{-\alpha}] = 2H_1 , \tag{1.124}$$

the metric is g_{ij} is just $\delta_{ij}/2$. This easily generalizes to our next example $SU(3) = A_2$. Here we have two commuting generators, H_1 and H_2. In the **3** representation, they are given by the diagonal Gell-Mann matrices

$$H_1 = \frac{1}{2}\begin{pmatrix} 1 & 0 & 0 \\ 0 & -1 & 0 \\ 0 & 0 & 0 \end{pmatrix} , \qquad H_2 = \frac{1}{2\sqrt{3}}\begin{pmatrix} 1 & 0 & 0 \\ 0 & 1 & 0 \\ 0 & 0 & -2 \end{pmatrix} .$$

The other generators are given by matrices which have only one non-zero entry in the off-diagonal position; there are six such generators, three represented by upper diagonal (nilpotent) matrices,

$$E_{\boldsymbol{\alpha}_1} = \begin{pmatrix} 0 & 1 & 0 \\ 0 & 0 & 0 \\ 0 & 0 & 0 \end{pmatrix} , \; E_{\boldsymbol{\alpha}_2} = \begin{pmatrix} 0 & 0 & 1 \\ 0 & 0 & 0 \\ 0 & 0 & 0 \end{pmatrix} , \; E_{\boldsymbol{\alpha}_3} = \begin{pmatrix} 0 & 0 & 0 \\ 0 & 0 & 1 \\ 0 & 0 & 0 \end{pmatrix} .$$

The other three, represented by lower diagonal matrices, correspond to the negative roots $E_{-\alpha_i}$. The commutation relations yield the three root vectors

$$\boldsymbol{\alpha}_1 = (1,0)\ , \qquad \boldsymbol{\alpha}_2 = (\frac{1}{2}, \frac{\sqrt{3}}{2})\ , \qquad \boldsymbol{\alpha}_3 = (-\frac{1}{2}, \frac{\sqrt{3}}{2})\ , \tag{1.125}$$

with the other three roots given by their negatives. A simple computation shows that the metric is also one-half times the unit matrix. It is instructive to plot these roots in the $H_1 - H_2$ plane

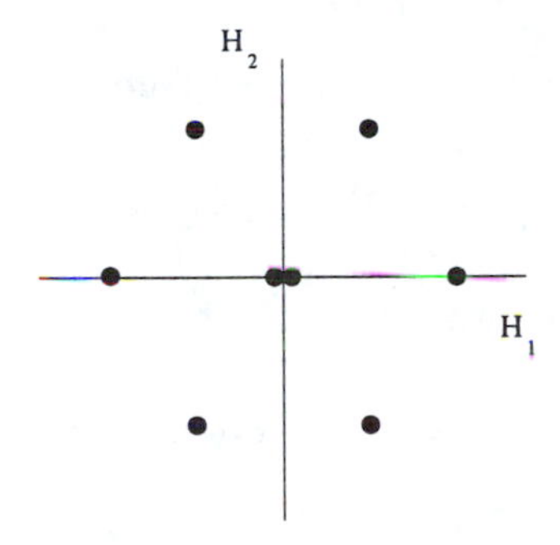

The choice of a convenient basis to express these roots is not obvious. Since half of the roots are the negative of the others, let us first split them in positive roots and negative roots. We define *positive roots* as those for which the first non-zero component is positive. Clearly the three positive roots $\boldsymbol{\alpha}_1$, $\boldsymbol{\alpha}_2$, and $-\boldsymbol{\alpha}_3$ are not all independent, and we can express

$$\boldsymbol{\alpha}_1 = \boldsymbol{\alpha}_2 + [-\boldsymbol{\alpha}_3]\ , \tag{1.126}$$

in such a way that the expansion coefficients are positive (and in this case equal) integers. The two roots $\boldsymbol{\alpha}_2$, and $-\boldsymbol{\alpha}_3$, are called the *simple* roots. There are as many simple roots as the rank of the algebra; they form a basis, called the α-basis, for the root space. Their properties characterize the algebra. In $SU(3)$, for example, the two simple roots have the same length, and from

$$g^{ij}(\alpha_2)_i(-\alpha_3)_j \equiv \boldsymbol{\alpha}_2 \cdot [-\boldsymbol{\alpha}_3] = -1\ , \tag{1.127}$$

the angle between them is 120°, as can easily be seen from the root diagram above. All this information about the simple roots is contained in the 2×2 *Cartan matrix*

$$A_{ij} \equiv 2\frac{\boldsymbol{\alpha}_i \cdot \boldsymbol{\alpha}_j}{\boldsymbol{\alpha}_i \cdot \boldsymbol{\alpha}_i} , \tag{1.128}$$

which for $SU(3)$ is just

$$\begin{pmatrix} 2 & -1 \\ -1 & 2 \end{pmatrix} .$$

This matrix is clearly independent of the basis in which we have expressed the simple roots. We can introduce another convenient basis, called the *ω-basis*, defined by the relations

$$2\frac{\boldsymbol{\omega}_i \cdot \boldsymbol{\alpha}_j}{\boldsymbol{\alpha}_i \cdot \boldsymbol{\alpha}_i} = \delta_{ij} . \tag{1.129}$$

In the case of $SU(3)$, the $\boldsymbol{\omega}$ vectors are given by

$$\boldsymbol{\omega}_1 = (\frac{1}{2}, \frac{1}{2\sqrt{3}}) , \qquad \boldsymbol{\omega}_2 = (0, \frac{1}{\sqrt{3}}) . \tag{1.130}$$

The relation between the two bases can be written in the form

$$\boldsymbol{\alpha}_1 = 2\boldsymbol{\omega}_1 - \boldsymbol{\omega}_1 , \qquad \boldsymbol{\alpha}_2 = -\boldsymbol{\omega}_1 + 2\boldsymbol{\omega}_2 , \tag{1.131}$$

so that the expansion coefficients are the rows of the Cartan matrix.

Finally we mention a third basis, called the *orthonormal basis*, in which the simple roots are conveniently expressed. Introduce three orthonormal vectors in a 3-dimensional space, $\mathbf{e}_i$, $i = 1, 2, 3$, such that $(\mathbf{e}_i, \mathbf{e}_j) = \delta_{ij}$. Then we have

$$\boldsymbol{\alpha}_1 = \mathbf{e}_1 - \mathbf{e}_2 , \qquad \boldsymbol{\alpha}_2 = \mathbf{e}_2 - \mathbf{e}_3 . \tag{1.132}$$

Let us apply this notation to the case of $SU(4) = A_3$. This rank-three algebra has three commuting generators, represented in its fundamental complex 4-dimensional representation by the traceless diagonal matrices,

$$H_1 = \frac{\text{diag}}{2}(1, -1, 0, 0) , \quad H_2 = \frac{\text{diag}}{2\sqrt{3}}(1, 1, -2, 0) , \quad H_3 = \frac{\text{diag}}{2\sqrt{6}}(1, 1, 1, -4) .$$

There are twelve ladder operators associated with the six root vectors and their negatives. The first three are the same as in $SU(3)$,

$$\boldsymbol{\alpha}_1 = (1, 0, 0) , \qquad \boldsymbol{\alpha}_2 = (\frac{1}{2}, \frac{\sqrt{3}}{2}, 0) , \qquad \boldsymbol{\alpha}_3 = (-\frac{1}{2}, \frac{\sqrt{3}}{2}, 0) ,$$

with the rest given by

$$\boldsymbol{\alpha}_4 = (\frac{1}{2}, \frac{1}{2\sqrt{3}}, \sqrt{\frac{2}{3}}) \, , \boldsymbol{\alpha}_5 = (-\frac{1}{2}, \frac{1}{2\sqrt{3}}, \sqrt{\frac{2}{3}}) \, , \boldsymbol{\alpha}_6 = (0, -\frac{1}{\sqrt{3}}, \sqrt{\frac{2}{3}}) \, .$$

The six positive roots are $\boldsymbol{\alpha}_1$, $\boldsymbol{\alpha}_2$, $-\boldsymbol{\alpha}_3$, $\boldsymbol{\alpha}_4$, $-\boldsymbol{\alpha}_5$, $-\boldsymbol{\alpha}_6$. Out of these we pick three simple roots

$$-\boldsymbol{\alpha}_3 \; , \; \boldsymbol{\alpha}_4 \; , -\boldsymbol{\alpha}_6 \; . \tag{1.133}$$

As for $SU(3)$, the remaining positive roots are linear combinations of these with positive (and in this case also equal) integer coefficients:

$$\boldsymbol{\alpha}_1 = [-\boldsymbol{\alpha}_3] + \boldsymbol{\alpha}_4 + [-\boldsymbol{\alpha}_6] \, , \boldsymbol{\alpha}_2 = \boldsymbol{\alpha}_4 + [-\boldsymbol{\alpha}_6] \, , [-\boldsymbol{\alpha}_5] = [-\boldsymbol{\alpha}_3] + [-\boldsymbol{\alpha}_6] \; .$$

Other properties are also similar: they have the same length, and the angle between $\boldsymbol{\alpha}_4$ and -$\boldsymbol{\alpha}_6$, $-\boldsymbol{\alpha}_3$ and $-\boldsymbol{\alpha}_6$ is 120°; however, the angle between $\boldsymbol{\alpha}_4$ and -$\boldsymbol{\alpha}_6$ is 90°. For $SU(4)$, the Cartan matrix is then

$$\begin{pmatrix} 2 & -1 & 0 \\ -1 & 2 & -1 \\ 0 & -1 & 2 \end{pmatrix} .$$

The ω-basis vectors are related to the α-basis by

$$\begin{aligned} \boldsymbol{\alpha}_1 &= 2\boldsymbol{\omega}_1 - \boldsymbol{\omega}_2 \; , \\ \boldsymbol{\alpha}_2 &= -\boldsymbol{\omega}_1 + 2\boldsymbol{\omega}_2 - \boldsymbol{\omega}_3 \; , \\ \boldsymbol{\alpha}_3 &= -\boldsymbol{\omega}_2 + 2\boldsymbol{\omega}_3 \; . \end{aligned} \tag{1.134}$$

The orthogonal basis spans a four-dimensional space with

$$\boldsymbol{\alpha}_1 = \mathbf{e}_1 - \mathbf{e}_2 \; , \qquad \boldsymbol{\alpha}_2 = \mathbf{e}_2 - \mathbf{e}_3 \; , \qquad \boldsymbol{\alpha}_3 = \mathbf{e}_3 - \mathbf{e}_4 \; . \tag{1.135}$$

By now the pattern should be obvious. The Cartan matrix contains a lot of information, and Dynkin devised a convenient way to describe its properties by means of a two-dimensional diagram. There, a simple root is entered as a dot. Two simple roots at 120° are joined by a single line. Dots not directly connected turn out to be at at 90° to one another. These rules produce the *Dynkin diagram* for the $SU(n+1)$ algebras, as a linear chain of dots connected by a single line.

o SU$_2$
1

o—o SU$_3$
1 2

o—o—o SU$_4$
1 2 3

The symmetric $n \times n$ Cartan matrix associated with the Lie algebra $SU(n+1)$ has integer or zero coefficients, given by

$$\begin{pmatrix} 2 & -1 & 0 & \cdot & \cdot & \cdot & 0 \\ -1 & 2 & -1 & \cdot & \cdot & \cdot & 0 \\ 0 & -1 & 2 & -1 & \cdot & \cdot & 0 \\ \cdot & \cdot & \cdot & \cdot & \cdot & \cdot & \cdot \\ 0 & 0 & 0 & \cdot & \cdot & -1 & 2 \end{pmatrix} .$$

Notice that most simple roots are perpendicular to one another.

Dynkin found that the same procedure could be generalized, allowing him to draw a specific Dynkin diagram for each Lie algebra.

This is made possible by the following facts about simple roots which we quote without proof:

– The angle between two simple roots can only be 90°, 120°, 135°, or 150°.

– Two simple roots at 120° have the same length.

– Two simple roots at 135° have a $1 : \sqrt{2}$ length ratio.

– Two simple roots at 150° have a $1 : \sqrt{3}$ their length ratio.

– All positive roots can be expressed as linear combinations of the simple roots with positive integer coefficients.

These facts about Lie algebras motivated Dynkin to associate to any Lie algebra, two-dimensional diagrams, according to the following rules:

–Associate with a short simple root a filled dot: •.

–Associate with a long simple root an empty dot: ∘.

–Connect the dots of two simple roots by by a single line if they are at 120° .

–Connect the dots of two simple roots by two lines if they are at 135°.

–Connect the dots of two simple roots by three lines if they are at 150°.

–Two simple roots not directly connected to one another are necessarily perpendicular.

This produces the following *Dynkin diagrams* for all the semi-simple Lie Algebras:

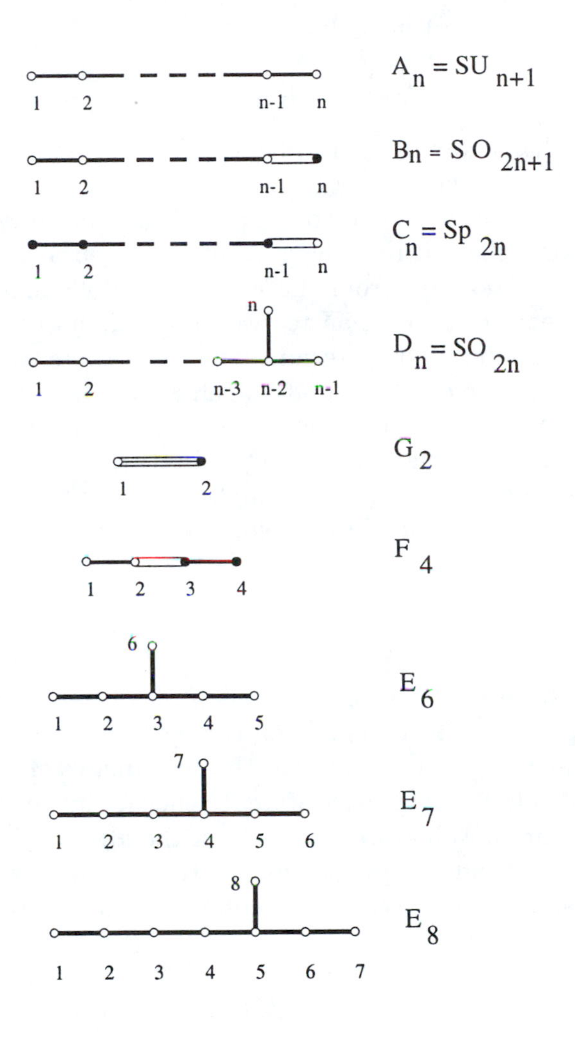

It is straightforward to simply read-off the Cartan matrix from the Dynkin diagram. Another immediate consequence is that there are many redundant diagrams for low rank algebras. In particular we see that there is only one algebra of rank one, since its Dynkin diagram is just one dot. This yields the following equivalences between Lie algebras

$$A_1 = B_1 = C_1 \ . \tag{1.136}$$

Although these Lie algebras are strictly equal, the groups they generate are not. For instance, the spinor representation of the rotation group is double valued, so that we say $SO_3 = SU_2/Z_2$. Finally for algebras of rank two and three, we have the equivalences

$$B_2 = C_2 \ ; \qquad D_2 = A_1 + A_1 \ ; \qquad A_3 = D_3 \ . \tag{1.137}$$

We also note that many diagrams display discrete symmetries; their significance will become clear later.

However, the information displayed by these diagrams does not allow us to reconstruct the full root diagram of the algebra. For instance, the Dynkin of $SU(3)$ tells us, from the number of dots that there are two roots of zero length, in addition to two simple roots of equal length at 120° from one another. The determine the other roots, we only have two facts: 1-) the negative of every root is itself a root, and 2-) the remaining positive roots are linear combinations of the two simple roots with integer coefficients of the same sign, but the Dynkin diagram does not tell us the values of these coefficients, nor the number of positive roots. Additional information must be provided to develop an algorithm to determine these coefficients.

Representations

Consider any state within a representation of a Lie algebra of rank l. It can be labelled by a set of l numbers, the eigenvalues of the commuting generators, which can be thought of as the coordinates of an l-dimensional vector, $\boldsymbol{\Lambda}$, called the *weight* of the state. In the adjoint representation, the weight vectors are nothing but the roots of the algebra.

The action of a ladder operator associated with a root $\boldsymbol{\alpha}$ on any state of weight $\boldsymbol{\Lambda}$ simply produces a different state labelled by a new weight vector

$$\boldsymbol{\Lambda}' = \boldsymbol{\Lambda} + \boldsymbol{\alpha}. \tag{1.138}$$

As a result, we can think of any representation as a collection of weights, differing from one another by the roots of the algebra. We can express the weights in any basis we choose. In terms of the basis of simple roots, we have

$$\boldsymbol{\Lambda} = 2\sum_{i=1}^{l} \overline{\lambda}_i \frac{\boldsymbol{\alpha}_i}{\boldsymbol{\alpha}_i \cdot \boldsymbol{\alpha}_i} , \tag{1.139}$$

where the $\overline{\lambda}_i$ describe the weight in the *dual* basis. The factor of 2 is traditional. Dynkin introduced a different basis with components

$$a_i = 2\frac{\boldsymbol{\Lambda} \cdot \boldsymbol{\alpha}_i}{\boldsymbol{\alpha}_i \cdot \boldsymbol{\alpha}_i} , \tag{1.140}$$

from which we see that

$$\boldsymbol{\Lambda} = \sum_{i=1}^{l} a_i \boldsymbol{\omega}_i , \tag{1.141}$$

which are the expansion coefficients of the weight in the ω basis. We also have

$$a_i = 2\sum_{i=1}^{l} \overline{\lambda}_j \frac{A_{ji}}{\boldsymbol{\alpha}_j \cdot \boldsymbol{\alpha}_j} . \tag{1.142}$$

For the jth simple root, $\overline{\lambda}_i = \delta_{ij}$: the components of the simple roots in the Dynkin basis are integers since they are just the rows of the Cartan matrix. The other roots, being combinations of these with integer coefficients, have themselves integer components in this basis. In the Dynkin basis, all weights in the adjoint representation have integer coordinates.

It can be shown that this is true for all representations, that is *all* weights have integer coefficients in the Dynkin basis. It follows that any state within a unitary irreducible representation can be labelled by a set of l integers, $(a_1, a_2, \ldots, a_k, \ldots, a_l)$: any weight can be represented by integers associated with the dots of the Dynkin diagram. We need only agree on a numbering convention to order these integers according to the sequences of the Dynkin diagrams.

This labelling does not tell us *which* representation the state belongs to; for example, the Dynkin label of states for spin $SU(2)$ is just twice the magnetic quantum number, and we know that states with Dynkin label (1), meaning $j_3 = 1/2$, occur in all half-integer spin representations. This example gives us the clue to identify the representation, as each representation of $SU(2)$ has only one state which has the maximum allowed value of the magnetic quantum number. It is the state $|j, j_3 = j\rangle$ that is annihilated by half the ladder operators. This feature generalizes to all unitary representations: each has exactly one state where the commuting generators have their maximum allowed eigenvalues, which are zero or positive . This state is called the *highest weight state*; it *uniquely* labels unitary irreducible representations with l *positive or zero* integers.

Conversely, any set of positive (or zero) integers uniquely labels a representation of the algebra. In the weight space, these weights are confined to being inside (or at the boundary if some highest weight components are zero) a region bounded by the l vectors $\boldsymbol{\omega}_i$. The inside of this region is called the *fundamental Weyl chamber* of the weight space. For $SU(3)$, it is the sliver bounded by the two vectors $(1/2, 1/2\sqrt{3})$ and $(0, 1/\sqrt{3})$. This region of weight space can be mapped into the whole weight space by the action of a discrete group of transformations, called the *Weyl* group. Each Lie algebra has its own characteristic Weyl group. For $SU(3)$, it is S_3, the permutation group on three objects. Its action can be described in terms of reflections about the ω-basis vectors. Any weight inside the fundamental Weyl chamber is called a *dominant* weight. Under the action of the Weyl group, it is taken into a finite number of weights all outside the fundamental chamber (since they are reflections about its boundaries). This set of weights is called the *Weyl orbit*, which is a closed in weight space. Thus we may view the whole weight space as an infinite number of non-intersecting Weyl orbits, each with one dominant weight in the fundamental chamber. For example the adjoint (octet) irrep of $SU(3)$ has highest weight state with Dynkin label $(1,1)$. Under S_3, this weight goes around its Weyl orbit of six weights. The other two states of the adjoint have Dynkin label $(0,0)$; they are at the origin of the Weyl cone, and are their own orbit. We can infer from this example the general anatomy of any irrep: it contains its highest weight state *once*, together with its Weyl orbit; it also contains a set of dominant weights, with their Weyl orbits; there can be several copies of the same dominant weight within one irrep. Thus in order to compute the content of the irrep, we need only know its dominant weights and their multiplicities. This method for describing irreps is favored in the mathematical literature, and we mention it to add to the (not inconsiderable) erudition of the reader.

We now proceed with our description of irreps along more conventional lines, essentially by acting the lowering operators on the highest weight state. This is done by subtracting from it the positive simple roots until one reaches either the negative of the highest weight for a real representation, or the negative of its conjugate for a complex representation. A weight that is obtained by k subtractions from the highest weight, is said to be at the kth level of the representation. The number of levels is called the *height* of the representation. The height of any representation can be computed, using the *level vector* $\mathbf{R} = (R_1, R_2, \ldots, R_l)$ which depends on the algebra, not the representation. It is given by

$$T = \sum_{i=1}^{l} R_i a_i \ , \tag{1.143}$$

where $(a_1\ a_2 \cdots\ a_l)$ is the highest weight, used to label a representation of a Lie algebra. The level vectors for the different Lie algebras are given by

$$\begin{aligned}
&A_n : [n, 2(n-1), \ldots, (n-2)3, (n-1)2, n] \ , \\
&B_n : [2n, 2(2n-1), \ldots, (n-1)(n+2), n(n+1)/2] \ , \\
&D_n : [2n-2, 2(2n-3), .., (n-2)(n+1), n(n-1)/2, n(n-1)/2] \ , \\
&C_n : [2n-1, 2(2n-2), \ldots, (n-1)(n+1), n^2] \ , \\
&G_2 : [10, 6] \ , \qquad F_4 : \ [22, 42, 30, 16] \ , \\
&E_6 : [16, 30, 42, 30, 16, 22] \ , \qquad E_7 : \ [34, 66, 96, 75, 52, 27, 49] \ , \\
&E_8 : [92, 182, 270, 220, 168, 114, 58, 136] \ .
\end{aligned} \tag{1.144}$$

Note that for $SO(2n)$ ($SO(2n+1)$), the last two (one) components break the pattern.

To make the construction of representations even simpler, Dynkin proved that the distribution of weights in any representation is spindle-shaped so that the number of weights at the kth level is the same as that of the $(T-k)$th level.

By definition, adding any positive simple root to the highest weight state does not produce another weight, while substracting a simple root does yield another state. For example the action of J_- on $|j, j\rangle$ yields, up to normalization, the state $|j, j-1\rangle$. In the Dynkin language, this translates as: the Cartan matrix, 2, gives the value of the one dimensional simple root, thus starting with the representation with highest weight (1), we obtain another weight within the representation by substracting the simple root, yielding (-1), the negative of the simple root; it is of course the state with $j_3 = -\frac{1}{2}$, and this yields the two-state spinor representation.

Starting with the representation $(2j)$, we generate the state $(2j - 2k)$, by substracting the simple root k times until we reach the negative $(-2j)$. The Dynkin label of the adjoint of $SU(2)$ is clearly (2).

To reconstruct the root system of any algebra, we need to know, besides the Dynkin diagram, the Dynkin label (the highest weight state) of its adjoint representation. The other states are then generated by subtracting the positive simple roots. The Dynkin labels (highest weight state) of the adjoint of each Lie algebras are:

$$
\begin{array}{lll}
\mathrm{A}_l : & (1\,0\,0\,\ldots\,0\,0\,1) & l \neq 1\,, \\
\mathrm{B}_l : & (0\,1\,0\,\ldots\,0\,0\,0) & \\
\mathrm{C}_l : & (2\,0\,0\,\ldots\,0\,0\,0) & \\
\mathrm{D}_l : & (0\,1\,0\,\ldots\,0\,0\,0) & \\
\mathrm{G}_2 : & (1\,0) & \\
\mathrm{F}_4 : & (1\,0\,0\,0) & \\
\mathrm{E}_6 : & (0\,0\;\;0\,0\,0\,1) & \\
\mathrm{E}_7 : & (1\,0\,0\,0\,0\,0\,0) & \\
\mathrm{E}_8 : & (0\,0\,0\,0\,0\,0\,1\,0) &
\end{array}
$$

We are now in a position to construct the states within a given representation. The crucial element is that any state with a weight with positive Dynkin coordinate in the ith entry, a_i, has a_i quanta of the ith root, which can then be subtracted that many times.

Consider the algebra $SU(4)$. Its simple roots are the rows of the Cartan matrix, $\boldsymbol{\alpha}_1 = (2,-1,0)$, and $\boldsymbol{\alpha}_2 = (-1,2,-1)$, $\boldsymbol{\alpha}_3 = (0,-1,2)$. Its level vector is $\mathbf{R} = (3,6,3)$. Let us start with its simplest representation with Dynkin label (100). Its level is therefore $T = 3$. The rule is that for any state of weight $(a_1 a_2 \ldots a_k \ldots a_l)$, the simple root $\boldsymbol{\alpha}_k$ can be subtracted k times, corresponding to the k-fold application of an annihilation operator, as long as a_k is positive. Hence we can subtract $\boldsymbol{\alpha}_1$ once from (100), obtaining $(-1\;1\;0)$, from which we can subtract $\boldsymbol{\alpha}_2$ once, obtaining $(0\,-1\;1)$, from which we can subtract $\boldsymbol{\alpha}_3$ once, yielding $(0\;0\,-1)$. Since there are no more positive Dynkin we are done. This representation has four states; it is the **4** of $SU(4)$. A similar procedure applied to the representation (001) yields another four-dimensional representation. It is the conjugate $\overline{\mathbf{4}}$.

Now for (101), the adjoint representation of $SU(4)$. It has six levels. We can subtract $\boldsymbol{\alpha}_{1,3}$ once, obtaining two states at the first level: $(-1\;1\;1)$ and $(1\;1{-}1)$. Similar subtractions yield at the next level the states $(0{-}1\;2)$, $(-1\;2-1)$, $(-1\;2-1)$, $(2-1\;0)$. These are the simple roots, but one of them appears twice. Clearly there is only one state associated with that simple root, and we only have three states at this level. The next level yields the three roots of zero length. Further subtraction reproduces the states but with negative entries. Hence we have $1+2+3+3+3+2+1 = 15$ states, corresponding to the generators of $SU(4)$.

The beauty of the Dynkin notation is that it applies equally well to a familiar algebra like $SU(4)$ as for an unfamiliar one like E_7.

There are many other advantages in describing representations by their Dynkin labels. Consider again the case of $SU_4 \approx SO_6$. We have seen that its lowest representation is the four dimensional complex vector, **4**, represented in Dynkinese by (100), while its complex conjugate $\overline{\mathbf{4}}$ is represented by (001). We note that complex conjugation corresponds to a symmetry of the Dynkin diagram: reflection about the vertical. The smallest real (self-conjugate) representation is therefore the (010). It is nothing but the six-dimensional vector representation of SO_6! In $SU(4)$ language, it corresponds to a second rank antisymmetric tensor, with the quantum numbers of the Kronecker product $\mathbf{4} \otimes \mathbf{4}_A = \mathbf{6}$.

Representations can be multiplied together to yield new representations. In general the product of two representations will contain the sum of several irreducible representations. However it will always contain a state of highest eight equal to the sum of the highest weights of the product representations. For the product of low-lying representations, that sum can be determined by the fact that the sum of their dimensions is equal to the product of the dimensions of the factor representations, and that the sum of their Dynkin indices is equal to the dimension of one factor representation times the Dynkin of the other, plus the other way around. In some cases, the product representation will be irreducible: the p-times antisymmetric product of the fundamental of SU_{n+1} always yields an irreducible representation, which can be written as a totally antisymmetric tensor of rank p. In Dynkinese, it is labelled with zeros except for a one at the pth position.

The product of two representations always contains one representation which is labeled by the sum of their Dynkin labels. For instance, in SU_4, the adjoint representation is just (101); note that it is symmetric under inversion and therefore real, as required.

There are two more types of Dynkin diagrams with interesting symmetries. The D_n Dynkin is symmetric under the interchange of the two dots at the n and $n-1$ positions. Representations with a one at one of these positions and zeros elsewhere are the spinor representations, and they are indeed complex; under conjugation, they flip into one another. For instance, in $D_5 = SO_{10}$, the complex spinor representation, **16** is represented by (00010), while its conjugate, $\overline{\mathbf{16}}$ is just (00001). On the other hand, the representation (10000) is manifestly real; it corresponds to the ten-dimensional vector. The representation (01000) represents the antisymmetric second rank tensor, and (00100) is the three-times antisymmetric tensor.

The upscale reader may have noticed that the Dynkin diagram for $SO_8 = D_4$ is just the Mercedes-Benz symbol which has a three-fold symmetry. There the (100) labels the eight-dimensional vector representation, while (010) and (001) label two eight-dimensional spinors. There is a special triality symmetry between these three representations, which has dramatic consequences in the formulation of superstring theories.

There is one more diagram which has a special symmetry. E_6 does indeed have complex representations. For instance, the complex **27** is written in Dynkinese as (100000), while its conjugate, $\overline{\mathbf{27}}$ is (000010). Groups with complex representations play a special role because they are the only ones that can describe the complex Weyl spinors that appear in the theory of weak interactions.

PROBLEMS

A. Using graphical methods, show that the Dynkin index of the symmetric product of two representations is equal to $(d_{\mathbf{r}} + 2)c_{\mathbf{r}}$.

B. Using graphical methods, show that the anomaly of the symmetric product of two representations is equal to $(d_{\mathbf{r}} + 4)A_{\mathbf{r}}$, where $d_{\mathbf{r}}$ and $A_{\mathbf{r}}$ are the dimension and anomaly of the representation. Deduce that
1-) the anomaly of the $\overline{\mathbf{3}}$ of $SU(3)$ is opposite that of the **3**.
2-) the anomaly of the sum of the two $SU(5)$ representations $\overline{\mathbf{5}} \oplus \mathbf{10}$ is free of anomalies, where **10** is the antisymmetric product of two **5**'s.

C. 1-) Using graphical methods, find the composition law for the trace of four representation matrices over a product of two representations.
2-) Explicitly construct the trace over four matrices in the fundamental and adjoint representation of $SU(2)$, and verify the composition law you have just derived.

D. Repeat the last part of problem C when the algebra is $SU(3)$.

E. Starting from its Dynkin diagram, find the Cartan matrix of G_2. Then work out and plot the roots of the algebra G_2, in the adjoint representation. The highest weight for the adjoint representation (10) (in the Dynkin basis), and the level vector has components $(10, 6)$.

F. There are many different ways to choose the two simple roots of $SU(3)$. Show that they are all related by the action of a discrete set of Weyl transformations, S_3, acting on the root diagram. Proceed to analyze the $(2, 0)$ irrep in terms of its Weyl orbits.

CHAPTER **2**

STANDARD MODEL

THE LAGRANGIAN

2.1 INTRODUCTION

The standard model of Elementary Particle Physics describes with amazing parsimony (only 19 parameters!) **all** known interactions over the scales that have been explored by experiments: from the Hubble radius of 10^{30} cm all the way down to scales of the order of 10^{-16} cm. Together with cosmological initial conditions when the universe was much tinier, the standard model is believed to encode the information necessary to deduce (in principle) all observed physical phenomena, life forms, etc...; its remarkable internal consistency is both a source of wonder and despair to those who seek to predict the future based on its flaws. Its structure does not run counter to any of our (rather incomplete) theoretical prejudices, and as a renormalizable model, it does not manifestly require new physics at a higher energy scale.

There are few obvious chinks in the armor of the standard model. Some are theoretical, suggested by the behavior of the standard model's parameters in the ultraviolet. Depending on the value of the mass of the Higgs particle, one can lose perturbative control in the Higgs self coupling close to experimental scales, suggesting the appearance of new interactions and/or new particles at those energies. The standard model also has Landau poles in its gauge and Yukawa couplings, but they typically occur at length scales much shorter than that at which quantum corrections to Einstein's theory of gravity are expected. In the absence of a satisfactory quantum theory of gravity, it is difficult to assess the meaning of these poles. On the other hand, the unification of the standard model to include gravity is the central conceptual problem of our time. Fortunately it need not concern us for the purposes of this book, as the scales of interest are at least seventeen orders of magnitudes smaller than the smallest scale hitherto explored! Only one candidate theory, superstring theory, addresses this question with any degree of success, but it only presents a qualitative picture which cannot yet be tested by experiment. The situation is reminiscent of the problem of the self-stress of the electron in classical electrodynamics: it tells you something is wrong with the theory, but it does not bring about an immediate confrontation with experiment.

The origin of different scales in nature is not answered by the standard model. There is but one fundamental constant with the dimension of length: Gravity comes with its own scale, the Planck length $\ell_{\rm Pl}$. All units of length should be scaled to it. In human terms, it corresponds to a very small distance indeed, 10^{-33} cm. In supernatural (Planck) units, the *Planckton?*, the length scales of the standard model are enormous:

$$\frac{\ell_{\rm N}}{\ell_{\rm Pl}} \sim \frac{10^{-13}}{10^{-33}} \sim 10^{20} \quad \text{and} \quad \frac{\ell_{\rm F}}{\ell_{\rm Pl}} \sim \frac{10^{-16}}{10^{-33}} \sim 10^{17} \ ,$$

where $\ell_{\rm N}, \ell_{\rm F}$ are the sizes associated with nuclear and weak interactions, respectively. It is a source of great intellectual worry that the standard model appears to be consistent at a scale that is so very different from the Planck scale. Naive expectations are for all phenomena to occur at their natural scale which, in fundamental physics, is the Planck scale. The understanding of these large numbers is probably the most fundamental question in microphysics today.

Today, there are no contradictions between particle physics experiments at accelerators and the standard model. Not only is it standing up to experiments in its gross features but also remarkably in numerous consistency checks that test its very quantum nature. A recent singular exception comes from non-accelerator experiments, which offers evidence for neutrino masses through flavor oscillations of neutrinos produced by cosmic rays. Further evidence from the measurements of neutrino fluxes from the Sun also suggest neutrinos masses through oscillations. This conflicts with the standard model in which the neutrinos which accompany the charged leptons in β decay are strictly massless. If these results are reproduced by newer neutrino detectors soon to be deployed, they provide the first true signal of physics beyond the standard model.

There may also be a possible conflict with cosmology. The standard model does contain the capability, through its anomaly, of generating a predominance of baryons over antibaryons in the universe. However, in its present formulation, it does not seem capable of reproducing the required amount of baryon asymmetry. This may be only a quantitative failure, tied to our incomplete understanding of the kinetics of the electroweak phase transition in the early universe.

Another arena of possible confrontation is the nature of dark matter. Observations indicate the presence of a dominant amount of non-luminous matter in the universe, so much so in fact that we should think of luminous matter as the foam that rides the crest of waves of dark matter in the cosmic ocean. The standard model, which contains no exotic matter, predicts all clustering neutral matter, dark and luminous, to be baryonic, while nucleosynthesis puts an upper bound on the amount of baryonic matter. As this bound seems to be exceeded by observation, the standard model must be extended to incorporate dark matter. This discussion does not include the prevailing theoretical prejudices based on inflation which predicts that the universe is at critical density, well above the nucleosynthesis bound. In this picture dark matter must be made up of type(s) of matter not in the standard model. While it is fair to say that the evidence for generalizing the standard model is most compelling in this domain, our knowledge of both the cosmological scenario before nucleosynthesis and of the late matter abundances in the universe are too nebulous to convincingly demonstrate the necessity of physics beyond the standard model.

Finally, in spite of its great success in confronting experiments, the standard model is not aesthetically satisfying: it contains particles of different spin, three disconnected gauge symmetries, two unexplained scales (nuclear size and neutron lifetime), as well as seventeen additional parameters, ranging over nine orders of magnitude.

Therefore it should not be too surprising if the standard model is not widely viewed as fundamental, but rather as the broken pieces of a more integrated structure. The hope of physicists is to glue its parts together into one theoretical construct with few if any undetermined parameters. Models for such a hypothetical structure abound in the theoretical literature, but they typically introduce many parameters, often more in fact than those they seek to explain! Although some of these models have definite theoretical appeal, their acceptance must await the verdict of experiments. While any proposal which reproduces even a *modicum* of already known facts can become an instant canditate for the Theory Of

Everything (TOE), we do not yet seem to have enough information to reconstruct the puzzle: the standard model is not sufficient to put Humpty-Dumpty back together again. It is tempting to think that such a unified structure may have emerged unscathed at the earliest moments after the Big Bang, only to be broken in the course of cosmological evolution into the chiral shards we observe today. Thus physicists who seek the TOE may be regarded as archeologists who probe our distant past when we were all in the same cosmic soup!

Since it is doubtful that direct experimental evidence can be obtained about physics at the Planck length, we have to restrain our theoretical ambitions. In this book, we consider extensions of the standard model which affect scales that are attainable by experiments in the near future. Hence, we present first the standard model, and then some of its minimal low-energy extensions, hoping to gather enough new shards to enable us to reconstruct the mother structure by the end of this book.

This chapter contains, after a short historical sketch, a self-contained summary of the standard model of Weak, Electromagnetic and Strong Interactions.

2.2 HISTORICAL PREAMBLE

Our presentation of the standard model does not follow historical lines, since its starting point is the standard model Lagrangian. Yet, its final form is the result of the inspired work of many experimentalists and theorists, over a period of seventy years or so. It is not possible to do justice to their contributions in the short description that follows. For detailed accounts, the reader is referred to various books and collection of reprints which are listed at the end of this book. Suffice it to say that the history of the standard model is remarkably rich as it mirrors the scientific effervescence of the natural sciences in this century.

The first third of the *XXth* century witnessed unparalleled scientific activities, spurred on by the dramatic experimental discoveries which resulted in the establishment of quantum mechanics, and general relativity. These set the stage for the intense bursts of experimental and theoretical breakthroughs, that led five decades later to the formulation of the standard model of the fundamental interactions. Below we present a decade by decade *esquisse* of these amazing developments.

• *1930's* In this decade many building blocks of the standard model were identified: its interactions, and its first family of elementary particles. At its onset, P. A. M. Dirac formulated one basic ingredient of the standard model, Quantum Electrodynamics, which describes the interaction of photons with matter. In 1933, E. Fermi generalized this work to include β decay, incorporating the neutrino, that had been postulated a few years earlier by W. Pauli, and the neutron, recently discovered by J. Chadwick. In 1935, H. Yukawa understood the short range of the strong interactions in terms of the exchange of a massive (elementary) particle. In 1937, a massive elementary particle was discovered in cosmic rays, but it interacted too weakly to be Yukawa's particle; it was the muon, the first member of the second family of elementary particles. The second half of the decade was spent confronting the theoretical difficulties associated with QED, but ended with O. Klein's remarkable anticipation of gauge theories as the root cause of β decay.

• *1940's* The computational rules of QED are understood by Feynman, Schwinger and Tomonaga, in terms of renormalization. Calculations and measurements of the anomalous magnetic moment of the electron and the Lamb shift provide spectacular evidence for their approach. A similar simplicity evades all formulations of the strong and weak interactions, although experiments begin to show impressive structures. Yukawa's true particle, the π meson is discovered in 1947, and soon after the first of a class of new particles, which we now call the K-mesons.

• *1950's* This is probably a spectacular decade in the history of the standard model. Its beginning saw the discovery of the Λ particle. These new strongly-interacting particles are always produced in pairs, with slow decay rates, features interpreted by M. Gell-Mann as evidence of *strangeness*, a new quantum number. The second half of the decade was just as impressive: neutrinos were detected in 1956 by C. Cowan and F. Reines. T. D. Lee and C. N. Yang proposed experiments to test parity violation by the weak interactions. Experimental verifications by C. S. Wu and by V. Telegdi soon followed. Soon thereafter, R. Marshak and E. C. G. Sudarshan and then R. Feynman and M. Gell-Mann identified the source of parity violation in terms of the $V - A$ vector and axial vector interactions. Spurred on by the 1954 work of C. N. Yang and R. L. Mills, theorists under the influence of J. Schwinger and others, soon realized the possibility that weak interactions could be caused by a massive vector particle, the W-boson.

• *1960's* This was the decade of great theoretical synthesis. First, the zoo of strongly interacting particles was organized in terms of $SU(3)$, by Y. Ne'eman and M. Gell-Mann at the beginning of the 1960's. A few years later, M. Gell-Mann and, independently G. Zweig, used this classification to postulate quarks as the building blocks of baryonic matter. The addition of spin to this synthesis by F. Gürsey and L. Radicati led to a new puzzle, interpreted by O. W. Greenberg in terms of parastatistics for hadrons. This led Y. Nambu in 1964 to propose that quarks interact with one another through a $SU(3)$ Yang-Mills theory. The beginning of this decade saw the emergence of the electroweak formulation of the standard model in its modern form, starting with the work of S.Glashow in 1961, then A. Salam and J.C. Ward, in 1964, and finally S. Weinberg in 1967. These works resulted in a model of the electroweak interactions of leptons, which after spontaneous breaking of the electroweak symmetry, reduces to QED. They predicted a new force, mediated by a neutral vector particle, the Z boson, and used ideas that originated in the BCS theory of superconductivity. Applied by Nambu to pion physics in 1961, and generalized by J. Goldstone, spontaneous breaking was shown by P. W. Higgs and R. Brout and F. Englert to generate massive gauge bosons from massless ones when the symmetry is gauged.

There were experimental surprises as well. The muon neutrino is discovered in 1962 by G. Danby et al in 1964, J. Cronin and V. Fitch discover CP violation. At the end of the decade a surprising scaling behavior (anticipated by J. Bjorken in 1966) is found in deep inelastic scattering experiments, implying the existence of hard constituents inside protons.

• *1970's* In 1970, S. Glashow, J. Iliopoulos and L. Maiani added theoretical evidence for a second family, by showing that a fourth *charm* quark could explain the absence of flavor-changing neutral interactions. This is followed by the theoretical breakthrough of G. 't Hooft, a student under the *aegis* of M. Veltman, who showed that the Glashow-Salam-Ward-Weinberg model was in fact renormalizable. C. Bouchiat, J. Iliopoulos and P. Meyer showed how quantum consistency required the existence of both quarks and leptons to cancel the Adler-Bell-Jackiw anomaly constraints. M. Gell-Mann and H. Fritzsch, interpreted the parastatistics as evidence for a new quantum number for quarks, *color*, and related it to the value of the π^0 lifetime. The modern version of quark dynamics, Quantum Chromodynamics, was finally formulated by M. Gell-Mann, H. Fritzsch and H. Leutwyler in 1973. G. 't Hooft, and then H. Politzer, and D. Gross and F. Wilczek showed that QCD was asymptotically-free, explaining the remarkable scaling results found by experiments in the late sixties. This showed that **all** interactions, strong, weak and electromagnetic were due to Yang-Mills vector exchange theories. In 1973, M. Kobayashi and T. Maskawa point out that a third family naturally implies CP-violation. In 1974, the J/Ψ charmed quark-antiquark bound state is discovered at

Brookhaven and SLAC, completing the first two families of elementary particles. But there is more: in 1975, M. Perl discovers the lepton member of the third family, the τ lepton. The *bottom* quark is soon thereafter discovered at FermiLab. Neutral current interactions are discovered in 1976 at CERN, and in 1978, the parity structure of the neutral current is determined to be that predicted in the *1960*'s. By 1979, the final formulation of the standard model was universally accepted as the paradigm of elementary particle physics, although many of its predictions still awaited discovery.

• *1980's* Compared with the *1970*'s, this decade seems anticlimactic, in spite of great experimental feats. It starts well enough with the discovery of the gauge bosons, the charged W and neutral Z bosons at CERN. Noteworthy are the measurements at CERN and SLAC of the width of the Z boson which limits the families to no more than the observed three. However, the rest of the decade is spent on further experimental verification of the standard model, and hunters for the three remaining particles predicted by the standard model, the top quark, the τ neutrino, and the Higgs boson, come up empty-handed, although the observed large $B - \bar{B}$ mixing hints at a very heavy top quark.

• *1990's* A systematic experimental study of the radiative structure of the standard model is begun at CERN and at SLAC. Agreement between experiment and theory continues to be spectacular. At FermiLab, the top quark is discovered in 1995, and strong kinematical evidence for the τ neutrino. at the end of this decade, the Higgs boson still eludes detection.

Today, the investigation of the standard model is almost over. Both theorists and experimentalist await with excitement the experimental exploration of scales at which the Higgs particle must manifest itself. The recent discovery of neutrino oscillations has put the first chink in its armor, tangibly indicating that there is more to the world than just the standard model.

2.3 THE LAGRANGIAN

The Lagrangian of the standard model is made up of four different parts, each named after illustrious scientists:

$$\mathcal{L}_{SM} = \mathcal{L}_{YM} + \mathcal{L}_{WD} + \mathcal{L}_{\text{Yu}} + \mathcal{L}_{H} \ . \tag{2.1}$$

The first is the Yang-Mills part $\mathcal{L}_{YM}$, which describe the low energy gauge groups of the standard model, SU_3 for color, SU_2 for weak isospin, and U_1 for hypercharge.

$$\begin{aligned}\mathcal{L}_{YM} &= \mathcal{L}_{QCD} + \mathcal{L}_{I_w} + \mathcal{L}_Y \ , \\ &= -\frac{1}{4g_3^2}\sum_{A=1}^{8} G^A_{\mu\nu}G^{\mu\nu A} - \frac{1}{4g_2^2}\sum_{a=1}^{3} F^a_{\mu\nu}F^{\mu\nu a} - \frac{1}{4g_1^2}B_{\mu\nu}B^{\mu\nu} \ , \end{aligned} \quad (2.2)$$

where g_1, g_2, g_3 are the dimensionless coupling constants, corresponding to color, weak isospin and hypercharge, respectively. The color field strengths are given by

$$G^A_{\mu\nu} = \partial_\mu A^A_\nu - \partial_\nu A^A_\mu - f^{ABC}A^B_\mu A^C_\nu, \qquad A, B, C = 1, ..., 8 \ , \quad (2.3)$$

with A^B_μ represent the eight gluon fields, and f^{ABC} are the structure functions of SU_3. The weak isospin field strengths

$$F^a_{\mu\nu} = \partial_\mu W^a_\nu - \partial_\nu W^a_\mu - \epsilon^{abc}W^b_\mu W^c_\nu, \qquad a, b, c = 1, 2, 3 \ , \quad (2.4)$$

are written in terms of the intermediate vector bosons W^a_μ and the SU_2 structure function ϵ^{abc}. Finally, the Abelian hypercharge field strength is

$$B_{\mu\nu} = \partial_\mu B_\nu - \partial_\nu B_\mu \ . \quad (2.5)$$

The second part of the Lagrangian, $\mathcal{L}_{WD}$, describes the fermion fields and their gauge interactions. The fermions in the standard model can be split into two categories, quarks which are triplets under the color gauge groups and leptons which have no color. Within each category some transform as weak doublets, some as weak singlets. Specifically, we have (all fermions are represented by two-component Weyl left-handed fields)

$$\text{weak doublet of leptons} : L_i = \begin{pmatrix} \nu_i \\ e_i \end{pmatrix}_L \sim (\mathbf{2}, \mathbf{1}^c)_{y_1} \ ,$$

$$\text{lepton weak singlets} : \bar{e}_{iL} \sim (\mathbf{1}, \mathbf{1}^c)_{y_2} \ ,$$

$$\text{quark weak doublet} : \mathbf{Q}_i = \begin{pmatrix} \mathbf{u}_i \\ \mathbf{d}_i \end{pmatrix}_L \sim (\mathbf{2}, \mathbf{3}^c)_{y_3} \ ,$$

$$\text{antiquark weak singlets} : \bar{\mathbf{u}}_{iL} \sim (\mathbf{1}, \bar{\mathbf{3}}^c)_{y_4} \ ,$$

$$\text{antiquark weak singlets} : \bar{\mathbf{d}}_{iL} \sim (\mathbf{1}, \bar{\mathbf{3}}^c)_{y_5} \ .$$

The notation denotes the color, isospin, and hypercharge assignments, in the form $(SU_2^W, SU_3^c)_Y$. The index i, ranges over the three families of chiral fermions. In terms of the matrices

$$\mathbf{W}_\mu = \frac{1}{2} W_\mu^a(x)\tau^a \ , \qquad \mathbf{A}_\mu = \frac{1}{2} A_\mu^A(x)\lambda^A \ ,$$

where τ^a and λ^A are the Pauli and Gell-Mann matrices for SU_2^W and SU_3^c, we can easily express the covariant derivatives on the various fermion fields

$$\mathcal{D}_\mu L_i = (\partial_\mu + i\mathbf{W}_\mu + \frac{i}{2} y_1 B_\mu) L_i \ ,$$

$$\mathcal{D}_\mu \bar{e}_i = (\partial_\mu + \frac{i}{2} y_2 B_\mu) \bar{e}_i \ ,$$

$$\mathcal{D}_\mu \mathbf{Q}_i = (\partial_\mu + i\mathbf{A}_\mu + i\mathbf{W}_\mu + \frac{i}{2} y_3 B_\mu) \mathbf{Q}_i \ ,$$

$$\mathcal{D}_\mu \bar{\mathbf{u}}_i = (\partial_\mu - i\mathbf{A}^*_\mu + \frac{i}{2} y_4 B_\mu) \bar{\mathbf{u}}_i \ ,$$

$$\mathcal{D}_\mu \bar{\mathbf{d}}_i = (\partial_\mu - i\mathbf{A}^*_\mu + \frac{i}{2} y_5 B_\mu) \bar{\mathbf{d}}_i \ .$$

The factor $\frac{1}{2}$ in front of the hypercharge field is conventional. This whole gauge structure is family independent. In terms of the above, the Weyl-Dirac Lagrangian is given by

$$\mathcal{L}_{WD} = \sum_i^3 \Big(L_i^\dagger \sigma^\mu \mathcal{D}_\mu L_i + \bar{e}_i^\dagger \sigma^\mu \mathcal{D}_\mu \bar{e}_i + \mathbf{Q}_i^\dagger \sigma^\mu \mathcal{D}_\mu \mathbf{Q}_i + \bar{\mathbf{u}}_i^\dagger \sigma^\mu \mathcal{D}_\mu \bar{\mathbf{u}}_i + \bar{\mathbf{d}}_i^\dagger \sigma^\mu \mathcal{D}_\mu \bar{\mathbf{d}}_i \Big) \ . \tag{2.6}$$

The partial Lagrangian $\mathcal{L}_{YM} + \mathcal{L}_{WD}$ displays large global symmetries. To start, there are several different family chiral symmetries, one for each set of fermions with the same quantum numbers. For example, the *global* transformation on the lepton doublet

$$L_i \to L_i' = U_{ij} L_j \ , \tag{2.7}$$

where U_{ij} is a 3×3 unitary matrix, which leaves the partial Lagrangian invariant. Since this invariance is true for all the different types of fermions we have the global family symmetry $U(3) \times U(3) \times U(3) \times U(3) \times U(3)$!

It is actually a wee bit smaller since two of the chiral U_1 currents are anomalous. We will address this question later on.

Much of this enormous global symmetry is explicitly broken by the Yukawa interactions of the standard model, by which fermion pairs interact with spinless particles. Let us therefore examine the quantum numbers of the two fermion operators and look for recurrences. The Lorentz scalar, color singlet, weak doublets occur in the combinations

$$\Delta I_{\mathrm{w}} = \frac{1}{2}: \qquad \widehat{\mathbf{Q}}\bar{\mathbf{u}}\ , \quad \widehat{\mathbf{Q}}\bar{\mathbf{d}}\ , \quad \widehat{L}\bar{e}\ , \tag{2.8}$$

where we have not shown the family indices, and have used the notation introduced in the previous chapter $\widehat{\eta} \equiv \eta^T \sigma_2$.

Classically, the hypercharge assignments are arbitrary, but at the quantum level they are constrained by the Adler-Bell-Jackiw anomaly, which spoils the renormalizability of the theory whenever it spoils the conservation of a gauged current. The Adler-Bardeen theorem asserts that the anomaly occurs only at $\mathcal{O}(\hbar)$ (one-loop order). It is therefore sufficient to demand that the relevant one loop triangle graphs vanish. The color group SU_3^c does not have any chiral anomaly, since there are as many (left-handed) quarks as antiquarks – it is vector-like. The weak isospin group SU_2^W is anomaly-free because of its structure: it has no d-coefficient. This leaves the hypercharge U_1^Y as the only possible candidate for gauged anomalies. These would appear in the following three triangle graphs (the internal lines represent left-handed fermions)

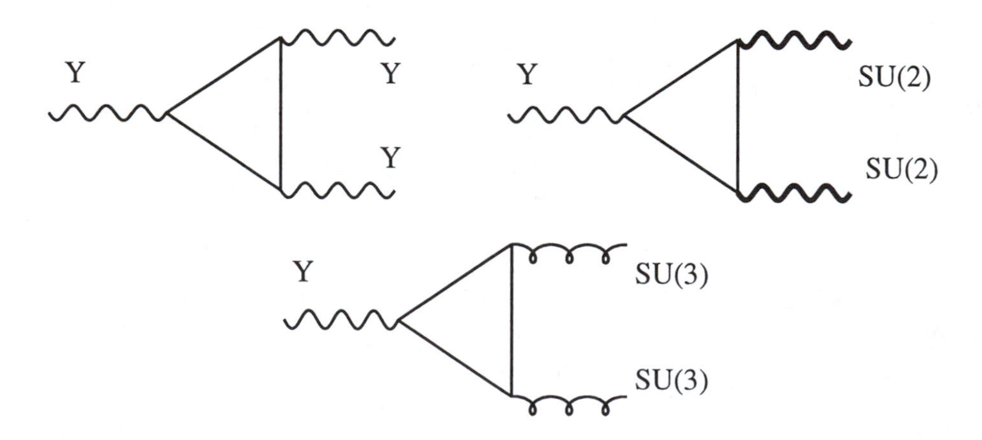

These vanish if the following equations are satisfied,

$$2y_3 + y_4 + y_5 = 0\ , \qquad y_1 + 3y_3 = 0\ ,$$

$$2y_1^3 + y_2^3 + 3(2y_3^3 + y_4^3 + y_5^3) = 0 .$$

They relate both quark and lepton hypercharges, providing the first hint of some sort of unification; two hypercharges are still undetermined, but one can be normalized arbitrarily by redefining the hypercharge coupling constant g_1. The vanishing of the mixed $SU(3) - U(1)$ anomaly condition requires the two quark bilinears to have opposite value of hypercharge.

We may consider another type of anomaly: the mixed gravitational anomaly which is generated by a graph of the form

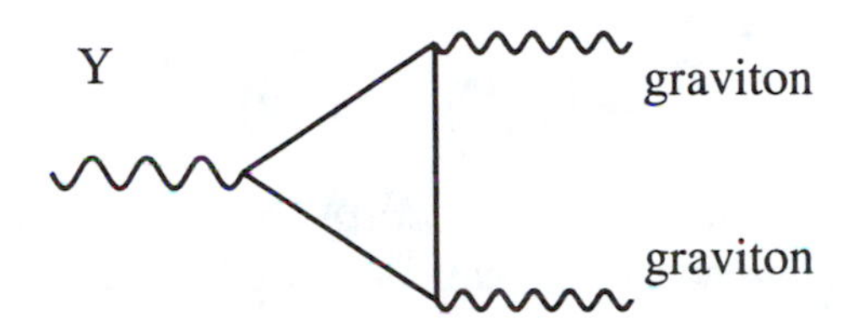

with one hypercharge gauge boson and two gravitons. Its vanishing results in the additional condition

$$2y_1 + y_2 + 6y_3 + 3y_4 + 3y_5 = 0 . \tag{2.9}$$

This anomaly condition is not on the same footing as the others since it brings in quantum gravity, which is not a renormalizable theory: it would affect the conservation of the hypercharge current only if graviton loops are taken seriously. Nevertheless it is likely to be present in a renormalizable theory of gravity. In the absence of Yukawa couplings, the mixed gravitational anomaly, together with the ABJ anomalies, yield the relation

$$y_2 = 6y_3 . \tag{2.10}$$

If we substitute this equation in the unmixed $U(1)$ anomaly condition, we obtain

$$18y_3(2y_3 - y_5)(4y_3 + y_5) = 0 . \tag{2.11}$$

It seems that we have three physically different solutions, but this is illusory: the second solution $y_5 = 2y_3$, and the third solution $y_5 = -4y_3$ can be seen to be the same with y_4 and y_5 interchanged. This is not surprising since these two fields are only distinguished by their hypercharges. We really have only two solutions, $y_3 = 0$ and $y_5 = 2y_3$. Of these, the simplest

$y_3 = 0$, is quite uninteresting: none of the three two fermion combinations have the same hypercharge. For the other solution, $y_5 = 2y_3$, $\widehat{L}\bar{e}$ and $\widehat{\mathbf{Q}}\bar{\mathbf{d}}$ have the same hypercharge. Which one does nature choose? It is the Yukawa couplings that settle the issue.

It is remarkable that one of these solutions is realized by the Yukawa couplings. Under the $SU_2^W \times U_1^Y$ gauge group, potential color singlet quark masses violate weak isospin by half units, *i.e.* $\Delta I_w = \frac{1}{2}$. In order to generate masses for the charged fermions, without violating the renormalizability of the theory, one introduces a spinless boson Higgs field H, transforming as a weak doublet, color singlet, and with hypercharge y_h:

$$H = \begin{pmatrix} H_1 \\ H_2 \end{pmatrix} \sim (\mathbf{1}^c, \mathbf{2})_{y_h} \ .$$

With only this field, all the gross features of the data can be explained. The following Yukawa couplings can be added

$$\mathcal{L}_{\text{Yu}} = i\widehat{L}_i \bar{e}_j H^* Y_{ij}^{[e]} + i\widehat{\mathbf{Q}}_i \bar{\mathbf{d}}_j H^* Y_{ij}^{[d]} + i\widehat{\mathbf{Q}}_i \bar{\mathbf{u}}_j \tau_2 H Y_{ij}^{[u]} + \text{c.c.} \ , \tag{2.12}$$

where τ_2 are the weak isospin indices and the Yukawa coupling constants are unknown complex 3×3 matrices $\mathbf{Y}^{[e]}$, $\mathbf{Y}^{[d]}$, and $\mathbf{Y}^{[u]}$. Invariance under hypercharge transformations implies the further relations

$$y_h = y_1 + y_2 = -(y_3 + y_4) = -(y_3 + y_5) \ .$$

Thess Yukawa couplings determine all the hypercharge assignments to be (in units of y_h),

$$y_1 = -1, \ y_2 = +2, \ y_3 = +\frac{1}{3}, \ y_4 = -\frac{4}{3}, \ y_5 = +\frac{2}{3} \ . \tag{2.13}$$

Hence, the absence of anomalies from the standard model requires hypercharge to be quantized. It is noteworthy that although the standard model does not include gravitational interactions, the mixed gravitational anomaly of the hypercharge happens to vanish.

The global symmetries of $\mathcal{L}_{YM} + \mathcal{L}_{WD}$, can be used to simplify the Yukawa couplings. We can write without loss of generality any matrix as the product of a unitary matrix times a real diagonal matrix times another unitary matrix, yielding for the lepton Yukawa matrix (not showing the family indices)

$$\mathbf{Y}^{[e]} = \mathbf{U}_e^t \mathbf{M}^{[e]} \mathbf{V}_e \ , \tag{2.14}$$

where $\mathbf{U}_e$ and $\mathbf{V}_e$ are 3×3 unitary matrices,

$$\mathbf{U}_e \mathbf{U}_e^\dagger = \mathbf{V}_e \mathbf{V}_e^\dagger = 1 \ , \tag{2.15}$$

and $\mathbf{M}^{[e]}$ is a real diagonal matrix. We can absorb the unitary matrices $\mathbf{U}_e$ and $\mathbf{V}_e$ through the field redefinitions

$$L' = \mathbf{U}_e L \ , \quad \bar{e}' = \mathbf{V}_e \bar{e} \ , \tag{2.16}$$

without affecting the rest of the Lagrangian $\mathcal{L}_{YM} + \mathcal{L}_{WD}$. Dropping the primes, the leptonic part of the Yukawa couplings becomes flavor-diagonal

$$i \sum_{i=1}^{3} \widehat{L}_i \bar{e}_i \ H^* y_{ii}^{[e]} \quad + \text{ c.c.} \ ,$$

where $y_{ii}^{[e]}$ are the diagonal elements of $\mathbf{M}^{[e]}$. The lepton Yukawa couplings therefore break the leptonic global symmetry $SU_3 \times SU_3 \times U_1 \times U_1$ down to three phase transformations,

$$L_i \to e^{i\alpha_i} L_i \ , \quad \bar{e}_i \to e^{-i\alpha_i} \bar{e}_i \ , \tag{2.17}$$

interpreted as the three lepton numbers, one for each family (electron-, muon-, and tau- numbers).

We can attempt similar simplifications in the quark Yukawa sector by setting

$$\mathbf{Y}^{[d]} = \mathbf{U}_d^t \mathbf{M}^{[d]} \mathbf{V}_d \ , \tag{2.18}$$

$$\mathbf{Y}^{[u]} = \mathbf{U}_u^t \mathbf{M}^{[u]} \mathbf{V}_u \ , \tag{2.19}$$

where $\mathbf{U}_{u,d}$ and $\mathbf{V}_{u,d}$ are unitary family matrices; $\mathbf{M}^{[d]}$ and $\mathbf{M}^{[u]}$ are diagonal and real.

We note immediately that the unitary matrices $\mathbf{V}_u$ and $\mathbf{V}_d$ can be absorbed in a redefinition of the fields

$$\bar{\mathbf{u}} \to \mathbf{V}_u \bar{\mathbf{u}} \ , \quad \bar{\mathbf{d}} \to \mathbf{V}_d \bar{\mathbf{d}} \ . \tag{2.20}$$

On the other hand, the matrices $\mathbf{U}_u$ and $\mathbf{U}_d$ cannot *both* be absorbed away by field redefinitions. Indeed, the disappearance of $\mathbf{U}_u$ in one Yukawa coupling forces its reappearance as $\mathbf{U}_u^\dagger$ in the other one. Similarly if $\mathbf{U}_d$ disappears from the second, it reappears as $\mathbf{U}_d^\dagger$ in the first. The best we can do is to rewrite the quark Yukawa couplings in the simplified form

$$i \ \widehat{\mathbf{Q}}_i \bar{\mathbf{d}}_i \ H^* y_{ii}^{[d]} + i \ \widehat{\mathbf{Q}}_i (\mathcal{V})_{ji} \bar{\mathbf{u}}_j \tau_2 H \ y_{jj}^{[u]} \ ,$$

where

$$\mathcal{V} = \mathbf{U}_u \mathbf{U}_d^\dagger \ , \tag{2.21}$$

is a unitary matrix. One can make a final modification by expurgating yet more redundant phases from $\mathcal{V}$. Indeed, any unitary matrix can be decomposed as

$$\mathcal{V} = \mathcal{P}^T \mathcal{U} \mathcal{P}' \ , \tag{2.22}$$

where $\mathcal{P}$ and $\mathcal{P}'$ are diagonal phase matrices generated by elements of the Cartan subalgebra, and $\mathcal{U}$ contains the remaining parameters. For a rotation matrix, this is exactly the Euler decomposition. In the physical case of three families, $\mathcal{V}$ depends on nine parameters, one of which is an overall phase. The other eight correspond to the generators of SU_3. Using this decomposition (called Iwasawa in mathematics), we see that $\mathcal{P}$ and $\mathcal{P}'$ each depend on two parameters, leaving the unitary $\mathcal{U}$ with four parameters. Three correspond to real rotations while the fourth one must be a phase, not in the $SO(3)$ subgroup of SU_3. Why is this decomposition useful? We see that $\mathcal{P}$ can be absorbed in $\bar{\mathbf{u}}$, and $\mathcal{P}'$ can be absorbed in $\mathbf{Q}$ and then, through the first coupling, transferred to $\bar{\mathbf{d}}$, where it finally disappears, leaving us with the final form of the Yukawa couplings of the standard model

$$\mathcal{L}_{\text{Yu}} = i\left(\widehat{L}_i y_{ii}^{[e]} \bar{e}_i + \widehat{\mathbf{Q}}_i y_{ii}^{[d]} \bar{\mathbf{d}}_i\right) H^* + i\ \widehat{\mathbf{Q}}_i \mathcal{U}_{ji} y_{jj}^{[u]} \bar{\mathbf{u}}_j \tau_2 H + \ \text{c.c.}\ . \tag{2.23}$$

Let us remark that if there had been only two families, the Euler decomposition would have left us with only one rotation angle, and no phase. With three families, the Yukawa sector of the standard model depend on thirteen (real) parameters: nine masses, three mixing angles and one phase. The matrix $\mathcal{U}$ is called the Cabibbo-Kobayashi-Maskawa (CKM) matrix. In the quark sector, the only remnant of the global symmetries of $\mathcal{L}_{WD}$ is a common phase transformation for all the quarks,

$$\mathbf{Q}_i \to e^{i\alpha}\mathbf{Q}_i \quad \bar{\mathbf{u}}_i \to e^{-i\alpha}\bar{\mathbf{u}}_i\ , \quad \bar{\mathbf{d}}_i \to e^{i\alpha}\bar{\mathbf{d}}_i\ , \tag{2.24}$$

and it corresponds to quark number (1/3 of baryon number). To summarize, the Yukawa couplings have reduced the global symmetries of the gauged kinetic terms to four phase symmetries, baryon number and the three lepton numbers. However it is easy to see (see problem) that of these one is anomalous, leaving only three truly conserved quantum numbers.

We now turn to the fourth and final part of the standard model Lagrangian which describes the Higgs doublet, its interactions with the gauge fields, and with itself. The gauge-covariant derivative acting on the Higgs takes the form

$$\mathcal{D}_\mu H = (\partial_\mu + i\mathbf{W}_\mu + \frac{i}{2} y_h B_\mu) H \ . \tag{2.25}$$

The Higgs Lagrangian is simply given by

$$\mathcal{L}_H = (\mathcal{D}_\mu H)^\dagger (\mathcal{D}^\mu H) - V(H) \ , \tag{2.26}$$

where V is the most general renormalizable potential invariant under $SU_2^W \times U_1^Y$; it is given by

$$V = -\mu^2 H^\dagger H + \lambda (H^\dagger H)^2 \ . \tag{2.27}$$

It can be shown that there is only one quartic polynomial with this invariance. The dimensionless coupling λ is taken to be positive to insure that V is bounded from below, and μ^2 is the only parameter with units of mass in the classical Lagrangian. The sign in front of the mass term is taken to be negative for reasons that will soon become obvious.

The alert reader has already noticed that $V(H)$ is invariant under a much larger symmetry, SO_4, acting on the four real components of the complex doublet. This symmetry is broken by other parts of the Lagrangian. Such a symmetry, shared only by part of the Lagrangian is sometimes called accidental. We have already encountered such symmetries in the Weyl-Dirac part of the Lagrangian. Since $SO_4 \sim SU_{2L} \times SU_{2R}$, the four vector made up of the four real components of the Higgs transforms as $(\mathbf{2},\mathbf{2})$. The first SU_{2L} is the gauged weak isospin. The diagonal sum of these two $SU(2)$s, is sometimes called "custodial"; it is broken both by the Yukawa couplings, and by the couplings to hypercharge.

We have now come to the end of our description of the Lagrangian of the standard model. We have seen that it depends on the three coupling constants g_1, g_2, g_3 of the gauge groups, on the 13 parameters of the Yukawa couplings (12 real + 1 phase), on the Higgs self coupling λ, and on one dimensionful "coupling" μ. Thus, the classical standard model depends on 18 parameters. In addition to being invariant under the gauge group $SU(3) \times SU(2) \times U(1)$, the standard model is (at the classical level) invariant under four global phase transformations, the three lepton numbers and baryon number.

We have stated that this counting is altered by quantum effects. Quantum anomalies break a linear combination of baryon and lepton numbers. As we shall investigate in detail, this also introduce another parameter that describes the QCD vacuum. These anomalies, however, do not alter the renormalizability of the theory since they affect only ungauged symmetries.

We conclude this description of the standard model Lagrangian by discussing its discrete symmetries, in particular its CP properties. Of course, the locality and reality of the Lagrangian implies the invariance of the standard model under the combined operation of CPT. The chiral character of weak isospin obviously breaks parity, but the CP analysis is more subtle. We start with the transformation of a generic fermion field under CP:

$$CP : \psi_L \to \sigma_2 \psi_L^* \ . \tag{2.28}$$

This transformation leaves the kinetic term invariant. Suppose ψ_L forms a representation of some gauge group. The fermion gauge interaction is

$$\mathcal{L}_{\text{int}} = ig\psi_L^\dagger \sigma_\mu \mathbf{T} \psi_L \cdot \mathbf{A}_\mu \ , \tag{2.29}$$

where $\mathbf{T}$ are the matrices which represent the group in the representation of ψ_L. Under CP, the currents transform as

$$i\psi_L^\dagger \mathbf{T} \psi_L \to i\psi_L^t \mathbf{T} \psi_L^* = -i\psi_L^\dagger \mathbf{T}^t \psi_L \ , \tag{2.30}$$

$$i\psi_L^\dagger \vec{\sigma} \mathbf{T} \psi_L \to i\psi_L^t \sigma_2 \vec{\sigma} \sigma_2 \mathbf{T} \psi_L^* = -i\psi_L^t \vec{\sigma}^t \vec{T} \psi_L^* = +i\psi_L^\dagger \vec{\sigma} \mathbf{T}^t \psi_L \ . \tag{2.31}$$

Thus invariance of this coupling is assured if we demand

$$A_0^S \to -A_0^S \ , \qquad A_i^S \to +A_i^S \ , \tag{2.32}$$

$$A_0^A \to +A_0^A \ , \qquad A_i^A \to -A_i^A \ , \tag{2.33}$$

where the superscripts S(A) denote the generators for which the $\mathbf{T}$ matrices are symmetric or antisymmetric. In SU_2 for example, the set S contains τ^3 and τ^1 while the A set contains τ^2. This is consistent with the naive expectation that charge conjugation changes the sign of i since the $\mathbf{T}$-matrices are hermitian. Also we recover the desired transformations under CP since the space components are odd under Parity. Further, under CP, the field strengths transform as

$$F_{0i}^S \to +F_{0i}^S \ , \qquad F_{ij}^S \to -F_{ij}^S \ , \tag{2.34}$$

$$F_{0i}^A \to -F_{0i}^A , \qquad F_{ij}^A \to +F_{ij}^A , \tag{2.35}$$

which leaves the Yang-Mills Lagrangian, $\vec{F}_{\mu\nu} \cdot \vec{F}^{\mu\nu}$, invariant. Note for future reference that the surface term

$$\epsilon^{\mu\nu\rho\sigma} \vec{F}_{\mu\nu} \cdot \vec{F}_{\rho\sigma} \sim \epsilon^{ijk} \vec{F}_{0i} \cdot \vec{F}_{jk} , \tag{2.36}$$

is clearly odd under CP and P.

A generic Yukawa contribution to the Lagrangian, involving two left handed field ψ and χ and a Higgs field ϕ, is given by

$$\mathcal{L}_{Yu} = y \widehat{\psi}_L \chi_L \phi - y^* \psi_L^\dagger \sigma_2 \chi_L^* \phi^* . \tag{2.37}$$

Under a CP transformation, it becomes

$$\begin{aligned} \mathcal{L}_{Yu} &\to y \psi_L^\dagger \sigma_2^T \sigma_2 \chi_L^* \phi^{CP} - y^* \psi_L^T \sigma_2 \sigma_2 \sigma_2^* \chi_L \phi^{*CP} \\ &= -y \psi_L^\dagger \sigma_2 \chi_L^* \phi^{CP} + y^* \psi_L^T \sigma_2 \chi_L \phi^{*CP} , \end{aligned} \tag{2.38}$$

where ϕ^{CP} is the CP transform of ϕ. It is invariant if

$$y \phi^{CP} = y^* \phi^* . \tag{2.39}$$

Since the CP transform of ϕ is the complex conjugate, we conclude that CP invariance is maintained if the Yukawa couplings are real, $y = y^*$. It follows that the standard model explicitly breaks CP because the three-family CKM matrix has just enough room to accomodate one complex phase. It also means that CP-violating effects will only appear when all three families are incorporated, since with two families, all CP-violating phases can be absorbed away. Thus we expect two possible manifestations of CP-violation. One where particles made up of the first two families are involved, in which case the CP-violation has to enter through loop diagrams involving the third family, the other at tree level with particles of all three families involved. Clearly, the study of CP-violation is a powerful probe into the structure of the standard model, where all CP-violating phenomena derive from only one parameter.

PROBLEMS

A. 1-) Show that with one complex Higgs doublet, one can form only one quartic invariant.
   2-) How many independent quartic invariants can be formed out of two doublets with the same hypercharge?

B. Consider all possible fermion bilinears which transform as a Lorentz vector. Sort them in terms of their baryon and lepton numbers. For each class, find the possible gauged quantum numbers of these combinations.

C. Repeat problem B for the fermion bilinears that are Lorentz invariants.

D. Suppose there were four chiral families of fermions. Find out the number of parameters needed to describe the Yukawa sector. How many CP-violating phases are there?

E. Using the triangle graphs, show that while neither quark nor lepton numbers are conserved, a certain linear combination survives.

F. Identify chiral combinations of fermions under the standard model gauge groups that satisfy the anomaly conditions. For reference, see P. Fishbane, S. Meshkov, and P. Ramond, *Phys. Lett.* **134B**, 81(1984).

CHAPTER **3**

STANDARD MODEL

VACUUM STRUCTURE

The standard model Lagrangian consists of two parts, $\mathcal{L}_{QCD}$ that describes the strong interactions, and $\mathcal{L}_{EW}$ the weak interactions, with the colored quarks linking the two. Both are invariant under color and Lorentz symmetries, and to the extent that experiments show no evidence of violation of either symmetry, the vacuum state of the standard model must be invariant as well. The presently available theoretical equipment is too primitive to probe for answers outside of perturbation theory, and these symmetries must be accepted as fact, required by observation. We infer that quark fields must vanish on the vacuum state of the standard modell. As a result, acting on the vacuum, the standard model Hamiltonian splits up into the sum of the electroweak and QCD parts, of the form $|\Omega_{QCD}\rangle|\Omega_{EW}\rangle$, where $|\Omega_{QCD}\rangle$ is the ground state of the QCD part of the model, and $|\Omega_{EW}\rangle$ is that of the electroweak part. We delay the discussion of the QCD vacuum state as it is determined by subtle quantum effects, and start with the electroweak vacuum $|\Omega_{EW}\rangle$.

3.1 THE ELECTROWEAK VACUUM

In this section we describe the lowest energy configuration of the different fields of the standard model. First, Lorentz invariance of the vacuum demands that we set all the fermion fields equal to zero. Second, since the $SU_2^W \times U_1^Y$ gauge fields give a positive definite contribution to the Hamiltonian, they also must vanish in the vacuum, corresponding to a pure gauge configurations of the gauge fields. In addition, as the gauged kinetic term for the Higgs fields also gives a positive contribution to the energy density, the covariant derivative of the Higgs field must vanish in the vacuum as well. It follows that in the electroweak vacuum, the electroweak potentials are gauge equivalent to zero, and the Higgs field is gauge equivalent to a constant.

The remaining part of the energy density is in the Higgs potential density, all other parts being zero in the vacuum configuration: the minimum of the Hamiltonian density occurs at the minimum of the potential, that is for the fields that satisfy

$$\frac{dV}{dH} = -\mu^2 H^\dagger + 2\lambda H^\dagger H H^\dagger = 0 \ , \tag{3.1}$$

which has two solutions, $H_{\rm vac} = 0$, and

$$H^\dagger_{\rm vac} H_{\rm vac} = \frac{\mu^2}{2\lambda} \equiv \frac{v^2}{2} \ . \tag{3.2}$$

The first solution, $H_{\rm vac} = 0$, is easily seen to corresponds to a maximum. The second solution leads to a closed continuous surface of minima parametrized by three angles

$$H_{\rm vac} = e^{i\theta_0} \mathbf{U} \begin{pmatrix} 0 \\ \frac{v}{\sqrt{2}} \end{pmatrix} \ , \tag{3.3}$$

where $\mathbf{U}$ is an element of the weak $SU(2)$,

$$\mathbf{U} = e^{i\frac{\vec{\tau}}{2}\cdot\vec{\theta}} \ . \tag{3.4}$$

This vacuum configuration can be written in its Euler form

$$\mathbf{U} = e^{i\frac{\tau_3}{2}\eta_1} e^{i\frac{\tau_2}{2}\eta_2} e^{i\frac{\tau_3}{2}\eta_3} \ . \tag{3.5}$$

Substitution in eq. (3.3) yields

$$H_{\rm vac} = \frac{v}{\sqrt{2}} \begin{pmatrix} e^{i(\theta' + \eta_1)} \sin\eta_2 \\ e^{i(\theta' - \eta_1)} \cos\eta_2 \end{pmatrix} \ , \tag{3.6}$$

where $\theta' = \theta_0 - \eta_3$. We have a continuous surface of minima, on which one moves by $SU_2^W \times U_1^Y$ gauge transformations, so that all points of the surface are gauge-equivalent.

In order to set up perturbation theory about this vacuum, we can use any one of these configurations as a starting point, so we start with the simplest

$$< H >= \frac{v}{\sqrt{2}}\begin{pmatrix}0\\1\end{pmatrix} . \tag{3.7}$$

We can understand why only three angles for the vacuum manifold by simple counting. By singling out one direction in weak isospin space, the isospin charge-changing transformations are "broken" in the sense that they no longer leave the vacuum configuration invariant, which accounts for two angles. Also, the lower spinor component has $I_{3\mathrm{w}} = -\frac{1}{2}$ and $y_h = 1$ (recall that setting $y_h = 1$ involves no loss of generality since any other value can be absorbed by redefining g_1). Its vacuum value breaks only one combination, which accounts for the third angle, and preserves the combination

$$\mathcal{Q} = I_{3\mathrm{w}} + \frac{Y}{2} , \tag{3.8}$$

which we later identify with the electric charge. This vacuum configuration is said to "break" the electroweak gauge group $SU_2^W \times U_1^Y$ down to U_1^γ, the gauge group of Maxwell's theory. The conserved charges of the various fermions of the theory are readily determined. For example the charge of each lepton doublet is given by

$$\mathcal{Q}\begin{pmatrix}\nu_e\\e\end{pmatrix}_L = \left[\begin{pmatrix}\frac{1}{2} & 0\\0 & -\frac{1}{2}\end{pmatrix} + \frac{(-1)}{2}\begin{pmatrix}1 & 0\\0 & 1\end{pmatrix}\right]\begin{pmatrix}\nu_e\\e\end{pmatrix}_L = \begin{pmatrix}0\\-e\end{pmatrix}_L . \tag{3.9}$$

Hence the upper lepton component is electrically neutral, and the lower negatively charged. The first family doublet components are identified with the left-handed electron neutrino and the left-handed part of the electron, and so on for the other two families. Similarly the electric charge of the left-handed anti-electron is

$$\mathcal{Q}\bar{e} = (0 + \frac{(2)}{2})\bar{e} = \bar{e} , \tag{3.10}$$

corresponding to a right-handed electron of charge -1. Both left- and right-handed charged leptons have the same electric charge. This charge assignment is called *vector-like.* One easily verifies that the quark charges are also vector-like.

The next step in identifying the degrees of freedom in the electroweak vacuum is to express the Lagrangian in terms of fields which vanish at the minimum. As we discussed, there are many $SU_2^W \times U_1^Y$ gauge-equivalent configurations. The simplest

$$H' = H - < H > , \tag{3.11}$$

is not very convenient, and we prefer to use the Kibble parametrization

$$H = e^{i\vec{\zeta}(x)\frac{\vec{\tau}}{v}} \begin{pmatrix} 0 \\ \frac{v+h(x)}{\sqrt{2}} \end{pmatrix} \equiv \mathbf{U}(x) H_0 \ ,$$

introducing the fields $\vec{\zeta}$ and h which vanish in the vacuum. The exponential term $\mathbf{U}(x)$ describes a field-valued $SU(2)$ unitary gauge transformation. Under gauge transformations along the three broken directions, $h(x)$ is invariant while the $\vec{\zeta}(x)$'s transform inhomogeneously

$$\zeta_a(x) \to \zeta_a(x) + \theta_a \ , \qquad a = 1, 2, 3 \ , \tag{3.12}$$

the way Nambu-Goldstone bosons are supposed to behave. To see that the real field $h(x)$ has zero value at minimum, we rewrite the potential as

$$V = \lambda \left(H^\dagger H - \frac{v^2}{2} \right)^2 - \lambda \frac{v^4}{4} \ , \tag{3.13}$$

and drop the last term (only gravity will be affected by this shift!). Substituting the Kibble form, we obtain

$$V = \lambda \left(\frac{1}{2}(v+h)^2 - \frac{v^2}{2} \right)^2 = \lambda v^2 h^2 + \lambda v h^3 + \frac{\lambda}{4} h^4 \ . \tag{3.14}$$

The three massless Nambu-Goldstone bosons associated with the three broken generators$\vec{\zeta}_a(x)$ have disappeared. There remains the field h which describes the Higgs particle. From the quadratic term in the potential, we obtain its mass

$$m_h = v\sqrt{2\lambda} = \mu\sqrt{2} \ . \tag{3.15}$$

To express the kinetic terms in the Kibble parametrization, we start with the covariant derivative

$$\mathcal{D}_\mu H = \frac{1}{\sqrt{2}} \left(\partial_\mu + i\mathbf{W}_\mu + \frac{i}{2} B_\mu \right) \mathbf{U} \begin{pmatrix} 0 \\ v + h(x) \end{pmatrix} \ . \tag{3.16}$$

The appropriate gauge transformations on $\mathbf{W}_\mu \to \mathbf{W}'_\mu$, allows us to rewrite the covariant derivative as

$$\mathcal{D}_\mu H = \frac{1}{\sqrt{2}}\mathbf{U}\left(\partial_\mu + i\mathbf{W}'_\mu + \frac{i}{2}B_\mu\right)\begin{pmatrix} 0 \\ v + h(x) \end{pmatrix} , \tag{3.17}$$

where

$$\mathbf{W}'_\mu = -i\mathbf{U}^\dagger\partial_\mu\mathbf{U} + \mathbf{U}^\dagger\mathbf{W}_\mu\mathbf{U} , \tag{3.18}$$

or in the useful operator notation,

$$\mathcal{D}_\mu = \mathbf{U}\mathcal{D}'_\mu\mathbf{U}^\dagger . \tag{3.19}$$

After a little more algebra and dropping the primes, we find

$$\mathcal{D}_\mu H = \frac{\mathbf{U}}{\sqrt{2}}\begin{pmatrix} \frac{i}{2}(W^1_\mu - iW^2_\mu)(v+h) \\ \partial_\mu h + \frac{i}{2}(B_\mu - W^3_\mu)(v+h) \end{pmatrix} . \tag{3.20}$$

Since $\mathbf{U}$ is unitary, it disappears completely from the Lagrangian, yielding

$$\mathcal{L}_H = \frac{1}{2}\partial_\mu h\partial^\mu h + \frac{1}{8}\left(B_\mu - W^3_\mu\right)\left(B^\mu - W^{3\mu}\right)(v+h)^2$$
$$+ \frac{1}{8}\left(W^1_\mu - iW^2_\mu\right)\left(W^{1\mu} + iW^{2\mu}\right)(v+h)^2 . \tag{3.21}$$

We recognize the standard kinetic term for the Higgs field h, and upon expansion of $(v+h)^2$, mass terms for some gauge fields. In our notation, the coupling constants have been hidden in the gauge potentials. We can reinstate them by the replacements

$$\mathbf{W}_\mu \to g_2\mathbf{W}_\mu , \quad B_\mu \to g_1 B_\mu , \tag{3.22}$$

where the new fields now have the usual kinetic term. The form of the (quadratic) kinetic terms is preserved by canonical transformations; on the real gauge fields, they reduce to orthogonal transformations. Thus $\mathcal{L}_H$ is seen to contain a quadratic mass for the canonical combination

$$Z_\mu = \cos\theta_\mathrm{w} W^3_\mu - \sin\theta_\mathrm{w} B_\mu , \tag{3.23}$$

written in terms of the Weinberg angle

$$\cos\theta_\mathrm{w} \equiv \frac{g_2}{\sqrt{g_1^2 + g_2^2}} . \tag{3.24}$$

Its orthogonal linear combination

$$A_\mu = \sin\theta_\mathrm{w} W^3_\mu + \cos\theta_\mathrm{w} B_\mu , \tag{3.25}$$

also corresponds to a canonical field. Since A_μ does not appear quadratically in the Lagrangian, it has no mass term, although it does have the standard Maxwell kinetic term coming from $\mathcal{L}_{YM}$. It is the vector potential associated with the massless photon.

The rest of the Lagrangian describes the couplings of the Higgs particle to the gauge bosons

$$\mathcal{L}_H = \frac{1}{2}\partial_\mu h \partial^\mu h + \frac{1}{2}m_Z^2 Z_\mu Z^\mu + m_W^2 W_\mu^+ W^{-\mu} + \frac{h}{v}(2 + \frac{h}{v})\left(\frac{1}{2}m_Z^2 Z_\mu Z^\mu + m_W^2 W_\mu^+ W^{-\mu}\right) , \tag{3.26}$$

where

$$W_\mu^\mp = \frac{1}{\sqrt{2}}(W_\mu^1 \pm iW_\mu^2) , \tag{3.27}$$

$$m_Z^2 = \frac{1}{4}v^2(g_1^2 + g_2^2) , \qquad m_W^2 = \frac{1}{4}v^2 g_2^2 . \tag{3.28}$$

The gauge potentials Z_μ and $W_\mu^\mp$, with canonical kinetic terms from the Yang-Mills part, are both massive, with masses related by the (tree-level) relation

$$\frac{m_W^2}{m_Z^2} = \cos^2\theta_{\rm w} . \tag{3.29}$$

It is also evident that the Higgs particle has no couplings to the photon; it is electrically neutral, although it couples to the massive gauge bosons proportionally to their masses squared.

These results can be neatly summarized: three symmetry operation were "broken"– the gauge fields corresponding to each broken symmetry become massive while that associated with the unbroken symmetry stays massless. By acquiring a mass, each gauge field has increased its degrees of freedom from two to three. These extra degrees of freedom are the $\zeta_a(x)$ of the Kibble decomposition, which were seen to disappear from the Lagrangian after judicious gauge transformations. This is the Higgs mechanism.

In this formulation, the Lagrangian contains only physical fields, since the Nambu-Goldstone bosons have disappeared (the ζ_a's have been eaten, and been reincarnated as the longitudinal degrees of freedom of the massive gauge bosons!). Called the *unitary gauge*, this parametrization is quite convenient for evaluating $\mathcal{L}_{EW}$ in terms of $h(x)$, although for performing calculations in the quantum field theory, another parametrization will prove more convenient.

To complete the story, we rewrite the couplings of the Higgs field to the fermions in the unitary gauge,

$$\mathcal{L}_{\text{Yu}}^{[e]} = i\widehat{L}_i \bar{e}_i H^* y_{ii}^{[e]} + \text{c.c.} \ , \tag{3.30}$$

$$= i\widehat{L}_i \bar{e}_i \mathbf{U}^* H_0 y_{ii}^{[e]} + \text{c.c.} \ . \tag{3.31}$$

Using the hermiticity of $\mathbf{U}$, $\mathbf{U}^* = (\mathbf{U}^\dagger)^t$, we set $L'_i = \mathbf{U}^\dagger L_i$, and rewrite the Yukawa part as

$$\mathcal{L}_{\text{Yu}}^{[e]} = i\widehat{L}'_i \bar{e}_i H_0 y_{ii}^{[e]} + \text{c.c.} \ , \tag{3.32}$$

$$= i\widehat{L}'_{2i} \bar{e}_i (\frac{v+h}{\sqrt{2}}) y_{ii}^{[e]} + \text{c.c.} \ . \tag{3.33}$$

The unitary matrix $\mathbf{U}$ disappears from the lepton doublet gauged kinetic term also by putting the gauge fields in the unitary gauge.

This form of the Yukawa coupling suggests that we interpret L_{2L} and $\bar{e}_L$ as the left-handed and right-handed parts of the charged massive leptons, e, μ, τ. Hence, setting $\bar{e}_{Li} = -\sigma_2 e^*_{Ri}$, and

$$e_{Ri} = (e_R, \mu_R, \tau_R) \ , \qquad L'_{2iL} = (e_L, \mu_L, \tau_L) \ , \tag{3.34}$$

we arrive at the simple form

$$\mathcal{L}_{\text{Yu}}^{[e]} = \frac{i}{\sqrt{2}}(v+h)(y_{11}^{[e]} e_R^\dagger e_L + y_{22}^{[e]} \mu_R^\dagger \mu_L + y_{33}^{[e]} \tau_R^\dagger \tau_L) + \text{c.c.} \ . \tag{3.35}$$

Remembering that both $y_{ii}^{[e]}$ and v are real, we can readily identify the lepton masses

$$\frac{v}{\sqrt{2}} y_{ii}^{[e]} = (m_e, m_\mu, m_\tau) \ . \tag{3.36}$$

In four-component Dirac notation, the Yukawa lepton couplings read simply

$$\mathcal{L}_{\text{Yu}}^{[e]} = i(1 + \frac{h}{v})[m_e \bar{e} e + m_\mu \bar{\mu} \mu + m_\tau \bar{\tau} \tau] \ . \tag{3.37}$$

These fields describe the *electron*, the *muon*, and the *tau* leptons. The Higgs doublet, by having a vacuum value, has generated the charged lepton masses. Note that the Higgs field couples to the leptons according to their masses.

The upper component of each lepton doublet has no mass terms, describing the massless neutrinos

$$L'_{1iL} = \nu_{iL} = (\nu_e, \nu_\mu, \nu_\tau)_L \ . \tag{3.38}$$

From here on we will use these symbols for the lepton fields. Not much is left of the $U(3) \times U(3)$ global symmetry of the lepton kinetic terms, only three global vectorial transformations, corresponding to the three lepton numbers.

The quark Yukawa couplings are re-expressed through a similar, albeit slightly more complicated analysis. For the down type quarks, the Higgs vacuum value translates into a mass for the lower components of the quark doublets, with the result

$$\mathcal{L}^{[d]}_{\text{Yu}} = i \ \widehat{\mathbf{Q}}_i \bar{\mathbf{d}}_i H^* \ y^{[d]}_{ii} + \text{c.c.} \tag{3.39}$$

$$= i(1 + \frac{h}{v})[m_d \bar{\mathbf{d}}\mathbf{d} + m_s \bar{\mathbf{s}}\mathbf{s} + m_b \bar{\mathbf{b}}\mathbf{b}] \ , \tag{3.40}$$

where we have made the following identifications

$$m_d = \frac{v}{\sqrt{2}} y^{[d]}_{11} \ , \quad m_s = \frac{v}{\sqrt{2}} y^{[d]}_{22} \ , \quad m_b = \frac{v}{\sqrt{2}} y^{[d]}_{33} \ , \tag{3.41}$$

$$\mathbf{d}_L = \mathbf{U}^\dagger \mathbf{Q}'_1 \ , \quad \mathbf{s}_L = \mathbf{U}^\dagger \mathbf{Q}'_2 \ , \quad \mathbf{b}_L = \mathbf{U}^\dagger \mathbf{Q}'_3 \ , \tag{3.42}$$

$$\mathbf{d}_R = \sigma_2 \bar{\mathbf{d}}^*_1 \ , \quad \mathbf{s}_R = \sigma_2 \bar{\mathbf{d}}^*_2 \ , \quad \mathbf{b}_R = \sigma_2 \bar{\mathbf{d}}^*_3 \ . \tag{3.43}$$

These eigenstates describe the *down*, *strange*, and *bottom* quarks.

The analysis of the remaining quark Yukawa couplings is a bit more involved. We write

$$\mathcal{L}^{[u]}_{\text{Yu}} = i \widehat{\mathbf{Q}}_i \mathcal{U}_{ji} \bar{\mathbf{u}}'_j y^{[u]}_{jj} \tau_2 H + \text{c.c.} \ , \tag{3.44}$$

$$= i \widehat{\mathbf{Q}}_i \mathcal{U}_{ji} \bar{\mathbf{u}}_j y^{[u]}_{jj} \tau_2 \mathbf{U} H_0 + \text{c.c.} \ . \tag{3.45}$$

Using $\tau_2 \vec{\tau} \tau_2 = -\vec{\tau}^t$, it follows that

$$\tau_2 \mathbf{U} = (\mathbf{U}^{-1})^t \tau_2 = (\mathbf{U}^\dagger)^t \tau_2 \ , \tag{3.46}$$

so that we may absorb $\mathbf{U}$ in the combination $\mathbf{Q}'_i = \mathbf{U}^\dagger \mathbf{Q}_i$; the result is

$$\mathcal{L}^{[u]}_{\rm Yu} = i\widehat{\mathbf{Q}}'_i \mathcal{U}_{ji} \bar{\mathbf{u}}_j y^{[u]}_{jj} \tau_2 H_0 + \text{c.c.} \ , \tag{3.47}$$

$$= \frac{i}{\sqrt{2}} \widehat{\mathbf{Q}}'_{1i} \mathcal{U}_{ji} \bar{\mathbf{u}}_j y^{[u]}_{jj} (v + h) + \text{c.c.} \ . \tag{3.48}$$

This unitary transformation on the quark doublet leaves its gauged kinetic term invariant as long as the fields are in the unitary gauge, and then $\mathbf{U}$ disappears totally from the Lagrangian. We define the *mass eigenstates*

$$\mathbf{u}_L = \mathcal{U}_{1i} \mathbf{Q}'_{1i}, \qquad \bar{\mathbf{u}}_1 = i\sigma_2 \mathbf{u}^*_R \ , \tag{3.49}$$

$$\mathbf{c}_L = \mathcal{U}_{2i} \mathbf{Q}'_{1i}, \qquad \bar{\mathbf{u}}_2 = i\sigma_2 \mathbf{c}^*_R \ , \tag{3.50}$$

$$\mathbf{t}_L = \mathcal{U}_{3i} \mathbf{Q}'_{1i}, \qquad \bar{\mathbf{u}}_3 = i\sigma_2 \mathbf{t}^*_R \ , \tag{3.51}$$

which describe the *up*, *charm*, and *top* quarks, with masses

$$\frac{v}{\sqrt{2}} y^{[u]}_{ii} = (m_u, m_c, m_t) \ . \tag{3.52}$$

This enables us to rewrite the Yukawa couplings in Dirac form as

$$\mathcal{L}^{[u]}_{\rm Yu} = i(1 + \frac{h}{v})(m_u \bar{\mathbf{u}}\mathbf{u} + m_c \bar{\mathbf{c}}\mathbf{c} + m_t \bar{\mathbf{t}}\mathbf{t}) \ . \tag{3.53}$$

We conclude that the Higgs field couples in a universal way to all the fermions, quarks and leptons, with a strength proportional to their masses, and to the gauge bosons according to the square of their masses; it does not couple to massless particles.

The couplings of the quarks and leptons to the gauge fields is determined by the covariant derivatives, but in this asymmetric vacuum, they are complicated, and must be rewritten in terms of the new mass eigenstates. Two important features occur: the mixing of W^3_μ and B_μ and the mixing among the left-handed quark mass eigenstates through the 3×3 family CKM matrix $\mathcal{U}$.

It is simplest to start with the leptons where the family indices play no role. The gauged kinetic term yields

$$
\begin{aligned}
&(\nu_e^\dagger\; e^\dagger)_L \sigma^\mu \partial_\mu \begin{pmatrix} \nu_e \\ e \end{pmatrix}_L \\
&\quad + \frac{i}{2}(\nu_e^\dagger\; e^\dagger)_L \begin{pmatrix} g_2 W_\mu^3 - g_1 B_\mu & g_2 W_\mu^1 - i g_2 W_\mu^2 \\ g_2 W_\mu^1 + i g_2 W_\mu^2 & -g_1 B_\mu - g_2 W_\mu^3 \end{pmatrix} \begin{pmatrix} \nu_e \\ e \end{pmatrix}_L
\end{aligned} \tag{3.54}
$$

(the **U** matrix is of course absorbed in the unitary gauge). We use the inverse relations

$$W_\mu^3 = \cos\theta_{\mathrm{w}} Z_\mu + \sin\theta_{\mathrm{w}} A_\mu \;, \tag{3.55}$$

$$B_\mu = -\sin\theta_{\mathrm{w}} Z_\mu + \cos\theta_{\mathrm{w}} A_\mu \;, \tag{3.56}$$

to rewrite the interaction terms in the form

$$
\begin{aligned}
&\frac{i}{2}\sqrt{g_1^2+g_2^2}\, Z_\mu \left(\nu_{eL}^\dagger \sigma^\mu \nu_{eL} + \frac{g_1^2 - g_2^2}{g_1^2 + g_2^2} e_L^\dagger \sigma^\mu e_L \right) - \frac{g_1 g_2}{\sqrt{g_1^2+g_2^2}} A_\mu e_L^\dagger \sigma^\mu e_L \\
&\quad + \frac{i}{2} g_2 [(W_\mu^1 - i W_\mu^2) \nu_{eL}^\dagger \sigma^\mu e_L + (W_\mu^1 + i W_\mu^2) e_L^\dagger \sigma^\mu \nu_{eL}] \;.
\end{aligned} \tag{3.57}
$$

The right-handed contributions to the lepton currents come from the lepton singlet gauged kinetic term, which is now rewritten as

$$e_R^\dagger \bar\sigma^\mu \partial_\mu e_R - i \frac{g_1 g_2}{\sqrt{g_1^2+g_2^2}} A_\mu e_R^\dagger \bar\sigma^\mu e_R + i \frac{g_1^2}{\sqrt{g_1^2+g_2^2}} Z_\mu e_R^\dagger \bar\sigma^\mu e_R \;.$$

Putting it all together, the lepton gauge interactions look like

$$
\begin{aligned}
&- i \frac{g_1 g_2}{\sqrt{g_1^2+g_2^2}} A_\mu (e_L^\dagger \sigma^\mu e_L + e_R^\dagger \bar\sigma^\mu e_R) \\
&+ \frac{i}{2}\sqrt{g_1^2+g_2^2}\, Z_\mu (\nu_{eL}^\dagger \sigma^\mu \nu_{eL} + \frac{g_1^2 - g_2^2}{g_1^2+g_2^2} e_L^\dagger \sigma^\mu e_L + 2\frac{g_1^2}{g_1^2+g_2^2} e_R^\dagger \bar\sigma^\mu e_R) \\
&+ \frac{i}{2} g_2 (W_\mu^1 - i W_\mu^2) \nu_{eL}^\dagger \sigma^\mu e_L + \frac{i}{2} g_2 (W_\mu^1 + i W_\mu^2) e_L^\dagger \sigma^\mu \nu_{eL} \;.
\end{aligned} \tag{3.58}
$$

We choose to rewrite the various couplings in terms of the electric charge and the Weinberg angle. The electric charge is clearly

$$e = \frac{g_1 g_2}{\sqrt{g_1^2 + g_2^2}} \ , \tag{3.59}$$

so that

$$\sqrt{g_1^2 + g_2^2} = \frac{e}{\cos\theta_{\rm w}\sin\theta_{\rm w}} \ , \qquad g_2 = \frac{e}{\sin\theta_{\rm w}} \ . \tag{3.60}$$

The interactions break up into the usual electromagnetic interaction,

$$-i\ e\ A_\mu(e_L^\dagger \sigma^\mu e_L + e_R^\dagger \bar\sigma^\mu e_R) \ , \tag{3.61}$$

the charged current interaction,

$$\frac{ie}{\sqrt{2}\sin\theta_{\rm w}}\left(W_\mu^- \nu_{eL}^\dagger \sigma^\mu e_L + W_\mu^+ e_L^\dagger \sigma^\mu \nu_{eL}\right) \ , \tag{3.62}$$

and the neutral current interaction

$$i\frac{eZ_\mu}{\cos\theta_{\rm w}\sin\theta_{\rm w}}\left(\frac{1}{2}\nu_{eL}^\dagger\sigma^\mu\nu_{eL} - \frac{1}{2}e_L^\dagger\sigma^\mu e_L + \sin^2\theta_{\rm w}(e_L^\dagger\sigma^\mu e_L + e_R^\dagger\bar\sigma^\mu e_R)\right) \ . \tag{3.63}$$

The same procedure applied to the quarks is complicated by the family mixing which results from the CKM matrix $\mathcal{U}$. Neglecting the QCD gluon interactions, we start with the quark doublets

$$\sum_{i=1}^{3} \mathbf{Q}_i^\dagger \sigma^\mu (ig_2 \mathbf{W}_\mu + \frac{i}{6} g_1 B_\mu)\mathbf{Q}_i \ . \tag{3.64}$$

The redefinitions of $\mathbf{Q}_i$ in terms of the mass eigenstates all involve unitary transformations. Thus both the gauged kinetic terms and the color gauge interactions are left invariant by these redefinitions. As for the electroweak part, we first rewrite it in terms of the gauge field mass eigenstates. A *modicum* of algebra gives the interaction terms

$$\begin{aligned}
&+ ieA_\mu\left(\frac{2}{3}\mathbf{Q}_{i1}^\dagger\sigma^\mu\mathbf{Q}_{i1} - \frac{1}{3}\mathbf{Q}_{i2}^\dagger\sigma^\mu\mathbf{Q}_{i2}\right) \\
&+ \frac{ieZ_\mu}{\cos\theta_{\rm w}\sin\theta_{\rm w}}\left((\frac{1}{2} - \frac{2}{3}\sin^2\theta_{\rm w})\mathbf{Q}_{1i}^\dagger\sigma^\mu\mathbf{Q}_{1i} - (\frac{1}{2} - \frac{1}{3}\sin^2\theta_{\rm w})\mathbf{Q}_{2i}^\dagger\sigma^\mu\mathbf{Q}_{2i}\right) \\
&+ \frac{ie}{\sqrt{2}\sin\theta_{\rm w}}W_\mu^+\mathbf{Q}_{1i}^\dagger\sigma^\mu\mathbf{Q}_{2i} + \frac{ie}{\sqrt{2}\sin\theta_{\rm w}}W_\mu^-\mathbf{Q}_{2i}^\dagger\sigma^\mu\mathbf{Q}_{1i} \ .
\end{aligned} \tag{3.65}$$

Similarly, the gauged kinetic terms for the right-handed parts yield the interactions

$$-ig_1\frac{B_\mu}{3}\sum_{i=1}^{3}\left(2\bar{\mathbf{u}}_i^\dagger\sigma^\mu\bar{\mathbf{u}}_i+\bar{\mathbf{d}}_i^\dagger\sigma^\mu\bar{\mathbf{d}}_i\right)=$$

$$\begin{aligned}&=-\frac{ie}{3}A_\mu\mathbf{d}_R^\dagger\bar{\sigma}^\mu\mathbf{d}_R+\frac{ie}{3}\sin^2\theta_{\mathrm{w}}Z_\mu\mathbf{d}_R^\dagger\bar{\sigma}^\mu\mathbf{d}_R\\&+\frac{2ie}{3}A_\mu\mathbf{u}_R^\dagger\bar{\sigma}^\mu\mathbf{u}_R-\frac{2ie}{3}\sin^2\theta_{\mathrm{w}}Z_\mu\mathbf{u}_R^\dagger\bar{\sigma}^\mu\mathbf{u}_R\\&+(\mathbf{u}_R\rightarrow\mathbf{c}_R\,,\mathbf{t}_R;\;\;\mathbf{d}_R\rightarrow\mathbf{s}_R\,,\mathbf{b}_R)\,.\end{aligned}\tag{3.66}$$

In both expressions, we have absorbed the family-diagonal $\mathbf{U}$ matrix by putting the $SU_2\times U_1$ gauge fields in the unitary gauge.

There remains to rewrite $\mathbf{Q}_{1i}$ and $\mathbf{Q}_{2i}$ in terms of the mass eigenstates. For the lower components things are pretty simple,

$$\mathbf{Q}_{2i}=\mathbf{d}_{iL}=(\mathbf{d}_L\,,\mathbf{s}_L\,,\mathbf{b}_L)\,;\tag{3.67}$$

On the other hand, the upper component's redefinition involve the CKM matrix in the form (we have dropped the primes of the unitary gauge)

$$\mathbf{Q}_{1iL}=(\mathcal{U}^\dagger)_{ij}\mathbf{u}_{jL}\,,\tag{3.68}$$

in terms of the previously identified mass eigenstates. In this redefinition the $\mathcal{U}$ matrix disappears from the A_μ and Z_μ couplings since it is unitary

$$\begin{aligned}\mathbf{Q}_{1i}^\dagger\sigma^\mu\mathbf{Q}_{1i}&=\mathbf{u}_{iL}^\dagger(\mathcal{U}\mathcal{U}^\dagger)_{ij}\sigma^\mu\mathbf{u}_{jL}\\&=\mathbf{u}_L^\dagger\sigma^\mu\mathbf{u}_L+\mathbf{c}_L^\dagger\sigma^\mu\mathbf{c}_L+\mathbf{t}_L^\dagger\sigma^\mu\mathbf{t}_L\,.\end{aligned}\tag{3.69}$$

Thus, because of the unitarity of the $\mathcal{U}$ matrix, we expect no tree-level transitions between quark flavors (mass eigenstates) caused by the neutral Z boson. We shall see later that this suppression is not exact in the full quantum theory, but kept small enough to agree with experiment. The mechanism by which it is sufficiently suppressed is called the GIM mechanism, named after S. Glashow, J. Iliopoulos and L. Maiani who predicted the *charm* quark to explain the suppression.

On the other hand, the charged current couplings do cause transitions between the quark flavors, since

$$\mathbf{Q}_{1i}^{\dagger}\sigma^{\mu}\mathbf{Q}_{2i} = \mathbf{u}_{jL}^{\dagger}\mathcal{U}_{ji}\sigma^{\mu}\mathbf{d}_{iL} \ . \tag{3.70}$$

We can now rewrite all the interactions between the fermions and the gauge fields in the simple form

$$\frac{ie}{\sqrt{2}\sin\theta_{\rm w}}(W_{\mu}^{+}J^{-\mu}+W_{\mu}^{-}J^{+\mu})+ieA_{\mu}J_{em}^{\mu}+\frac{ieZ_{\mu}}{\sin\theta_{\rm w}\cos\theta_{\rm w}}[J^{3\mu}-\sin^{2}\theta_{\rm w}J_{em}^{\mu}] , \tag{3.71}$$

where the charged, neutral and electromagnetic currents are, respectively,

$$J_{\mu}^{-} = \mathbf{u}_{jL}^{\dagger}\mathcal{U}_{ji}\sigma_{\mu}\mathbf{d}_{iL} \ , \quad J_{\mu}^{+} = \mathbf{d}_{jL}^{\dagger}(\mathcal{U}^{\dagger})_{ji}\sigma_{\mu}\mathbf{u}_{iL} \ , \tag{3.72}$$

$$J_{\mu}^{3} = \frac{1}{2}(\mathbf{u}_{jL}^{\dagger}\sigma_{\mu}\mathbf{u}_{jL} - \mathbf{d}_{jL}^{\dagger}\sigma_{\mu}\mathbf{d}_{jL}) \ , \tag{3.73}$$

$$J_{em}^{\mu} = \frac{2}{3}(\mathbf{u}_{jL}^{\dagger}\sigma^{\mu}\mathbf{u}_{jL} + \mathbf{u}_{jR}^{\dagger}\bar{\sigma}^{\mu}\mathbf{u}_{jR}) - \frac{1}{3}(\mathbf{d}_{jL}^{\dagger}\sigma^{\mu}\mathbf{d}_{jL} + \mathbf{d}_{jR}^{\dagger}\bar{\sigma}^{\mu}\mathbf{d}_{jR}) - (e_{jL}^{\dagger}\sigma^{\mu}e_{jL} + e_{jR}^{\dagger}\bar{\sigma}^{\mu}e_{jR}) \ . \tag{3.74}$$

It is clear from the above that the up-type and down-type quarks have charge $\frac{2}{3}$ and $-\frac{1}{3}$, respectively, and the charged leptons have charge -1. The neutrinos which do not couple to the photon, are electrically neutral.

To complete the writing of the standard model Lagrangian in the unitary gauge, there remains the Yang-Mills part which contains the self-interactions of the gauge fields. The QCD part remains unaltered, since all the quark field redefinitions we have performed are unitary. It reads

$$\mathcal{L}_{QCD} + ig_{3}A_{\mu}^{A}(x)J^{A\mu}(x) \ ,$$

where the gluon currents are given by

$$J_{\mu}^{A} = \mathbf{u}_{L}^{\dagger}\sigma_{\mu}\frac{\lambda^{A}}{2}\mathbf{u}_{L} + \mathbf{u}_{R}^{\dagger}\bar{\sigma}_{\mu}\frac{\lambda^{A}}{2}\mathbf{u}_{R} + \mathbf{d}_{L}^{\dagger}\sigma_{\mu}\frac{\lambda^{A}}{2}\mathbf{d}_{L} + \mathbf{d}_{R}^{\dagger}\bar{\sigma}_{\mu}\frac{\lambda^{A}}{2}\mathbf{d}_{R} \ , \tag{3.75}$$

or in Dirac form

$$J_{\mu}^{A} = \bar{\mathbf{u}}\gamma_{\mu}\frac{\lambda^{A}}{2}\mathbf{u} + \bar{\mathbf{d}}\gamma_{\mu}\frac{\lambda^{A}}{2}\mathbf{d} \ , \tag{3.76}$$

where we have summed over family indices.

In the electroweak sector, all the gauge potential redefinitions are also unitary (canonical), and their kinetic terms are not altered, taking the expected form

$$-\frac{1}{4}(\partial_\mu A_\nu - \partial_\nu A_\mu)^2 - \frac{1}{4}(\partial_\mu Z_\nu - \partial_\nu Z_\mu)^2 \tag{3.77}$$

$$-\frac{1}{2}(\partial^\mu W^{+\nu} - \partial^\nu W^{+\mu})(\partial_\mu W^-_\nu - \partial_\nu W^-_\mu)\ . \tag{3.78}$$

After some serious algebra and integration by parts of the cross-term in $(F^3_{\mu\nu})^2$, the interactions linear in the couplings are just given by

$$i(eA^\mu + \frac{e}{\tan\theta_\mathrm{w}} Z^\mu)\left[W^{+\nu}(\partial_\nu W^-_\mu - \partial_\mu W^-_\nu) - W^{-\nu}(\partial_\nu W^+_\mu - \partial_\mu W^+_\nu)\right]\ , \tag{3.79}$$

where we have used the convenient decompositions

$$F^\pm_{\mu\nu} = \frac{1}{\sqrt{2}}(F^1_{\mu\nu} \mp i F^2_{\mu\nu}) = \partial_\mu W^\pm_\nu - \partial_\nu W^\pm_\mu \mp i g_2 (W^3_\nu W^\pm_\mu - W^3_\mu W^\pm_\nu)\ , \tag{3.80}$$

and

$$F^3_{\mu\nu} = \partial_\mu W^3_\nu - \partial_\nu W^3_\mu - i g_2 (W^-_\mu W^+_\nu - W^+_\mu W^-_\nu)\ . \tag{3.81}$$

The quartic self-interactions yield the complicated terms

$$\frac{1}{2}\frac{e^2}{\sin^2\theta_\mathrm{w}}[(W^-_\mu W^{-\mu})(W^+_\nu W^{+\nu}) - (W^-_\mu W^{+\mu})^2]$$
$$- \frac{e^2}{\sin^2\theta_\mathrm{w}}[(\cos\theta_\mathrm{w} Z_\mu + \sin\theta_\mathrm{w} A_\mu)(\cos\theta_\mathrm{w} Z^\mu + \sin\theta_\mathrm{w} A^\mu)(W^{+\rho} W^-_\rho)$$
$$- (\cos\theta_\mathrm{w} Z_\mu + \sin\theta_\mathrm{w} A_\mu) W^{+\mu}(\cos\theta_\mathrm{w} Z_\rho + \sin\theta_\mathrm{w} A_\rho) W^{-\rho}]\ . \tag{3.82}$$

This concludes the description of the standard model Lagrangian in the unitary gauge. This gauge choice is useful for identifying the physical degrees of freedom in the theory, but ackward for calculations.

To summarize, the standard model contains:

- **Gauge Particles:**

  - Eight massless vector gluons which mediate QCD interactions among quarks.
  - One massless photon which mediates the QED interactions of the charged particles.
  - One massive neutral spin one boson, the Z boson, with mass

$$M_Z = \frac{v}{2}\frac{e}{\sin\theta_w \cos\theta_w} , \tag{3.83}$$

  - One massive singly charged spin one boson, W^+ (and its antiparticle W^-), with mass

$$M_W = \frac{v}{2}\frac{e}{\sin\theta_w} . \tag{3.84}$$

- **Fermions:**

  - Three families of quarks and leptons, each containing one massless left-handed neutrino, one lepton of charge -1, and quarks of charge $2/3$ and $-1/3$.

- **Scalar Boson:**

  - One neutral Higgs spinless field of mass

$$M_H = v\sqrt{2\lambda} \qquad (v = \frac{\mu^2}{\lambda}) . \tag{3.85}$$

Interactions among these particles display some noteworthy features:

- **Interactions**

  - All charged particles interact with the photon with a strength proportional to their electric charge. The QED interactions are diagonal, preserve parity (P), and charge conjugation (C).
  - Only quarks interact with the eight QCD gluons. Their QCD interactions are universal and diagonal, and are, at the classical level, parity (P) and charge conjugation (C) invariant. These invariances are broken by subtle non-perturbative quantum effects associated with quantum anomalies.

- Charge-changing gauge interactions between different left-handed fermions mediated by the $W^{\pm}$ bosons. These preserve chirality, but violate P and C. They are family-diagonal and preserve CP on the leptons; for quarks, they generate transitions between different families, and explicitly break CP invariance.
- Charge-conserving interactions between the same left- and right-handed quarks and leptons, mediated by the Z-boson.
- Chirality-changing Yukawa interactions among massive fermions mediated by the neutral Higgs particle, to which they couple according to their mass.
- Cubic interactions between the Z's, the $W^{\pm}$ and the Higgs boson, with strength proportional to their mass squared.
- Cubic and quartic self-interactions of the Higgs particle.
- Quartic Interactions among the massive gauge bosons, and the photon.
- Quartic Yang-Mills interactions among the massless gluons.

These interactions respect four global quantum numbers that do not correspond to gauge symmetries: quark number, and the three lepton numbers, electron number, muon number, and tau number. A discussion, to come later, of the non-perturbative aspects of the vacuum state of the standard model, will alter these invariances.

PROBLEMS

A. Use the charge operator (3.8), to obtain the values of the left and right-handed quark charges.

B. Suppose we add to the standard model a weak isotriplet Higgs field with unit hypercharge. Write the covariant derivative acting on this new field. Assume that it gets a vacuum value v', distinct from that of the usual Higgs doublet. Find the new tree-level relation between the W and Z boson masses, and deduce a bound on its *vev*, using the experimental values of the masses of the gauge bosons.

C. Couple the Higgs triplet of the previous problem to the fermions of the standard model. Determine which fermions get a mass as a result of this coupling. Infer a bound on v'.

D. Consider the extension of the standard model with two Higgs doublets with opposite hypercharge, H_u and H_d. Assume the first couples to the charge 2/3 quarks, the second to the charge -1/3 quarks and to the charged leptons. Derive the most general potential $V(H_u, H_d)$ that is consistent with the symmetries of the original model with one Higgs doublet.

E. Analyze the vacuum structure of the two Higgs model of the previous model, and identify the new physical particles it contains.

3.2 CALCULATING IN THE STANDARD MODEL

The classical Lagrangian of the previous section provides the input for evaluating the quantum transition amplitudes. These are generated by the vacuum to vacuum transition amplitude in the presence of external sources which couple to the fields of the theory. Let us quickly remind the reader how this is done.

For a theory with classical action functional $S[\varphi]$, the transition ampitude is given by the path integral

$$\langle\Omega|\Omega\rangle_J = \int \mathcal{D}\varphi(x) \exp\left\{\frac{i}{\hbar}S[\varphi] + \frac{i}{\hbar}\int d^4x J(x)\varphi(x)\right\} . \tag{3.86}$$

In order to handle the oscillatory character of the integrand, one may introduce a convergence factor of the form

$$\exp\left\{-\frac{\epsilon}{2}\int d^4x\varphi^2(x)\right\} , \tag{3.87}$$

for each field, where ϵ is a positive small number which is taken to zero at the end of all calculations. This is the same as shifting the mass squared of each boson field by $-i\epsilon$; it corresponds to a particular choice of boundary condition for the propagator of the field.

Another way is to start from the path integral in Euclidean space, and do all the calculations there, and when a physical answer is required, continue it to Minkowski space. The procedure is rather simple. We intoduce Euclidean coordinates, $\bar{x}_\mu$ as follows

$$\bar{x}_\mu = (ix_0, \vec{x}) , \tag{3.88}$$

where the unbarred coordinates are in Minkowski space. Correspondingly, the Minkowski measure is given by

$$d^4x = -id^4\bar{x} . \tag{3.89}$$

It follows that

$$\partial_\mu \varphi \partial^\mu \varphi = -\bar{\partial}_\mu \varphi \bar{\partial}_\mu \varphi \ , \tag{3.90}$$

where

$$\bar{\partial}_\mu = \frac{\partial}{\partial \bar{x}_\mu} \ . \tag{3.91}$$

Similarly, the Euclidean time component of the vector potential is purely imaginary. We also note that there is no distinction between upper and lower indices.

For a generic field theory, the Euclidean space integrand of the path integral is then

$$\exp\left\{ - \int d^4\bar{x} \left[\frac{1}{2}(\bar{\partial}_\mu \varphi \bar{\partial}_\mu \varphi + m^2 \varphi^2) + V^{cl}_{int}(\varphi) + J(x)\varphi(x) \right] \right\} \ . \tag{3.92}$$

The integral for the free part is Gaussian, and can be evaluated, at least when the interaction terms are small.

It is sometimes useful to consider the transition amplitude as a functional of the Legendre transform of the source J. If we set in Euclidean space

$$\langle \Omega | \Omega \rangle_J \equiv e^{-Z[J]} \ , \tag{3.93}$$

and introduce the "classical field"

$$\varphi_{cl} \equiv \frac{\delta Z[J]}{\delta J} \ , \tag{3.94}$$

then the "action functional", defined through

$$\Gamma_{eff}[\varphi_{cl}] \equiv Z[J] - \int d^4\bar{x} J(\bar{x}) \varphi_{cl}(\bar{x}) \ , \tag{3.95}$$

takes on a particularly transparent form. For the generic field theory above, it has the same form as the classical action, augmented by its quantum corrections

$$\Gamma_{eff}[\varphi_{cl}] = \int d^4\bar{x} \{ \frac{1}{2} F_1(\varphi_{cl}) \bar{\partial}_\mu \varphi_{cl} \bar{\partial}_\mu \varphi_{cl} + V_{eff}(\varphi_{cl}) + F_2(\varphi_{cl}) [\bar{\partial}_\mu \varphi_{cl} \bar{\partial}_\mu \varphi_{cl}]^2 + \cdots \} \ . \tag{3.96}$$

where

$$F_1(\varphi_{cl}) = 1 + \mathcal{O}(\hbar) \ , \qquad F_2(\varphi_{cl}) = \mathcal{O}(\hbar) \ , \tag{3.97}$$

$$V_{eff}(\varphi_{cl}) = m^2_{phys}\varphi^2_{cl} + V^{cl}_{int}(\varphi_{cl}) + \mathcal{O}(\hbar) \ . \tag{3.98}$$

Here m_{phys} is the physical mass of the particle described by the field; V_{eff} is the effective potential. Hence Γ_{eff} reduces to the classical action in the classical limit ($\hbar \to 0$), while encapsulating *all* the quantum corrections to the classical action. Unlike the renormalizable classical Lagrangian which contains only up to fourth power of the field, and second power of derivatives, the quantum action contains arbitrarily high powers of both the derivatives and the fields.

In the following, we set up the necessary tools to calculate Γ in perturbation theory. The process is much more complicated in the standard model than in this generic field theory, but the idea is the same: calculate the effective action.

The standard model contains two types of gauge theories, QCD with the unbroken SU^c_3 color gauge group, and the Electroweak theory based on $SU^W_2 \times U^Y_1$, spontaneously broken to Maxwell's U^γ_1. Perturbative calculations in spontaneously broken gauge theories contain extra subtelties of their own on top of those already present in unbroken gauge theories.

In the previous section, we presented the Lagrangian in a special gauge called the unitary gauge where all the Nambu-Goldstone bosons have disappeared to become the longitudinal modes of the massive gauge fields. This gauge provides a very clear discussion of the classical Lagrangian, but it is not the most convenient for calculating quantum corrections. We digress for a moment to remind the reader of the different ways both broken and unbroken gauges are fixed in order to perform perturbative calculations.

Gauge fixing in unbroken gauge theories

In the following, we only summarize the salient facts. Gauge fixing is necessary because gauge invariance is realized on more fields than there are degrees of freedom. At a working level, the problem arises through the non-invertability of the propagator of the gauge field. The most elegant formulation of the problem is due to Faddeev and Popov who present the whole problem as one of redefining the path integral measure. The path integration over the gauge potentials contains an infinite number of redundant variables, since the integrand is invariant under gauge transformations. One must start from a reduced measure, the Haar measure, which does not involve the redundant integrations. It necessarily depends on the choice of gauge for the potentials.

In Abelian gauge theory, with the Fermi gauge choice $\bar{\partial}_\mu A^\mu \approx 0$, the net result of the Faddeev-Popov procedure is to replace the Euclidean space classical Lagrangian by

$$\mathcal{L}_{\rm cl} = \frac{1}{4g^2} F_{\mu\nu} F_{\mu\nu} \ , \quad F_{\mu\nu} = \bar{\partial}_\mu A_\nu - \bar{\partial}_\nu A_\mu \ , \tag{3.99}$$

by the Faddev-Popov Lagrangian

$$\mathcal{L}_{\rm FP} = \frac{1}{4g^2} F_{\mu\nu} F_{\mu\nu} + \frac{1}{2\alpha g^2} (\bar{\partial} \cdot A)^2 + i \bar{\partial}_\mu \eta^* \bar{\partial}_\mu \eta \ , \tag{3.100}$$

where $\eta(x)$ are complex anticommuting classical Grassmann fields (called ghosts), and α is a gauge parameter. The transition amplitude is obtained by path-integrating over the gauge potential A_μ and the ghosts

$$\langle \Omega | \Omega \rangle_{J_\mu, \xi, \xi^*} = \int \mathcal{D} A_\mu \mathcal{D} \eta \mathcal{D} \eta^* \exp \left\{ \int d^4 \bar{x} \left[\mathcal{L}_{\rm FP} + J_\mu A_\mu + i \eta \xi + i \eta^* \xi^* \right] \right\} \ . \tag{3.101}$$

Clearly the ghosts play no role in this case, but we have left them in order to point out an important invariance of this path integral. Although this Lagrangian is no longer invariant under arbitrary gauge transformations, it still displays invariance under a special class of gauge transformations of the form

$$\delta A_\mu = \bar{\partial}_\mu (\zeta^* \eta + \eta^* \zeta) \ , \tag{3.102}$$

$$\delta \eta = \frac{i}{g^2 \alpha} (\bar{\partial} \cdot A) \zeta \ , \qquad \delta \eta^* = - \frac{i}{g^2 \alpha} (\bar{\partial} \cdot A) \zeta^* \ , \tag{3.103}$$

where ζ and ζ^* are global Grassmann variables. This invariance of the gauge fixed Lagrangian is called BRST, after the names of its inventors, serves as the quantum version of the classical gauge invariance.

It is easy to check that the BRST transformations are nilpotent, that is $\delta\delta = 0$. The effective Lagrangian can be rewritten in the elegant form

$$\mathcal{L}_{\rm FP} = \frac{1}{4g^2} F_{\mu\nu} F_{\mu\nu} + \delta \left(\bar{\partial} \cdot A \eta^* \right) \ , \tag{3.104}$$

which shows it to be invariant, given the nilpotency of the BRST transformations.

This invariance, the quantum remnant of the classical invariance, is sufficient to derive the necessary miracles (Ward identities) that insure renormalizability. Note that the ghosts do not couple in the Abelian theory. In this gauge, the gauge field propagator is just

$$\frac{1}{\bar{p}^2}\left(\delta_{\mu\nu} - (1-\alpha)\frac{\bar{p}_\mu \bar{p}_\nu}{\bar{p}^2}\right)$$

The special choice $\alpha = 0$ corresponds to the Landau gauge, while $\alpha = 1$ yields the Feynman gauge. It is sometimes convenient to keep α in the calculations, as a check, since it should disappear from physical answers.

The situation generalizes to non-Abelian gauge theories. The only difference is that the ghosts couple to the physical fields and thus must be included in physical calculations. If $\mathcal{G}^B = 0$ is the gauge fixing condition, the classical Yang Mills Lagrangian is altered to

$$\mathcal{L}^{\rm FP}_{YM} = \mathcal{L}^{\rm cl}_{YM} + \frac{1}{2\alpha}\mathcal{G}^B\mathcal{G}^B - i\eta^{*A}\frac{\delta\mathcal{G}^A}{\delta\omega^B}\eta^B \ , \tag{3.105}$$

where η^{*A}, η^A are the Grassmann ghosts, transforming as the members of the adjoint representation of the gauge algebra, and $\frac{\delta\mathcal{G}^B}{\delta\omega^C}$ denotes the change of the gauge function under a gauge transformation with parameters ω^C. As in the Abelian case, this Lagrangian is also invariant under a nilpotent BRST transformation.

If we choose $\mathcal{G}^B = \bar{\partial}\cdot A^B$, the (Euclidean) gauge propagator is given by

A B

$$\frac{\delta^{AB}}{\bar{p}^2}\left(\delta_{\mu\nu} - (1-\alpha)\frac{\bar{p}_\mu \bar{p}_\nu}{\bar{p}^2}\right)$$

The ghost propagator is given by

A B

$$-i\frac{\delta^{AB}}{\bar{p}^2}$$

Since we have

$$\frac{\delta\mathcal{G}^A}{\delta\omega^B} = -\bar{\partial}_\mu\bar{\partial}_\mu\delta^{AB} - f^{ABC}\bar{\partial}_\mu A^C_\mu \ , \tag{3.106}$$

the ghosts interact with the gauge potential.

These procedures are well understood and detailed in many field theory books. In the standard model, they apply to the unbroken gauge groups, $SU_3^c \times U_1^\gamma$. The broken $SU_2^W \times U_1^Y$ presents new challenges which alter this situation somewhat, and makes a detailed discussion compulsory.

Gauge fixing in broken gauge theories

Before discussing the standard model, consider the simpler model of a spontaneously broken Abelian gauge theory. Its Minkowski space Lagrangian is

$$\mathcal{L}_{cl} = -\frac{1}{4g^2}F^{\mu\nu}F_{\mu\nu} + (\mathcal{D}_\mu\phi)^*(\mathcal{D}^\mu\phi) + \mu^2\phi^*\phi - \lambda(\phi^*\phi)^2 \,, \tag{3.107}$$

where ϕ has unit charge

$$\mathcal{D}_\mu\phi = (\partial_\mu + iA_\mu)\phi \,. \tag{3.108}$$

The field configuration with minimum energy density is such that

$$\phi_0^*\phi_0 = \frac{\mu^2}{2\lambda} \equiv \frac{v^2}{2} \,. \tag{3.109}$$

The covariant kinetic term generates a mass term for the gauge field $M = gv$. Rather than proceeding to the unitary gauge, we simply express the Lagrangian in terms of the most general decomposition of the fields which have zero vacuum value:

$$\phi = \frac{1}{\sqrt{2}}(\varphi_1 + i\varphi_2) + \frac{v}{\sqrt{2}} \,, \tag{3.110}$$

choosing v to be real (students: what if v is not real?). A little bit of algebra enables us to rewrite the quadratic part of the Euclidean space Lagrangian

$$\begin{aligned}\mathcal{L}_{\rm cl} =& \frac{1}{4}(\bar\partial_\mu A_\nu - \bar\partial_\nu A_\mu)^2 + \frac{1}{2}(\bar\partial_\mu\varphi_1\bar\partial_\mu\varphi_1 + \bar\partial_\mu\varphi_2\bar\partial_\mu\varphi_2)\\ &+ \frac{1}{2}M^2A_\mu A_\mu + MA_\mu\bar\partial_\mu\varphi_2 + \frac{1}{2}m_H^2\varphi_1^2 \,,\end{aligned} \tag{3.111}$$

where $m_H^2 = 2\lambda v^2$. Let us contrast this expression with that obtained by writing the Higgs field in the Kibble form

$$\phi = \frac{1}{\sqrt{2}} e^{i\frac{\theta(x)}{v}} (\rho(x) + v) \ .$$

In the unitary gauge, $\theta(x)$ is absorbed by an appropriate gauge transformation. The quadratic part of the Lagrangian then looks much nicer

$$\mathcal{L}_{\rm cl} = \frac{1}{4}(\bar{\partial}_\mu A_\nu - \bar{\partial}_\nu A_\mu)^2 + \frac{1}{2}\bar{\partial}_\mu \rho \bar{\partial}_\mu \rho + \frac{1}{2} M^2 A_\mu A_\mu \ . \qquad (3.112)$$

The Kibble form is a special case of the general decomposition, obtained from the first by setting

$$\varphi_1 = \frac{1}{\sqrt{2}}(\rho(x) + v) \ , \quad \varphi_2 = 0 \ ,$$

which has actually frozen A_μ in a particular gauge. These two parameterizations correspond to different gauge choices.

There remains to see how to characterize these different gauge choices. In the unitary gauge, we note that the combination $\phi_0^* \phi$ is real $\phi_0^* \phi = \frac{1}{2} v(v + \rho(x))$, suggesting the gauge condition

$$(\phi_0^* \phi - \phi^* \phi_0) \approx 0 \qquad \text{(unitary gauge)} \ , \qquad (3.113)$$

which does not depend on the gauge fields.

When the Lagrangian is expressed in terms of $\varphi_1(x)$ and $\varphi_2(x)$, there is a quadratic term $M A_\mu \bar{\partial}_\mu \varphi_2$, which mixes the gauge field and the imaginary part of the scalar field. After integration by parts, we rewrite it as $-M\varphi_2 \bar{\partial} \cdot A$, suggestive of a gauge since it involves $\bar{\partial} \cdot A$ which is like a naive gauge function. By comparing it with the old fashioned Fermi term $\frac{1}{2\alpha}(\bar{\partial} \cdot A)^2$, it can be viewed as a cross term between $\partial \cdot A$ and $\alpha M \varphi_2$. Thus, following 't Hooft, we change the name of the gauge parameter, and choose the gauge function to be

$$\mathcal{G} = (\bar{\partial} \cdot A + \xi M \varphi_2) = \bar{\partial} \cdot A - i\xi(\phi_0^* \phi - \phi_0 \phi^*) \ . \qquad (3.114)$$

This gauge is called the R_ξ gauge. We now see that the addition of the term

$$\frac{1}{2\xi} \mathcal{G}^2 \qquad (3.115)$$

to $\mathcal{L}_{\rm cl}$ exactly cancels the troublesome cross-term, and creates a mass term for the φ_2 field. We now apply the same procedure as in the unbroken case, but with this gauge function. The Faddeev-Popov improved Lagrangian in Euclidean space is now

$$\mathcal{L}_{\rm FP} = \frac{1}{4}(\bar{\partial}_\mu A_\nu - \bar{\partial}_\nu A_\mu)^2 + \frac{1}{2}(\bar{\partial}_\mu \varphi_1 \bar{\partial}_\mu \varphi_1 + \bar{\partial}_\mu \varphi_2 \bar{\partial}_\mu \varphi_2) + \mathcal{L}_{\rm ghost}$$
$$+ \frac{1}{2\xi}(\bar{\partial} \cdot A)^2 + \frac{1}{2}M^2 A_\mu A_\mu + \frac{1}{2}m_H^2 \varphi_1^2 + \frac{1}{2}\xi M^2 \varphi_2^2 \, . \qquad (3.116)$$

The ghost part is obtained from

$$\mathcal{L}_{\rm ghost} = -i\eta^* \frac{\delta \mathcal{G}}{\delta \omega}\eta = i\eta^* \left(\frac{1}{g}\bar{\partial}_\mu \bar{\partial}_\mu - \xi M(\varphi_1 + v) \right) \eta \, , \qquad (3.117)$$

is the change of the gauge function under an infinitesimal gauge transformation. After absorbing $\frac{1}{g}$ in the ghost fields, we obtain

$$\mathcal{L}_{\rm ghost} = -i\bar{\partial}_\mu \eta^* \bar{\partial}_\mu \eta - i\xi M^2 \eta^* \eta - i\xi g M \eta^* \eta \varphi_1 \, . \qquad (3.118)$$

Noote that in this gauge, even in the Abelian case, the ghosts interact and must be taken into account.

We now can read-off the various propagators from the full effective Lagrangian

$$A_\mu : \quad \frac{1}{\bar{p}^2 + M^2}[\delta_{\mu\nu} - (1-\xi)\frac{\bar{p}_\mu \bar{p}_\nu}{\bar{p}^2 + \xi M^2}] \, ,$$

$$\varphi_1 : \quad \frac{1}{\bar{p}^2 + m_H^2} \, ,$$

$$\varphi_2 : \quad \frac{1}{\bar{p}^2 + \xi M^2} \, ,$$

$$\eta : \quad -i\frac{1}{\bar{p}^2 + \xi M^2} \, .$$

We note that both the unphysical field φ_2 and the ghosts have a ξ-dependent pole. Note that as $\xi \to \infty$, φ_2 becomes infinitely massive, and the gauge propagator reduces to

$$\frac{1}{\bar{p}^2 + M^2}\left(\delta_{\mu\nu} + \frac{\bar{p}_\mu \bar{p}_\nu}{M^2}\right) .$$

We now understand why the unitary gauge may not be so convenient as the gauge propagator contains terms along its momentum. When $\xi = 1$, the propagator is the simplest, in fact the same as in the Feynman gauge. This is an example of S. Fubini's "law of conservation of trouble": simplicity in the gauge propagator means a complicated ghost and Higgs structure, and vice-versa. As all good calculators have found, it is advisable to keep ξ until all calculations are done, since the absence of ξ in physical answers provides a partial check to calculations.

Gauge Fixing in the Electroweak Model

Gauge fixing in the electroweak part of the standard model follows along the same lines, albeit with some complications; in the R_ξ gauge, new ghosts appear due to the non-Abelian nature of SU_2, there are more than one gauge conditions to consider, and there are additional subtelties due to the presence of the unbroken electromagnetic subgroup.

It is instructive to first characterize the unitary gauge ($\xi = \infty$). In that gauge, the Higgs field is taken to be

$$H = \frac{1}{\sqrt{2}}\begin{pmatrix} 0 \\ h(x) + v \end{pmatrix} , \tag{3.119}$$

where the Nambu-Goldstone bosons have been absorbed by an appropriate SU_2^W gauge transformation. This particular form of H can be obtained by demanding that, given

$$H_0 = \frac{1}{\sqrt{2}}\begin{pmatrix} 0 \\ v \end{pmatrix} , \tag{3.120}$$

we impose the condition

$$\mathrm{Im}\,(H_0^\dagger \vec{\tau} H) = 0 . \tag{3.121}$$

This condition characterizes the unitary gauge. In general, we start from the parametrization

$$H(x) = \begin{bmatrix} \varphi^+(x) \\ \frac{1}{\sqrt{2}}(\varphi_1(x) + i\varphi_2(x) + v) \end{bmatrix} , \tag{3.122}$$

which produces in the Lagrangian the cross-terms

$$\frac{i}{2}g_2v(W_\mu^+\bar{\partial}_\mu\varphi^- - W_\mu^-\bar{\partial}_\mu\varphi^+) + i\frac{v}{2}(g_1B_\mu - g_2W_\mu^3)\bar{\partial}_\mu\varphi_2 ,$$

which must be cancelled by the appropriate gauge choices. For the two gauge groups, they are

$$\mathcal{G}^a = \bar{\partial}_\mu W_\mu^a - \frac{i}{2}\xi g_2(H_0^\dagger\tau^a H - H^\dagger\tau^a H_0) , \tag{3.123}$$

$$\mathcal{G} = \bar{\partial}_\mu B_\mu - i\frac{\xi}{2}g_1(H_0^\dagger H - H^\dagger H_0) . \tag{3.124}$$

Note that the same gauge parameter appears in both (What happens if one choose ξ_1 and ξ_2 for each?). These particular choices are sufficient for our purposes. They lead to the following extra terms in the Lagrangian

$$\frac{1}{2\xi}\left[\mathcal{G}^a\mathcal{G}^a + \mathcal{G}\mathcal{G} + (\bar{\partial}_\mu W_\mu^a)^2 + (\bar{\partial}_\mu B_\mu)^2\right] + \xi M_W^2\varphi^+\varphi^- + \frac{1}{2}\xi M_Z^2\varphi_2^2 .$$

In this general gauge, there are ghost fields accompanying both SU_2 and U_1 gauge particles. Let η^a , $a = 1, 2, 3$ be the SU_2 ghosts, χ the U_1 ghost. Then, they appear in the Lagrangian in the form

$$-i(\eta^{*a}\chi^*)\begin{pmatrix} \frac{\delta\mathcal{G}^a}{\delta\omega^b} & \frac{\delta\mathcal{G}^a}{\delta\omega} \\ \frac{\delta\mathcal{G}}{\delta\omega^b} & \frac{\delta\mathcal{G}}{\delta\omega} \end{pmatrix}\begin{pmatrix} \eta^b \\ \chi \end{pmatrix} .$$

Under the gauge transformations

$$\delta W_\mu^a = -\frac{1}{g_2}\bar{\partial}_\mu\omega^a + \epsilon^{abc}W_\mu^b\omega^c , \tag{3.125}$$

$$\delta H = i\frac{\tau^a}{2}\omega^a H + i\omega H , \tag{3.126}$$

$$\delta B_\mu = -\frac{2}{g_1}\bar{\partial}_\mu\omega , \tag{3.127}$$

we find

$$\frac{\delta\mathcal{G}}{\delta\omega} = \frac{2}{g_1}[-\bar{\partial}_\mu\bar{\partial}_\mu + \frac{\xi}{4}g_1^2v(\varphi_1 + v)] ,$$

$$\frac{\delta \mathcal{G}}{\delta \omega^a} = \frac{\xi g_1}{4\sqrt{2}} \begin{cases} -v\sqrt{2}(\varphi_1 + v) & \text{a=3} , \\ v(\varphi^+ + \varphi^-) & \text{a=1} , \\ iv(\varphi^+ - \varphi^-) & \text{a=2} , \end{cases}$$

$$\frac{\delta \mathcal{G}^a}{\delta \omega} = \frac{\xi g_2}{2\sqrt{2}} \begin{cases} -v(\varphi_1 + v) & \text{a=3} , \\ v(\varphi^+ + \varphi^-) & \text{a=1} , \\ iv(\varphi^+ - \varphi^-) & \text{a=2} , \end{cases}$$

$$\frac{\delta \mathcal{G}^a}{\delta \omega^c} = \frac{1}{g_2}[-\bar{\partial}_\mu \bar{\partial}_\mu \delta^{ac} + g_2 \epsilon^{abc} \bar{\partial}_\mu W^b_\mu + \frac{\xi}{4} g_2^2 v(\varphi_1 + v)\delta^{ac}]$$

$$+ \frac{i\xi g_2}{4\sqrt{2}} \epsilon^{abc} \begin{cases} 0 & \text{b=3} , \\ v(\varphi^+ - \varphi^-) & \text{b=1} , \\ iv(\varphi^+ + \varphi^-) & \text{b=2} . \end{cases}$$

We now absorb $\frac{2}{g_1}$ into ω and $\frac{1}{g_2}$ into ω^a, and introduce the combinations

$$\chi_z = \cos\theta_{\mathrm{W}} \chi - \sin\theta_{\mathrm{W}} \eta^3, \tag{3.128}$$

$$\chi_\gamma = \sin\theta_{\mathrm{W}} \chi + \cos\theta_{\mathrm{W}} \eta^3 \tag{3.129}$$

$$\eta^\pm = \frac{1}{\sqrt{2}}(\eta^1 \mp i\eta^2) , \tag{3.130}$$

and obtain canonical kinetic and mass terms

$$i\{(\bar{\partial}_\mu \bar{\chi}_z \bar{\partial}_\mu \chi_z + \bar{\partial}_\mu \bar{\chi}_\gamma \bar{\partial}_\mu \chi_\gamma + \bar{\partial}_\mu \bar{\eta}^+ \bar{\partial}_\mu \eta^- + \bar{\partial}_\mu \bar{\eta}^- \bar{\partial}_\mu \eta^+$$
$$+ \xi M_Z^2 \bar{\chi}_z \chi_z + \xi M_W^2 (\bar{\eta}^+ \eta^- + \bar{\eta}^- \eta^+)\} . \tag{3.131}$$

Note that there are two charged ghosts and two charged antighosts.

It is straightforward to obtain the Euclidean space propagators for the gauge and scalar fields:

$$Z_\mu : \quad \frac{1}{\bar{p}^2 + M_Z^2}\left[\delta_{\mu\nu} - (1-\xi)\frac{\bar{p}_\mu \bar{p}_\nu}{\bar{p}^2 + \xi M_Z^2}\right] ,$$

$$W_\mu^\pm : \quad \frac{1}{\bar{p}^2 + M_W^2}\left[\delta_{\mu\nu} - (1-\xi)\frac{\bar{p}_\mu \bar{p}_\nu}{\bar{p}^2 + \xi M_W^2}\right] ,$$

$$A_\mu : \quad \frac{1}{\bar{p}^2}\left[\delta_{\mu\nu} - (1-\xi)\frac{\bar{p}_\mu \bar{p}_\nu}{\bar{p}^2}\right] ,$$

$$\varphi_2 : \quad \frac{1}{\bar{p}^2 + \xi M_Z^2} \qquad \text{(fictitious)} ,$$

$$\varphi^\pm : \quad \frac{1}{\bar{p}^2 + \xi M_W^2} \qquad \text{(fictitious)} ,$$

$$\varphi_1 : \quad \frac{1}{\bar{p}^2 + M_H^2} \qquad \text{(Higgs)} ,$$

as well as the ghost propagators

$$\chi_Z : \quad \frac{-i}{\bar{p}^2 + \xi M_Z^2} ,$$

$$\chi_\gamma : \quad \frac{-i}{\bar{p}^2} ,$$

$$\eta^\pm : \quad \frac{-i}{\bar{p}^2 + \xi M_W^2} .$$

The interactions between the ghosts and the gauge fields occur through the term

$$-g_2\epsilon^{abc}\bar{\eta}^a\epsilon^{abc}\bar{\partial}_\mu(W^b_\mu\eta^c) \ ,$$

which is rewritten as (recognize the A_μ coupling as the electromagnetic current for the charged ghosts ($g_2 \sin\theta_w = e$))

$$\begin{aligned}
&- ig_2(\cos\theta_w Z_\mu + \sin\theta_w A_\mu)(\bar{\partial}_\mu\bar{\eta}^+\eta^- - \bar{\partial}_\mu\bar{\eta}^-\eta^+) \\
&+ ig_2(\cos\theta_w\bar{\chi}_\gamma - \sin\theta_w\bar{\chi}_z)(\bar{\partial}_\mu W^+_\mu\eta^- - \bar{\partial}_\mu W^-_\mu\eta^+) \\
&- ig_2(\bar{\partial}_\mu W^+_\mu\bar{\eta}^- - \bar{\partial}_\mu W^-_\mu\bar{\eta}^+)(\cos\theta_w\chi_\gamma - \sin\theta_w\chi_z) \ . \qquad (3.132)
\end{aligned}$$

The interaction of the Higgs field φ_1 with the ghosts is obtained by replacing all the ghosts masses $\xi M^2_{W,Z}$ by the factor $(1+\frac{\varphi_1}{v})$, leading to the interaction terms

$$\frac{\xi}{v}\varphi_1\left\{M^2_Z\bar{\chi}_z\chi_z + M^2_W(\bar{\eta}^+\eta^- + \bar{\eta}^-\eta^+)\right\} \ . \qquad (3.133)$$

Finally, the ghost interactions with the Nambu-Goldstone bosons are given by

$$\begin{aligned}
&\frac{\xi}{2}\varphi^+\left\{\frac{eM_Z}{\sin\theta_w}\bar{\chi}_z\eta^- + 2eM_W\bar{\eta}^-(\chi_\gamma + \frac{1}{\tan 2\theta_w}\chi_z)\right\} \\
&+\frac{\xi}{2}\varphi^-\left\{\frac{eM_Z}{\sin\theta_w}\bar{\chi}_z\eta^+ + 2eM_W\bar{\eta}^+(\chi_\gamma + \frac{1}{\tan 2\theta_w}\chi_z)\right\} \ .
\end{aligned}$$

We leave it as an exercise to extract Feynman rules from these expressions. For completeness, we note that the QCD part of the standard model, being an unbroken gauge theory, has simpler Feynman rules, with its own gauge parameters. We do not dwell on this point since it is adequately covered in the many QFT primers to be found in good libraries.

We have assembled all the ingredients necessary for computing perturbative quantum corrections. In the electroweak sector, the tree level interactions do reproduce the gross features of the interactions among leptons. However interactions involving quarks at large distances must be reinterpreted.

An essential element in comparing any quantum field theory in detail with experiment is the quantum renormalization group. It endows all the input parameters in the classical Lagrangian with a dependence on the scale at which they are measured; coupling constants, masses, and CKM angles all depend on scale, in a precise way which can be calculated in the context of perturbation theory.

For QCD, this dependence tells us that its coupling constant grows large with distance, and decreases at shorter distances (asymptotic freedom). As a result, quarks and gluons cannot exist as isolated asymptotic states, far away from other colored objects. Only colorless combinations of three quarks, or a quark and an antiquark can escape this infernal force. Baryons are identified with the three-quark combination, mesons with the quark-antiquark pair. Under these assumptions, the quarks inside a baryon may be considered as quasi-free as long as they do not stray too far away from their fellow constituents. In this case, "far" means a nuclear size $\sim$ fermi. Thus we will treat each quark as a quasifree *parton*, each with a fraction of the baryon's momentum. We will assume this picture when treating the electroweak interactions of quarks.

PROBLEMS

A. In the $U(1)$ toy model, start from the unitary gauge condition

$$\mathcal{G} = (\phi_0^* \phi - \phi^* \phi_0) ,$$

and derive the expressions for all the Euclidean space propagators. Verify that they correspond to those derived in the text in the limit $\xi \to \infty$.

B. Derive the Feynman rules for the three and four-point vertices among physical particles of the standard model.

C. Verify the Feynman rules for the standard model gauge bosons and ghost propagators in the text in the R_ξ gauge.

D. Derive the Feynman rules involving the ghosts of the standard model in the R_ξ gauge.

E. Derive the form of the BRST transformation in the standard model.

F. Discuss the reason why one uses only one gauge parameter for the gauge conditions (3.123) and (3.124). Can you think of any situation where it might be useful to assign different parameters?

G. Derive the form of the Faddeev-Popov Lagrangian when there are two Higgs doublets. Use your result to read-off the propagators for the new Higgs particles.

CHAPTER 4

STANDARD MODEL

TREE-LEVEL PROCESSES

This chapter describes in some detail the tree-level computations of some elementary standard model processes. Both leptons and quarks are treated on the same footing, ignoring for the most part that quarks do not exist as free states. Quantum corrections to these results are relegated to a later chapter.

4.1 TREE-LEVEL PARAMETERS

In chapter 2 we described the classical standard model Lagrangian in terms of eighteen parameters: three gauge coupling constants, three charged lepton masses, six quark masses, three flavor-mixing angles, one CP-violating phase, the Higgs mass and the vacuum value of the Higgs field which is fixed by the masses of the W and Z gauge bosons. All these parameters have been measured to varying degrees of accuracy, except the Higgs mass for which there is an upper limit of 95 GeV (LEP, 1999).

All parameters of the standard model run (some only crawl) with the scale at which they are measured. Lepton masses which hardly run with scale can be measured directly through kinematics. To distinguish the running from the physical fermion mass, we denote the latter in capital letters. The experimental values of the charged lepton masses are

$$M_e = .511 \text{ MeV} \ , \quad M_\mu = 105.66 \text{ MeV} \ , \quad M_\tau = 1777.05 \text{ MeV} \ . \quad (4.1)$$

Since quarks are subject to the strong QCD force, they do not exist as asymptotic states. This makes the measurements of their masses at best indirect, and subject to qualifications. The masses of the three lightest quarks are extracted from the effective low energy chiral Lagrangian (see next chapter), with the range of values

$$M_u = 2 - 8 \text{ MeV} \ , \quad M_d = 5 - 15 \text{ MeV} \ , \quad M_s = 100 - 300 \text{ MeV} \ . \quad (4.2)$$

For heavier quarks, one may adopt a universal operational definition for their physical mass as say, half the energy needed for pair production of the quark-antiquark bound state. The other three quarks have masses which are all above the QCD scale, making their extraction from the data more direct, with values ranging over

$$M_c = 1.0 - 1.6 \text{ GeV} \ , \ M_b = 4.1 - 4.5 \text{ GeV} \ , \ M_t = 175.5 \pm 5.5 \text{ GeV} \ . \ (4.3)$$

The masses and widths of the W and Z bosons have been measured to remarkable accuracies

$$M_W = 80.37 \pm 0.09 \text{ GeV} \ , \qquad \Gamma_W = 2.06 \pm 0.06 \text{ GeV} \ , \qquad (4.4)$$

$$M_Z = 91.187 \pm 0.007 \text{ GeV} \ , \qquad \Gamma_Z = 2.496 \pm 0.0027 \text{ GeV} \ . \qquad (4.5)$$

The rest of the parameters appear in the Cabibbo-Kobayashi-Maskawa matrix, traditionally represented as

$$\mathcal{U} = \begin{pmatrix} V_{ud} & V_{us} & V_{ub} \\ V_{cd} & V_{cs} & V_{cb} \\ V_{tu} & V_{ts} & V_{tb} \end{pmatrix} \ . \qquad (4.6)$$

Not all of its nine matrix elements have been measured directly, but all can be inferred from the unitarity relations

$$\sum_{k=d,s,b} V^*_{ki} V_{kj} = \delta_{ij} \ , \ i,j = u,c,t \ . \qquad (4.7)$$

Since the rotation group is generated by the three family Gell-Mann matrices $\boldsymbol{\lambda}_2$, $\boldsymbol{\lambda}_5$, $\boldsymbol{\lambda}_7$, we can write the CKM matrix *à la* Euler

$$\mathcal{U} = e^{ia_4\boldsymbol{\lambda}_4}e^{ia_5\boldsymbol{\lambda}_5}e^{ia_7\boldsymbol{\lambda}_7}e^{ia_2\boldsymbol{\lambda}_2} , \tag{4.8}$$

where the CP-violating phase is set along $\boldsymbol{\lambda}_4$. Following L. Wolfenstein, we express its parameters as a power series in the Cabibbo angle,

$$a_2 = \lambda \ ; \ a_7 = A\lambda^2 \ ; \ a_5 = A\rho\lambda^3 \ ; \ a_4 = -A\eta\lambda^3 , \tag{4.9}$$

where λ is the Cabibbo angle. The first three parametrize the rotations, and $A\eta$ denotes the CP violating phase. In matrix form,

$$\mathcal{U} = \begin{pmatrix} 1-\frac{\lambda^2}{2} & \lambda & A\lambda^3(\rho - i\eta) \\ -\lambda & 1-\frac{\lambda^2}{2} & A\lambda^2 \\ A\lambda^3(1-\rho-i\eta) & -A\lambda^2 & 1 \end{pmatrix} , \tag{4.10}$$

where A, ρ, and η are of order one. Their (*1999*) values

$$\lambda \approx .2205 \pm .0018 \ ; \ A \approx 0.81 \pm 0.06 , \tag{4.11}$$

are obtained from direct measurements of V_{us} and V_{cb}, respectively. From the branching ratio of B-meson decay into charm and up quarks, we get

$$\sqrt{\rho^2 + \eta^2} = 0.36 \ \pm 0.09 . \tag{4.12}$$

The extraction of the phase from data is more indirect, as it necessarily involves loops with the third family, since CP-violation in B-decay has not yet been observed. One can at best constrain its value, either by deducing V_{td} from $B - \overline{B}$ mixing, or $K - \overline{K}$ mixing. One can quote the central values

$$\rho \approx 0.05 , \qquad \eta \approx 0.35 . \tag{4.13}$$

Finally we note that since the CKM matrix is unitary, the first column times the complex conjugate of the third column must vanish, yielding

$$A\lambda^3(\rho + i\eta) - A\lambda^3 + A\lambda^3(1 - \rho - i\eta) = 0. \tag{4.14}$$

With this parametrization, this is obviously tautological, but we may view the three factors, $\rho + i\eta$, -1, and $1 - \rho - i\eta$ as the sides of a triangle in the complex (ρ, η) plane, called the *unitarity triangle*. Its sides and its angles (CP-violation) can be independently determined from experiments. If after these measurements, the triangle does not close, there must be sources of CP-violation beyond the standard model.

4.2 W-DECAY

As a first example of a tree-level application of the standard model, we calculate the decay rate of the W-boson into an electron and its associated antineutrino,

$$W^- \to e^- + \bar{\nu}_e \ .$$

The decay proceeds at tree-level through the interaction

$$\mathcal{L}_{int} = \frac{ig_2}{\sqrt{2}} W_\mu^+(x) J^{-\mu}(x) \ , \tag{4.15}$$

using only the electron part of the charged current. This process is represented by the Feynman diagram

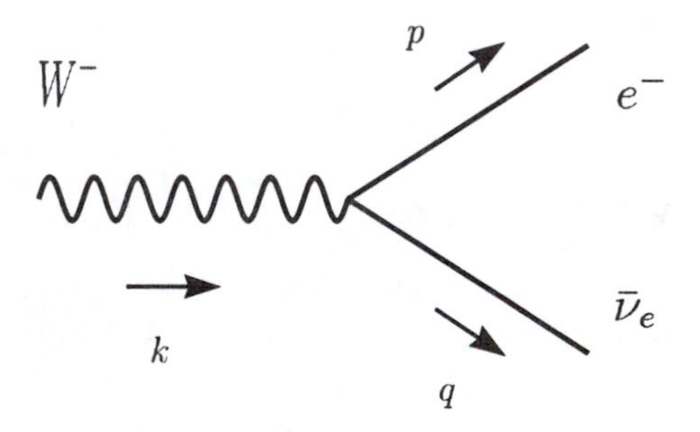

It corresponds to the matrix element

$$T_{fi} = \frac{ig_2}{\sqrt{2}} \int d^4x \langle W^{-(\lambda)}(k) | W_\mu^+(x) \bar{\nu}_e(x) \gamma^\mu L \ e(x) | e^{(a)}(p) \ \bar{\nu}^{(b)}(q) \rangle \ , \tag{4.16}$$

where we have rewritten the current in terms of four-component Dirac fields to facilitate the computations to come; L is the projection operator onto left-handed states

$$L = \frac{1 + \gamma_5}{2} \ . \tag{4.17}$$

Assuming that the plane wave states are normalized in a box of volume V, we have

$$\langle W^{-(\lambda)}(k) | W_\mu^+(x) | 0 \rangle = \frac{1}{\sqrt{V}} \frac{e^{-ik\cdot x}}{\sqrt{2E_W}} \epsilon_\mu^{*(\lambda)}(k) \ , \tag{4.18}$$

where $\epsilon_\mu^{(\lambda)}$ denote the plane wave solutions to the massive Klein-Gordon equation with polarization $\lambda = 1, 2, 3$; they obey

$$k^\mu \epsilon_\mu^{(\lambda)}(k) = 0 \; . \tag{4.19}$$

Similarly, the fermion fields acting on the plane wave states yield

$$\langle 0|e(x)|e^{(a)}(p)\rangle = \frac{1}{\sqrt{V}} \frac{1}{\sqrt{2E_e}} e^{ip\cdot x} u^{(a)}(p) \; , \tag{4.20}$$

$$\langle 0|\bar{\nu}_e(x)|\bar{\nu}_e^{(b)}(q)\rangle = \frac{1}{\sqrt{V}} \frac{1}{\sqrt{2E_{\nu_e}}} e^{iq\cdot x} \bar{v}^{(b)}(q) \; , \tag{4.21}$$

where a, b are the fermion polarizations which run over two values. The four-component covariant plane wave states, $u^{(a)}$ and $v^{(a)}$, are normalized to

$$\bar{u}^{(a)}(p) u^{(b)}(p) = 2m_e \delta_{ab} \; , \qquad \bar{v}^{(a)}(p) v^{(b)}(p) = -2m_{\nu_e} \delta^{ab} \; , \tag{4.22}$$

where the bar denotes the Pauli adjoint. Putting it all together, and integrating over x, we obtain the matrix element

$$T_{fi} = \frac{ig_2 (2\pi)^4}{4V^{3/2}} \frac{\delta^{(4)}(k - p - q)}{\sqrt{E_W E_e E_{\nu_e}}} \epsilon_\mu^{*(\lambda)}(k) \bar{v}^{(a)}(q) \gamma^\mu L u^{(b)}(p) \; . \tag{4.23}$$

Its absolute square gives the transition probability over all of space and time. One of the two δ-functions (times $(2\pi)^4$) is interpreted as the volume of space-time, VT.

The differential decay rate is the transition probability per unit of time, multiplied by the number of final states; with our normalization of the plane wave states, for each final state with three momentum $\vec{p}$, the number of available states is

$$\frac{V}{(2\pi)^3} d^3p \; , \tag{4.24}$$

leading to the differential decay rate

$$\begin{aligned} d\Gamma = & \frac{g_2^2}{16(2\pi)^2 E_W} \delta^{(4)}(k - p - q) \frac{d^3p}{E_e} \frac{d^3q}{E_{\nu_e}} \times \\ & \epsilon_\mu^{*(\lambda)}(k) \epsilon_\rho^{(\lambda)}(k) \bar{v}^{(a)}(q) \gamma^\mu L u^{(b)}(p) \bar{u}^{(b)}(p) \gamma^\rho L v^{(a)}(q) \; . \end{aligned} \tag{4.25}$$

The sums over the neutrino and electron polarizations, using

$$\sum_a v^{(a)}(q)\bar{v}^{(a)}(q) = \not{q} - m_{\nu_e} \; , \qquad \sum_a u^{(a)}(p)\bar{u}^{(a)}(p) = \not{p} + m_e \; , \tag{4.26}$$

yield the trace

$$\mathrm{Tr}\,(\gamma_\mu L[\not{p} + m_e]\gamma_\rho L[\not{q} - m_{\nu_e}]) = \mathrm{Tr}\,(\gamma_\mu \not{p}\, \gamma_\rho L \not{q}) \; , \tag{4.27}$$

since $L\gamma_\rho L = 0$. To obtain the total decay rate, we average over the three initial polarizations of the W-boson, using

$$\sum_\lambda \epsilon_\mu^{*(\lambda)}(k)\epsilon_\rho^{(\lambda)}(k) = -(g_{\mu\rho} - \frac{k_\mu k_\rho}{M_W^2}). \tag{4.28}$$

Substituting in the differential decay rate, and dividing by the number of polarization states of the W, we obtain

$$d\Gamma = \frac{-g_2^2\delta^{(4)}(k-p-q)}{48(2\pi)^2 E_W}\frac{d^3p}{E_e}\frac{d^3q}{E_{\nu_e}}(g^{\mu\rho} - \frac{k^\mu k^\rho}{M_W^2})\mathrm{Tr}(\gamma_\mu \not{p}\gamma_\rho L\not{q}) \; . \tag{4.29}$$

To evaluate the trace, we use the identity

$$\gamma_\mu\gamma_\nu\gamma_\rho = g_{\mu\nu}\gamma_\rho + g_{\nu\rho}\gamma_\mu - g_{\mu\rho}\gamma_\nu + i\epsilon_{\mu\nu\rho\sigma}\gamma_5\gamma^\sigma \; , \tag{4.30}$$

where $\epsilon_{0123} = 1$, with the result

$$\mathrm{Tr}\,(\gamma_\mu\gamma_\nu\gamma_\rho\gamma_\sigma L) = 2(g_{\mu\nu}g_{\rho\sigma} + g_{\mu\sigma}g_{\rho\nu} - g_{\mu\rho}g_{\sigma\nu} + i\epsilon_{\mu\nu\rho\sigma}) \; . \tag{4.31}$$

Then, by using the kinematical relations ($q \cdot p \equiv q_0 p_0 - \vec{q}\cdot\vec{p}$)

$$p\cdot k = m_e^2 + p\cdot q \; , \quad q\cdot k = p\cdot q \; , \; p\cdot q = \frac{1}{2}(M_W^2 - m_e^2) \; , \tag{4.32}$$

we find, in the rest frame of the W-boson,

$$d\Gamma = \frac{g_2^2 M_W}{24(2\pi)^2}\left(1 - \frac{m_e^2}{M_W^2}\right)\left(1 + \frac{m_e^2}{2M_W^2}\right)\delta^{(4)}(k-p-q)\frac{d^3p}{E_e}\frac{d^3q}{E_{\nu_e}} \; . \tag{4.33}$$

Use of the generic phase space integral formula

$$I \equiv \int \frac{d^3p}{p_0}\frac{d^3q}{q_0}\delta^{(4)}(k-p-q) \; ,$$
$$= \frac{2\pi}{k^2}\sqrt{[k^2-(m_1-m_2)^2][k^2-(m_1+m_2)^2]} \; , \tag{4.34}$$

where

$$q_0 = \sqrt{m_1^2 + \vec{q}\cdot\vec{q}} \; , \qquad p_0 = \sqrt{m_2^2 + \vec{p}\cdot\vec{p}} \; .$$

Setting $m_1 = 0$ and $m_2 = m_e$, it yields the final result

$$\Gamma(W^- \to e^- + \bar{\nu}_e) = \frac{G_F M_W^3}{6\pi\sqrt{2}}\left(1 - \frac{m_e^2}{M_W^2}\right)^2\left(1 + \frac{m_e^2}{2M_W^2}\right) \; , \tag{4.35}$$

written in terms of the Fermi coupling constant, G_F,

$$\frac{G_F}{\sqrt{2}} = \frac{g_2^2}{8M_W^2} \; , \tag{4.36}$$

which parametrizes the strength of the current-current interaction in the effective Hamiltonian

$$\mathcal{H}_{int} = 4\frac{G_F}{\sqrt{2}}J_\mu^+ J^{-\mu} \; . \tag{4.37}$$

The factor of 4 is to account for the L in the charged current (Fermi did not envisage Parity violation). Note the weak dependence of the decay rate on the outgoing particle masses, as long as they are small compared to the W mass: the decay rate of the W into an electron, a muon, and a tau lepton are nearly equal. This lepton universality is experimentally verified: the decay leptonic branching ratios of the W into electron, muon, and tau leptons are $10.9 \pm .04\%$, $10.2 \pm 0.05\%$, and $11.3 \pm 0.08\%$, respectively.

The W boson also decays into a quark and an antiquark of different flavors. Since quarks do not exist as asymptotic states, the computation of the decay width into a particular combination of quarks and antiquarks, say $\bar{\mathbf{u}}\mathbf{d}$, has no direct physical meaning. As the quark and antiquark leave the point of interaction and fly apart from one another, the strong QCD force grows with their separation. It abates only when color neutral combinations of quarks and antiquarks, such as pions, have been created. This is achieved by means of sparking of the QCD vacuum through quark-antiquark pair creation. At present, physicists do not have the tools necessary to compute in detail this "hadronization" process. We only know is that it takes place 100% of the time.

There is hope since as long as we do not seek too many details of the hadronic final state, we can use unitarity to finesse these subtelties. For example, the probability that a W decays into hadrons, without specifying anything about the hadrons, is the sum of the probability that it decays into orthogonal quark-antiquark pairs

$$\Gamma(W^- \to \text{ Hadrons}) = \sum_{d_i=d,s,b} \Gamma(W^- \to \bar{\mathbf{u}}\mathbf{d}_i) + \sum_{d_i=d,s,b} \Gamma(W^- \to \bar{\mathbf{c}}\mathbf{d}_i) \ . \tag{4.38}$$

Note that we have not included the top quark, since it is too heavy for this decay to proceed, by which we mean that hadrons that contain the top quark are more massive than the W.

We can compute these decay widths the same way as for the leptonic decay. Since the quarks are all much lighter than the W, we can treat them all as massless. In this case, the calculation has already been done. It is easy to see that, for instance,

$$\Gamma(W^- \to \bar{\mathbf{u}}\mathbf{d}) = 3|V_{ud}|^2\Gamma(W^- \to e^- + \bar{\nu}_e) \ , \tag{4.39}$$

where the factor of three is for the three colors of quarks, and V_{ud} is the relevant CKM matrix element. Thus, we find that

$$\begin{aligned}\Gamma(W^- \to \text{ Hadrons}) &= 3 \sum_{\substack{i=u,c\\ j=d,s,b}} |V_{ij}|^2\Gamma(W^- \to e^- + \bar{\nu}_e) \ , \\ &= 6\Gamma(W^- \to e^- + \bar{\nu}_e) \ . \end{aligned} \tag{4.40}$$

In the last we have used the unitarity of the CKM matrix. Thus we expect that the W will decay into hadrons 2/3 of the time, and into each charged lepton 1/9 of the time.

PROBLEMS

A. Verify the generic phase space formula Eq. (4.34).

B. Find the fraction of the time the W decays into strange hadrons.

C. Identify the main decay channel of the top quark, and calculate its lifetime.

D. 1-) Suppose there is a fourth chiral family of quarks and leptons, but that only its charged lepton and the massless associated neutrino are lighter than the W. Using the latest from the Particle Data Book, what is the lightest possible value of the charged lepton mass?
2-) Assuming that the fourth neutrino happens to have a (Dirac) mass, calculate the fourth family leptonic contribution to the W decay rate. What can you say about the possible values of the masses of these fourth family leptons?

4.3 Z-DECAY

Using the results of the previous section, it is straightforward to compute the width of the Z boson in the standard model. The only difference with the previous section is that the interaction now proceeds via

$$\mathcal{L}_{int} = \frac{ieZ^\mu}{\sin\theta_{\rm w}\cos\theta_{\rm w}}[J^3_\mu - \sin^2\theta_{\rm w} J^{em}_\mu] \ . \tag{4.41}$$

For the neutrino of each species, we just use the formula (4.22) with $m_e = 0$, replace M_W by M_Z, and

$$\frac{g_2}{\sqrt{2}} \to \frac{g_2}{\cos\theta_W} \ .$$

The result is

$$\Gamma(Z^0 \to \nu_e + \bar{\nu}_e) = \left(\frac{1}{4}\right)\frac{g_2^2 M_Z}{24\pi\cos^2\theta_W} = \frac{G_F M_Z^3}{12\pi\sqrt{2}} \ . \tag{4.42}$$

Of course, this width can never be measured directly, since we do not have the means to detect the neutrinos; it can only be measured by default, that is, by independently measuring the width into visible products, and then the total width.

The decay width into electron-positron pair is given by

$$\Gamma(Z^0 \to e^+ + e^-) = (a_L + a_R)\frac{G_F M_Z^3}{3\pi\sqrt{2}} \ , \tag{4.43}$$

where we have neglected the mass of the electron, and thus added incoherently the left and right contributions, with

$$a_L = \frac{1}{4} - \sin^2\theta_W + \sin^4\theta_W \ ; \qquad a_R = \sin^4\theta_W \ . \tag{4.44}$$

By using unitarity arguments similar to those the previous section, we can obtain the total rate into hadrons. The tree-level result is, neglecting the masses of the quarks,

$$\Gamma(Z^0 \to \text{Hadrons}) = \left(\frac{15}{4} - 7\sin^2\theta_W + \frac{22}{3}\sin^4\theta_W\right)\frac{G_F M_Z^3}{3\pi\sqrt{2}} \ . \tag{4.45}$$

These formulae can be used to extract the value of $\sin^2\theta_W$ from experiments.

Finally, let us calculate polarized Z decay; we start from Eq. (4.41), but without summing over the Z polarizations. It is easier to do the calculations in the Z rest frame where the polarization vector is purely space-like with components

$$\epsilon_\mu^{(3)} = (0,0,0,1) \ , \tag{4.46}$$

$$\epsilon_\mu^{(\pm)} = \frac{1}{\sqrt{2}}(0,1,\mp i,0) \ , \tag{4.47}$$

Since we neglect the mass of the electron, the spin of the outgoing electron points away from its momentum (if it is left-handed), and that of the positron along its momentum. By conservation of angular momentum, the spin vector of the Z is aligned with the direction of the outgoing positron.

The contribution to the decay rate from a transversely polarized Z into a left-handed electron-positron pair is then

$$d\Gamma_T = \frac{a_L g_2^2 \delta^{(4)}(k-p-q)}{24(2\pi)^2\cos^2\theta_W M_Z}\frac{d^3p}{E_e}\frac{d^3q}{E_{\bar{e}}}\epsilon_\mu^{*(+)}\epsilon_\rho^{(+)}\mathrm{Tr}(\gamma^\mu \not{p}\gamma^\rho L \not{q}). \tag{4.48}$$

A similar equation holds for the decay width into a right-handed pair, but with the L projection operator replaced by

$$R = \frac{1-\gamma_5}{2} \ , \tag{4.49}$$

and a_L replaced by a_R. Using Eq. (4.47), and evaluating the trace, we find

$$d\Gamma_T = \frac{a_L g_2^2 \delta^{(4)}(k-p-q)}{24(2\pi)^2 \cos^2\theta_W M_Z} \frac{d^3p}{E_e} \frac{d^3q}{E_{\bar{e}}} (q_0 p_0 - q_3 p_3 - q_0 p_3 + q_3 p_0) \ . \quad (4.50)$$

In the rest frame of the Z, we have (neglecting the electron mass),

$$p_0 = q_0 = \frac{M_Z}{2} \ ; \qquad \vec{p} = -\vec{q} = \frac{M_Z}{2}\hat{\mathbf{n}}, \quad (4.51)$$

where $\hat{\mathbf{n}}$ is a unit vector. After performing the δ-function integrations, we find

$$d\Gamma_T = \frac{a_L g_2^2 M_Z}{48(2\pi)^2 \cos^2\theta_W} d\Omega \left(1 + \hat{n}_3^2 - 2\hat{n}_3\right) . \quad (4.52)$$

This leads to the decay angular distribution

$$\frac{d\Gamma_T}{d\cos\theta} = \frac{G_F M_Z^3}{12\pi\sqrt{2}} a_L (1-\cos\theta)^2 \ , \quad (4.53)$$

where θ is the direction of the electron in the Z rest frame, relative its polarization. This angular dependence is connected with the amplitude for a Z polarized along one axis to be found with polarization at an angle θ from it. It vanishes at $\theta = 0$, when the electron-positron spin is anti-aligned with that of the Z. A similar formula holds for the right handed electrons with a change of sign in $\cos\theta$, and a_L replaced by a_R.

One can show that the decay of the longitudinal Z leads to the formula

$$\frac{d\Gamma_{\text{Long}}}{d\cos\theta} = \frac{G_F M_Z^3}{12\pi\sqrt{2}} 2a_L \sin^2\theta \ . \quad (4.54)$$

PROBLEMS

A. Compute the branching ratio

$$\frac{\Gamma(Z^0 \to \mathbf{b}\bar{\mathbf{b}})}{\Gamma(Z^0 \to \text{hadrons})}$$

and compare with the latest experimental value.

B. Compute the total Z decay rate into τ pair, without neglecting the τ masses.

C. Verify Eq. (4.54) for the angular probability distribution for an electron in the decay of a longitudinally polarized Z at rest. Discuss its physical interpretation.

D. Consider the process $e^+e^- \to \mu^+\mu^-$ at the Z resonance with longitudinally polarized electrons in the initial state. Derive the expression for the longitudinal polarization asymmetry in this collision. Discuss how it is measured and its advantages in the extraction of the value of $\sin^2\theta_W$ over other methods. For reference, see B.W. Lynn and C. Verzegnassi, *Phys. Rev.* **D 35**, 3326(1987).

E. Show that, using present data, the hypothesis of a fourth family with a massless neutrino is ruled out. Then assume that the neutrino of the fourth family has a mass. Together with the results of problem D of last section, what are the allowed ranges of masses for the hypothetical fourth neutrino and charged lepton?

4.4 MUON DECAY

In the standard model, the muon decays into a muon neutrino, an electron and its associated antineutrino

$$\mu^- \to \nu_\mu + e^- + \bar{\nu}_e \ .$$

The interaction responsible for this decay is the same as that for W and Z decays, namely

$$\mathcal{L}_{int} = \frac{ig_2}{\sqrt{2}} W^{+\alpha} \left[\nu_{eL}^\dagger \sigma_\alpha e_L + \nu_{\mu L}^\dagger \sigma_\alpha \mu_L + \cdots \right] \ . \tag{4.55}$$

Since the W-boson is eight hundred times more massive than the muon, the muon cannot decay into a physical W, but this is quantum field theory: the W can appear as an intermediate state which can be off its mass-shell; the emitted W can then in turn decay into an electron and its associated neutrino. This is shown by the following Feynman diagram

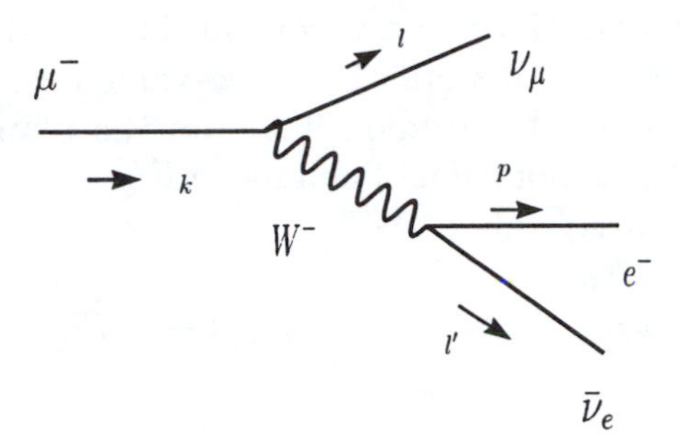

It corresponds to the effective interaction Hamiltonian

$$H_{int} = \frac{-g_2^2}{2} \int d^4x \int d^4y \Delta_F^{\alpha\beta}(x-y) \nu_{\mu L}^\dagger(x) \sigma_\alpha \mu_L(x) e_L^\dagger(y) \sigma_\beta \nu_{eL}(y) , \tag{4.56}$$

where

$$\Delta_F^{\alpha\beta}(x-y) = g^{\alpha\beta} \frac{1}{(2\pi)^4} \int d^4q \frac{e^{-iq\cdot(x-y)}}{q^2 - M_W^2 + i\epsilon} , \tag{4.57}$$

is the Feynman propagator of the W-boson in the Feynman gauge ($\xi = 1$). We are using the Minkowski representation of the propagator for this problem. We drop the $i\epsilon$ term since it is only relevant when the W is on its mass-shell, which is clearly not the case here.

We use the simple Fierz identity

$$\nu_{eL} \nu_{\mu L}^\dagger = -\frac{1}{2} \sigma_\rho \nu_{\mu L}^\dagger \sigma^\rho \nu_{eL} , \tag{4.58}$$

as well as

$$\sigma^\alpha \sigma_\rho \sigma_\alpha = -2\sigma_\rho , \tag{4.59}$$

to rewrite the interaction in the form

$$H_{int} = \frac{-g_2^2}{2} \int d^4x \int d^4y \Delta_F(x-y) \nu_{\mu L}^\dagger(x) \sigma^\rho \nu_{eL}(y) e_L^\dagger(y) \sigma_\rho \mu_L(x) . \tag{4.60}$$

The matrix element between the states shown in the Feynman diagram yields the transition amplitude over the whole of space-time

$$T_{fi} = \langle \mu^{(a)}(k) | H_{int} | e^{(b)}(p); \bar{\nu}_e^{(c)}(l'); \nu_\mu^{(d)}(l) \rangle ,$$

where the states denote the plane wave solutions of the Dirac equation. The superscripts a, b, c, d denote their two-valued polarizations. When evaluating the action of the fermion fields on the plane wave states, we switch over to the Dirac notation. Then using Eqs (4.6-7), and integrating over x, y, and q, we arrive at

$$T_{fi} = (2\pi)^4 \delta^{(4)}(k - p - l - l')\hat{T}_{fi} , \tag{4.61}$$

where

$$\hat{T}_{fi} = \frac{-g_2^2}{8V^2} \frac{1}{\sqrt{E_\mu E_e E_{\nu_e} E_{\nu_\mu}}} \frac{\bar{u}^{(d)}(l)\gamma^\rho L v^{(c)}(l')\bar{u}^{(b)}(p)\gamma_\rho L u^{(a)}(k)}{(k-l)^2 - M_W^2} . \tag{4.62}$$

By now the steps to obtain the differential decay rate should be familiar. First we take the absolute square of the transition amplitude over space-time, interpret $(2\pi)^4$ times one δ function as the volume of space-time, then divide by the time interval, and multiply by the number of available final states. The V's cancel, and we are left with

$$\begin{aligned} d\Gamma = & \frac{g_2^4}{64(2\pi)^5 E_\mu} \frac{\delta^{(4)}(k-p-l-l')}{\left[(k-l)^2 - M_W^2\right]^2} \frac{d^3p}{E_e} \frac{d^3l'}{E_{\nu_e}} \frac{d^3l}{E_{\nu_\mu}} \\ & \times |\bar{u}^{(d)}(l)\gamma^\rho L v^{(c)}(l')\bar{u}^{(b)}(p)\gamma_\rho L u^{(a)}(k)|^2 . \end{aligned} \tag{4.63}$$

The neutrinos are not detected, and the sum over their polarizations yields the trace

$$\mathrm{Tr}\,(\gamma_\rho \not{l}' \gamma_\sigma L \not{l}) = 2\left(l_\rho l'_\sigma + l'_\rho l_\sigma - g_{\rho\sigma} l \cdot l' - i\epsilon_{\rho\sigma\alpha\beta} l^\alpha l'^\beta\right) . \tag{4.64}$$

In the limit of zero electron mass, the electron comes out purely left-handed. Summing over the electron's polarizations, we find

$$\begin{aligned} d\Gamma = & \frac{g_2^4}{32(2\pi)^5 E_\mu} \frac{\delta^{(4)}(k-p-l-l')}{\left[(k-l)^2 - M_W^2\right]^2} \frac{d^3p}{E_e} \frac{d^3l'}{E_{\nu_\mu}} \frac{d^3l}{E_{\nu_\mu}} \\ & \times \left(l_\rho l'_\sigma + l'_\rho l_\sigma - g_{\rho\sigma} l \cdot l' - i\epsilon_{\rho\sigma\alpha\beta} l^\alpha l'^\beta\right) \bar{u}^{(a)}(k)\gamma_\rho \not{p} \gamma_\sigma L u^{(a)}(k) . \end{aligned} \tag{4.65}$$

If the polarization of the decaying muon is not measured, we average over its polarization states to obtain the decay rate, obtaining

$$
\begin{aligned}
d\Gamma =& \frac{g_2^4}{32(2\pi)^5 E_\mu} \frac{\delta^{(4)}(k-p-l-l')}{\left[(k-l)^2 - M_W^2\right]^2} \frac{d^3p}{E_e} \frac{d^3l'}{E_{\nu_e}} \frac{d^3l}{E_{\nu_\mu}} \\
& \times \left(l_\rho l'_\sigma + l'_\rho l_\sigma - g_{\rho\sigma} l \cdot l' - i\epsilon_{\rho\sigma\alpha\beta} l^\alpha l'^\beta\right) \\
& \times \left(p^\rho k^\sigma + k^\rho p^\sigma - g^{\rho\sigma} p \cdot k - i\epsilon^{\rho\sigma}{}_{\tau\lambda} p^\tau k^\lambda\right) . \qquad (4.66)
\end{aligned}
$$

With the help of the identity

$$
\epsilon_{\rho\sigma\alpha\beta}\epsilon^{\rho\sigma}{}_{\tau\lambda} = -2(g_{\alpha\tau}g_{\beta\lambda} - g_{\alpha\lambda}g_{\beta\tau}), \qquad (4.67)
$$

we arrive at the simpler expression

$$
d\Gamma = \frac{g_2^4}{8(2\pi)^5 E_\mu} \frac{\delta^{(4)}(k-p-l-l')}{((k-l)^2 - M_W^2)^2} \frac{d^3p}{E_e} \frac{d^3l'}{E_{\nu_e}} \frac{d^3l}{E_{\nu_\mu}} (p \cdot l)(k \cdot l') .
$$

The phase space integral over the momenta of the two neutrinos

$$
I_{\alpha\beta}(k,p) \equiv \int \frac{d^3l'}{E_{\nu_e}} \frac{d^3l}{E_{\nu_\mu}} \delta^{(4)}(k-p-l-l') \frac{l_\alpha l'_\beta}{\left[(k-l)^2 - M_W^2\right]^2} ,
$$

can be simplified since, to $\mathcal{O}(\frac{m_{e,\mu}}{M_W})$ we can neglect the momentum transfer in the propagator, leaving

$$
I_{\alpha\beta}(k-p) = \int \frac{d^3l'}{E_{\nu_e}} \frac{d^3l}{E_{\nu_\mu}} \delta^{(4)}(k-p-l-l') l_\alpha l'_\beta .
$$

We now use another generic phase space integral formula

$$
\begin{aligned}
I_{\alpha\beta} &\equiv \int \frac{d^3p}{\sqrt{m_2^2 + \vec{p}\cdot\vec{p}}} \frac{d^3q}{\sqrt{m_1^2 + \vec{q}\cdot\vec{q}}} \delta^{(4)}(k-p-q) p_\alpha q_\beta , \\
&= \frac{I}{12k^4}(k^2[k^2 - (m_1 - m_2)^2][k^2 - (m_1+m_2)^2] g_{\alpha\beta} \\
&\qquad + 2[k^4 + k^2(m_1^2 + m_2^2) - 2(m_1^2 - m_2^2)] k_\alpha k_\beta) , \qquad (4.68)
\end{aligned}
$$

where I was defined in the previous section. Applying it to our case, $m_1 = 0$ and $m_2 = m_e \approx 0$, we find

$$I_{\alpha\beta}(k-p) \approx \frac{\pi}{6}\left[(p-k)^2 g_{\alpha\beta} + 2(p-k)_\alpha (p-k)_\beta\right] \ . \tag{4.69}$$

Now we write

$$d^3p = E_e^2 dE_e d\Omega \ , \tag{4.70}$$

and introduce the dimensionless variable

$$x = \frac{2E_e}{m_\mu} \ , \tag{4.71}$$

which varies between zero and one. Integrating over the angles, we obtain the differential decay rate in the rest frame of the muon

$$\frac{d\Gamma}{dx} = \frac{g_2^4 m_\mu^5}{32 \cdot 2^5 \pi^3 M_W^4} x^2 \left(1 - \frac{2x}{3}\right) \ . \tag{4.72}$$

The total decay rate is

$$\begin{aligned} \Gamma &= \frac{g_2^4 m_\mu^5}{32 \cdot 2^5 \pi^3 M_W^4} \int_0^1 x^2 dx \left(1 - \frac{2x}{3}\right) \ , \\ &= \frac{e^4 m_\mu^5}{192 \cdot 2^5 \pi^3 \sin^4\theta_W M_W^4} \ , \end{aligned} \tag{4.73}$$

that is

$$\Gamma(\mu^- \to \nu_\mu + e^- + \bar{\nu}_e) = \frac{G_F \mu^5}{192\pi^3} \ . \tag{4.74}$$

PROBLEMS

A. Find the maximum kinematically allowed value of $(k-l)^2$. Estimate the error in the rate of τ decay from neglecting the momentum transfer term in the propagator

B. Verify the phase space formula Eq. (4.68).

C. When the decaying muon is at rest and polarized, calculate the dependence of the decay rate on the direction of the electron.

CHAPTER **5**

QCD AT LARGE DISTANCES

Quantum Chromodynamics, the Yang-Mills theory based on the color group $SU(3)$, is the most complicated part of the standard model because it is not perturbative, except at very high energies. Its description is deceptively simple: eight massless vector gluons interacting with one another, and with the quarks through vector currents. The vector gluons have no electroweak quantum numbers of their own, although they affect electroweak interactions through the quarks which share both color and electroweak quantum numbers. In this chapter we present some of the perturbative and non-perturbative aspects of QCD.

The gluon self interactions cause the effective QCD coupling constant to grow at large distances (infrared slavery) and decrease at short distances (asymptotic freedom). At short distances, where the quark-gluon coupling is weaker, the quarks appear as quasi-independent degrees of freedom, and perturbation theory in the quark-gluon coupling is applicable. This is the domain of perturbative QCD, which we discuss only to the extent that it applies to the electroweak interactions. We have only primitive theoretical tools to investigate the behavior of the quark couplings at low energies, but the wealth of low energy Strong Interaction experimental data suggests a plausible picture: only color-neutral states, those immune to the QCD forces, can exist as asymptotic states. At large distances, the quark language ceases to be useful, to be replaced by an alternate description of the strong interactions in terms of these color singlet states, which we discuss in this chapter.

We then consider the vacuum state of QCD. It consists of all field configurations for which the energy density vanishes. The chromo-electric and chromo-magnetic fields vanish with the potential in pure gauge configurations. In a non-Abelian theory, such configurations can be split into inequivalent classes, globally distinct from one another. The existence of tunnelling transitions between these distinct sectors which, for strong coupling are not suppressed, results in the appearance of CP violation by the strong interactions, and in the breaking of a global flavor-independent chiral symmetry.

5.1 DIMENSIONAL TRANSMUTATION

Consider a Yang-Mills theory invariant under a compact group generated by any of the semisimple Lie algebras. Assume that the theory includes all possible renormalizable interactions with Weyl fermions $\psi_L^{[i]}$ belonging to unitary irreducible representations $\mathbf{s}_i$ and a set of spinless fields $\phi^{[a]}$ belonging to the representations $\mathbf{r}_a$.

Quantum effects cause the gauge coupling "constant" of this theory to become scale-dependent according to the renormalization group equation, shown here to $\mathcal{O}(\hbar)$ (one loop),

$$\frac{d\alpha}{dt} = -\frac{1}{2\pi}\alpha^2 \left(\frac{11}{3} C_{\text{adj}} - \frac{2}{3}\sum_i C_{\mathbf{s}_i} - \frac{1}{6}\sum_a C_{\mathbf{r}_a} \right) + \mathcal{O}(\alpha^3) \ , \qquad (5.1)$$

where $\alpha = \frac{g^2}{4\pi}$, $t = \ln(\mu/\mu_0)$, with μ_0 as an arbitrary reference energy, and μ is any energy. The group-theoretic numbers C_{adj}, $C_{\mathbf{r},\mathbf{s}}$ are the "Dynkin indices" associated with the adjoint and the $\mathbf{r}, \mathbf{s}$ representations, respectively. With the Lie algebra represented by the matrices $\mathbf{T}_r^A$ in the representation $\mathbf{r}$, the Dynkin index is simply

$$\text{Tr}\,(\mathbf{T}_r^A \mathbf{T}_r^B) = C_{\mathbf{r}} \delta^{AB} \ . \qquad (5.2)$$

The fields contribute to this equation only at energies above their masses. Specializing to $SU(3)$, the QCD coupling obeys the equation

$$\begin{aligned} \frac{d\alpha_{QCD}}{dt} &= -\frac{1}{2\pi}\alpha_{QCD}^2 \left(\frac{11}{3}\cdot 3 - \frac{2}{3}(\frac{1}{2} + \frac{1}{2})n_q \right) \ , \\ &= -\frac{1}{2\pi}\alpha_{QCD}^2 (11 - \frac{2}{3} n_q) \ , \end{aligned} \qquad (5.3)$$

where n_q is the number of quark flavors (in QCD there are no spinless fields with color). This equation is easily integrated to yield

$$\alpha_{QCD}(t) = \frac{2\pi}{(11 - \frac{2}{3}n_q)} \frac{1}{t} \ . \tag{5.4}$$

It follows that $\alpha_{QCD}(t)$ decreases at large energies (short distances), a phenomenon known as *asymptotic freedom*. The integration constant has been chosen so that the coupling blows up when $\mu = \mu_0$.

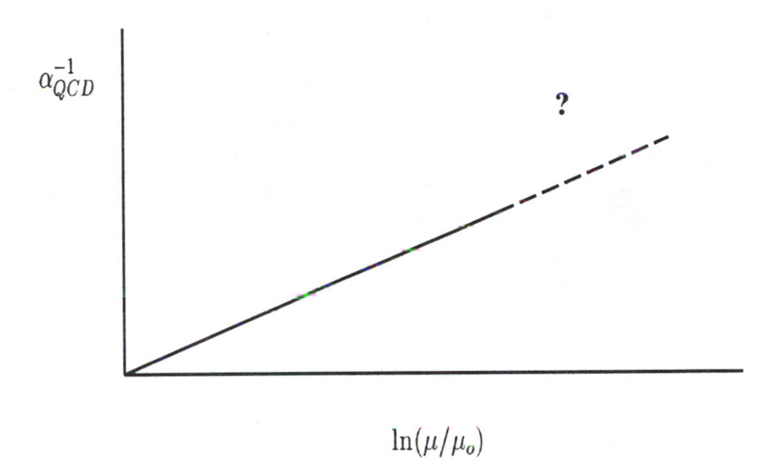

This picture is valid as long as the coupling is perturbative. We can only infer that, if small in the ultraviolet, α_{QCD} increases towards the infrared, eventually leaving the domain of perturbation theory. Perturbative control of QCD is achieved only at distances that are very small compared to some reference length, above which perturbation theory fails. An outstanding unsolved theoretical problem is to derive from the QCD Lagrangian its infrared effective theory, where perturbative methods using quarks and gluons are inapplicable.

It is fortunate that our lack of theoretical control is offset by the enormous amount of experimental information about the Strong Interactions in the static limit. As a result, we have developed a compelling intuitive picture: in the infrared limit, the QCD force is so strong that only color-neutral combinations can escape its cauldron and exist as free particles; these bound states are the nucleons, the mesons, and the resonances encountered in the study of Strong Interactions. These interactions have a natural length scale, their range, associated with the inverse pion mass, as well as other scales such as the radius of a proton, neutron, etc... .

The classical QCD Lagrangian contains no scale, only a dimensionless coupling constant. However in the quantum theory, because of the variation of the coupling with scale, we can exchange the measurement of the dimensionless coupling for the scale at which it is measured, *via* "dimensional transmutation", a term coined by S. Coleman. It is connected with the critical energy below which the coupling is no longer perturbative. At higher energies, we expect the quark picture to emerge, while at lower energies the quark picture fades out to be replaced by interactions among the bound states.

The strong coupling constant has been measured at many different scales, functions of the particulars of each experiment. Since the standard model predicts how the coupling runs with scale, one can in principle check the consistency among all these values. On the other hand, since there is but one physical scale associated with QCD, we can agree on a convention and extract that one scale from experiments conducted at different energies. It is instructive to see how one may define such a scale. The guiding principle is that all physical quantities must be renormalization group-invariant. As a first step, consider the combination

$$\Lambda = \mu e^{-t} = A\mu \ \exp\left(-\frac{2\pi}{(11-\frac{2}{3}n_q)\alpha_{QCD}(t)}\right) , \tag{5.5}$$

where A is an arbitrary integration constant. In the one-loop approximation, it is renormalization group-invariant, since it does not depend on the scale μ. For measurements where the coupling is very tiny, this expression may serve as *the* physical scale associated with QCD. Its numerical value is determined by experiment. We can invert the equation and rewrite the coupling in terms of the physical scale as

$$\alpha_{QCD}(t_{\text{phys}}) = \frac{1}{(11-\frac{2}{3}n_q)}\frac{4\pi}{t_{\text{phys}}} , \tag{5.6}$$

where now

$$t_{\text{phys}} = \ln\left(\frac{\mu^2}{\Lambda^2}\right) . \tag{5.7}$$

In this formula, Λ is the scale at which the coupling blows up. Extracting the physical scale of QCD from experiment appears straightforward: first determine the value of α at some very large scale above the six-quark threshold, and then compute the value of Λ, having beforehand agreed on a convention that fixes the integration constant A.

This would be a great formula to use if the measurements were performed around 10^{16} GeV where the QCD gauge coupling is tiny, and higher order effects can be neglected. Alas, life is not so simple: experiments are carried out at much lower energies, where α is not small enough to neglect higher loop corrections. Also the number of quarks which appear in this formula changes over the experimentally accessible range.

The necessity to go beyond one loop brings in more subtelties. Computing in any QFT involves the choice of a renormalization prescription to fix the finite part of the counterterms. Prescription dependence of the β function sets in at the three-loop order. In dimensional regularization, the most convenient prescription is the one in which the counterterms contain only the pole terms needed to absorb the divergences. It is called the MS prescription. In this prescription, the bare coupling g_0 is related to the renormalized coupling g_R through the Laurent series

$$g_0 = \mu^{2\epsilon}\left[g_R + \sum_{k=1}^{\infty}\frac{a_k}{\epsilon^k}\right] , \tag{5.8}$$

where $\epsilon = 2 - \frac{d}{2}$ (d = number of space-time dimensions); the coefficients a_k, which depend on g_R, are calculated in perturbation theory. Differentiating both sides with respect to the scale μ, we find

$$\frac{dg_R}{dt} = -2(1 - g_R\frac{d}{dg_R})a_1(g_R) , \tag{5.9}$$

yielding the renormalization group equation. Quark thresholds are treated most simply in this prescription: at a given scale μ, one includes only those quarks with masses less than μ – the reason is that the effect of massive quark pairs is suppressed at lesser energies. The result at one loop is discontinuous jumps in the coefficients as mass thresholds are crossed, corresponding to kinks in the curve for $\alpha(t)$. At two loops and higher, one must include matching functions which smooth-out these kinks. Away from the thresholds one only needs to insert the appropriate number of quarks.

In the MS scheme, the β-function has been calculated to three loops,

$$\frac{d}{dt}\alpha_{QCD}(t) \equiv \beta_1\alpha_{QCD}^2 + \beta_2\alpha_{QCD}^3 + \beta_3\alpha_{QCD}^4 + \cdots , \tag{5.10}$$

where

$$\beta_1 = -\frac{1}{2\pi}[11 - \frac{2}{3}n_q] , \; \beta_2 = -\frac{1}{4\pi^2}[51 - \frac{19}{3}n_q] ,$$
$$\beta_3 = -\frac{1}{64\pi^3}[2857 - \frac{5033}{9}n_q + \frac{325}{27}n_q^2] , \tag{5.11}$$

in which n_q is the number of quark flavors lighter than the energy scale considered. All these coefficients depend on the number of quarks, and only the first two are independent of the renormalization prescription.

This equation can be integrated with slightly less trivial results, but the physics has not changed– we still need one measured data point to fix the value of α_{QCD} at some scale μ. We can now define a better renormalization group invariant scale

$$\Lambda_{MS}^{[n_q]}(A) \equiv A\mu \exp\left[\frac{1}{\beta_1\alpha(\mu)} + \frac{\beta_2}{\beta_1^2}\ln\left(\frac{-\beta_1\alpha(\mu)}{2}\right) + \right. \\ \left. + \frac{\beta_3\beta_1 - \beta_2^2}{\beta_1^3}\alpha(\mu) + \mathcal{O}(\alpha^2)\right] , \quad (5.12)$$

where A is an arbitrary integration constant. In order to adopt a universally accepted value of this scale, we need to agree on the number of quark flavors, and on the value of the integration constant. Fixing the unknown integration constant is just a convention. The dependence on the number of quark flavors depends on the scale at which the measurement of α is performed. As long as no quark thresholds are crossed, one obtains the same value for the renormalization group independent scale.

We may now turn this equation around and solve iteratively for α_{QCD} as a function of the agreed-upon scale factor

$$t_{MS} \equiv \ln\left(\frac{\mu^2}{\Lambda_{MS}^2}\right) . \quad (5.13)$$

This procedure yields the analytic form

$$\alpha_{QCD}(t_{MS}) = \frac{-2}{\beta_1 t_{MS}}\left\{1 + \frac{2\beta_2}{\beta_1^2}\frac{\ln t_{MS}}{t_{MS}} + \frac{4\beta_2^2}{\beta_1^4 t_{MS}^2}(\ln t_{MS} - \frac{1}{2})^2 + \right. \\ \left. + \frac{\beta_3\beta_1}{\beta_2^2} - \frac{5}{4} + \mathcal{O}(\frac{1}{t_{MS}^3})\right\} , \quad (5.14)$$

where we have set $A = 1$. The physical scale is again where the coupling blows up, indicative only of the order of magnitude, since this formula is valid where the coupling is perturbative. We can massage this equation to express the coupling at some scale in terms of it at another in the form

$$\frac{1}{\alpha(t)} = \frac{1}{\alpha(0)} - \beta_1 t - \frac{\beta_2}{\beta_1}\ln\left[\frac{\alpha(t)}{\alpha(0)}\right] - \frac{1}{\beta_1^2}(\beta_3\beta_1 - \beta_2^2)(\alpha(t) - \alpha(0)) + O(\alpha^2) . \quad (5.15)$$

The only subtelty in the extrapolation occurs when crossing quark thresholds. Although it bypasses the definition of a fundamental scale, this process involves a complicated iterative procedure.

It is more convenient to use a slight technical modification of the MS prescription, designed to absorb the floatsam which accompanies the poles in dimensional regularization; it is called the $\overline{MS}$ prescription, and it affects the definition of the scale through

$$\Lambda_{\overline{MS}} = \Lambda_{MS}(4\pi)^{-\gamma/2} , \tag{5.16}$$

where γ is the Euler-Mascheroni constant. The β function is not affected by this change.

The measurements of α_{QCD} can be put in two categories. Low energy determinations from charm and bottom physics, as well as the hadronic τ decay rate, and higher energy measurements, notably in the decay of the Z boson into hadrons. All experimental values obtained at these different scales turn out to be within one standard deviation of one another, after renormalization group corrections are taken into account. A recent (and welcome) custom is to use the renormalization group to run all measurements of α_{QCD} to the Z mass for comparison purposes. An alternative would be to extract from each measurement a value for a previously agreed upon Λ_{QCD}.

The Υ-system provides an excellent laboratory to determine the value of α_{QCD} at low energies. The Υ-system describes bound $\mathbf{b}\overline{\mathbf{b}}$ states, starting with the $1s$ state with a mass of 9.460 GeV. This scale is sufficiently large to allow for a perturbative determination of α_{QCD} (provided one believes the non-relativistic approximation for treating $\mathbf{b}\overline{\mathbf{b}}$ dynamics). For instance calculations yield the ratio

$$\frac{\Gamma(\Upsilon \to \gamma + 2 \text{ gluons})}{\Gamma(\Upsilon \to \mu^+\mu^-)} = \frac{8(\pi^2 - 9)}{9\pi\alpha_{QED}} \alpha^2_{QCD}(M_\Upsilon)[1 + \mathcal{O}(\alpha_{QCD})] , \tag{5.17}$$

which is measured to be 2.75 to within .15%. Yet another determination comes from the ratio

$$\frac{\Gamma(\Upsilon \to 3 \text{ gluons})}{\Gamma(\Upsilon \to \mu^+\mu^-)} = \frac{10(\pi^2 - 9)}{9\pi} \frac{\alpha^3_{QCD}(M_\Upsilon)}{\alpha^2_{QED}} [1 + \mathcal{O}(\alpha_{QCD})] . \tag{5.18}$$

These yield the strong coupling at different scales. We then use Eq. (5.15) to obtain α_{QCD} at higher energy, particularly at the Z mass. This procedure yields (CLEO Collaboration, *Phys. Rev.* **D55**, 5273(1997))

$$\alpha_{QCD}(M_Z) = 0.110 \pm 0.001 \pm 0.007 \ , \tag{5.19}$$

with both statistical and systematic errors shown.

There are many other ways to extract α_{QCD} from experiment. One is by analyzing the departure of the deep inelastic structure functions from their parton behavior (Bjorken scaling); another is to consider $e^+ \ e^- \to$ jets, for which extraction of the strong coupling constant value is very complicated, as these experiments are conducted at energies where the effective coupling is rather large. Hence, higher order effects must be included when comparing theory with data. To make matters worse, by working to a finite order in perturbation theory, one necessarily introduces an artificial dependence on the renormalization scale. This induces errors in the value of α_{QCD} extracted from these processes.

One can also measure the total annihilation cross section of electron-positron into hadrons at lower energies, with the result

$$\alpha_{QCD}(34 \text{ Gev}) = 0.146^{+0.031}_{-0.026} \ , \tag{5.20}$$

but the most direct way to measure the value of the QCD coupling is through a careful measurement of $e^+e^- \to$ hadrons at the Z pole where most strong interaction effects are minimal. This yields

$$\alpha_{QCD}(M_Z) = 0.122 \pm 0.004 \ . \tag{5.21}$$

By taking an average over all measurements, one obtains a world average at the Z mass

$$\alpha_{QCD}(M_Z) = 0.1190 \pm 0.0058 \ . \tag{5.22}$$

This corresponds to the five-quark value

$$\Lambda^{(5)}_{\overline{MS}} \approx 220^{+78}_{-63} \text{ MeV} \ . \tag{5.23}$$

This is how dimensional transmutation works! From a more fundamental point of view, it would be far better to specify the value of the QCD coupling at the universal reference scale, the Planck scale, or even the GUT scale, but since no experiments are performed at those scales, we have to suffer through this cumbersome extraction procedure. Finally we note that lattice simulations of QCD yield comparable values for the strong coupling constant.

PROBLEMS

A. Suppose that measurements of α_S at the Z mass, and in Υ decay had indicated a 30% discrepancy in their value of $\Lambda_{\overline{MS}}$. Discuss quantitatively possible theoretical explanations. Which of the two values of $\Lambda_{\overline{MS}}$ you expect to be larger?

B. Verify that Λ_{MS} as defined in the text is renormalization group invariant to three loops.

C. Derive Eq. (5.15), starting from the anatytic form for α_{QCD}.

D. Calculate the ratios (5.17) and (5.18) and the scale at which each measures the strong coupling constant.

E. Assume the existence of another unbroken Yang-Mills theory with the following characteristics: it is locally invariant under $SU(4)$; it couples to new fermions transforming as **4** of the gauge group. One left-handed quartet transforms as an electroweak doublet, with hypercharge y_1, and two right-handed electroweak singlets with hypercharges y_2 and y_3, respectively.
1-) Determine the values of the hypercharges which avoid anomalies.
2-) Assume that its gauge coupling is the same as the QCD coupling at the Planck scale. What is the value of its $\Lambda'_{\overline{MS}}$?

5.2 THE CHIRAL LAGRANGIAN APPROXIMATION

Having ascertained that α_{QCD} becomes large at small energies, we can now proceed to discuss QCD in the static limit. We do not expect the structure of QCD at large distances to be easily described in terms of quarks and gluons, since, when last seen, their coupling constant was climbing out of perturbative range. The "Confinement Hypothesis" asserts that only those objects which are impervious to the color force may be identified with free but strongly interacting particles. This hypothesis, besides sounding plausible, has the virtue of being in accord with experiment: no isolated quark has ever been found (except by countless generations of freshmen duplicating Millikan's oil drop experiment!), and all the known strongly interacting particles can be interpreted in terms of colorless bound states.

It is an exercise in group theory to form color singlet combinations out of the quark fields. The simplest such combinations are made up of quarks, the $q\bar{q}$ quark-antiquark bound states, corresponding to $\mathbf{3} \otimes \bar{\mathbf{3}} = \mathbf{1} \oplus \mathbf{8}$, as well as the totally antisymmetric combination of three quarks, corresponding to $(\mathbf{3}\otimes\mathbf{3}\otimes\mathbf{3})_A = \mathbf{1}$. These bound states are identified with the mesons which mediate the strong force, and the nucleons, respectively. They are easily recognized because of their electroweak quantum numbers. QCD also allows for colorless combinations of gluons, called glue-balls, corresponding to the symmetric combinations of two and three gluons according to $(\mathbf{8} \otimes \mathbf{8})_S = (\mathbf{1} \oplus \cdots)$, and $(\mathbf{8} \otimes \mathbf{8} \otimes \mathbf{8})_S = (\mathbf{1} \oplus \cdots)$. These bound states are extremely hard to find, which is to be expected since they have no electroweak quantum numbers.

It would be too simplistic to think of these bounds states as being made up solely of these combinations; color singlets can be formed by including arbitrary number of gluons and quark-antiquark pairs.

Experiments show that baryons have the same electroweak quantum numbers as those of three (symmetrized) quarks, and mesons have the electroweak quantum numbers of quark-antiquark pairs. However one can obtain the same electroweak combinations by adding an arbitrary number of gluons, or by a judicious addition of quark-antiquark pairs. Hence, we must think of the physical mesons and baryons as highly complicated combinations of multiquark-multiglue states

$$\text{Mesons :} \qquad |\, \mathbf{q}^i\bar{\mathbf{q}}^j(1 + \mathbf{q}^k\bar{\mathbf{q}}^k + \mathbf{glue} + \cdots) > ,$$

$$\text{Baryons :} \qquad |\, \mathbf{q}^i\mathbf{q}^j\mathbf{q}^k(1 + \mathbf{q}^l\bar{\mathbf{q}}^l + \mathbf{glue} + \cdots) > .$$

Here i, j, k, l, are flavor indices; the color indices have been summed over and are not shown. The mesons have integer spin, starting with spin zero for the quark antiquark pair in an s-wave configuration, and spins antialigned. The baryons have half integer spins starting with spin one-half. We have used a highly schematic notation to emphasize that any colorless combination is actually a linear superposition of all possible color singlets with the same quantum numbers. Similarly, glue-balls contain electroweak singlet p-wave combinations of quarks and antiquarks (see problem).

Although consistent with experiment, the confinement hypothesis has yet to be proven from first principles. It is not our purpose to worry about such things, especially since experiments provide strong circumstantial evidence for it. The actual proof remains an outstanding unsolved problem in theoretical physics; its solution is sure to advance our knowledge of non-perturbative field theory in a dramatic way.

How then do we relate QCD to the mundane world of the nuclear force, of $\pi - N$ scattering, etc...? The key observation is that the electroweak quantum numbers which are shared by quarks, and their composites, mesons and baryons allow us to perceive a remarkable phenomenon, namely that some of the symmetry of the QCD Lagrangian is spontaneously broken, presumably by the color force, which is itself left unbroken. This open the way for a beautiful description of the static limit of QCD. Historically, Nambu, reasoning by analogy to the BCS theory of superconductivity, identified the pion as the harbinger of that spontaneous breaking. This of course predates quarks, QCD and gluons.

QCD is the fundamental theory of strong interactions. We may think of the two descriptions, the quark-gluon phase at short distances, and the baryon-meson phase at large distances, as two different phases of the same theory. The scale which comes with QCD, Λ_{QCD}, provides a quantitative way to differentiate between these two phases. For instance, quark masses which appear in the QCD Lagrangian can be put in two different classes, depending on their value with respect to that scale. If a quark has a mass larger than Λ_{QCD}, it may be vieved as a quasi-free particle in its rest frame. However if the quark is light with respect to Λ_{QCD}, it is always strongly interacting in its rest frame.

The two phases of QCD share some symmetries, for example, charge invariance. Other symmetries may be treated differently in the two phases: in the quark-gluon phase, some symmetries are realized *linearly*, while the same symmetries in the baryon-meson phase are realized *non-linearly*, in the particular way that is characteristic of spontaneously broken symmetries.

While we do not have the tools to derive the meson-baryon phase from first principles, we can go a long way just by understanding the way the symmetries of the QCD Lagrangian are realized on mesons and baryons.

To make things as simple as possible we (temporarily) turn off the weak interactions, and consider the QCD Lagrangian, with non-zero quark masses. We further segregate these masses in two classes; those lighter than Λ_{QCD}, and those heavier. There are two quarks u, d, which are thought to have masses definitely much smaller than Λ_{QCD}, and one, the s quark with a mass comparable to Λ_{QCD}. The other three quarks, c, b, and t have masses larger than Λ_{QCD}, and are not expected to play a role in this large distance description of QCD.

To start, we set the masses of the three "light" quarks to zero. In this limit, called the chiral limit, the light quark QCD Lagrangian is just

$$\mathcal{L}_{QCD} = -\frac{1}{4}\,\mathrm{Tr}\,(\mathbf{G}_{\mu\nu}\mathbf{G}^{\mu\nu}) + \sum_{i=1}^{3} \bar{\mathbf{q}}_i \gamma_\mu \mathcal{D}^\mu \mathbf{q}_i \ . \tag{5.24}$$

By setting these masses to zero, we have gained a very large global symmetry: $SU_L(3) \times SU_R(3) \times U(1)$, linearly realized on the three light quarks, u, d, s:

$$SU_L(3) \ : \ \mathbf{q}_L \to U_L \mathbf{q}_L \ ; \qquad SU_R(3) \ : \ \mathbf{q}_R \to U_R \mathbf{q}_R \ , \tag{5.25}$$

$$U(1) \ : \ \mathbf{q}_{L,R} \to e^{i\alpha} \mathbf{q}_{L,R} \ . \tag{5.26}$$

We have not included the flavor-blind chiral $U(1)$

$$\mathbf{q}_L \to e^{i\beta} \mathbf{q}_L \qquad \mathbf{q}_R \to e^{-i\beta} \mathbf{q}_R \ , \tag{5.27}$$

because it is not conserved due to quantum effects we have yet to discuss.

What happens to these chiral symmetries in the Meson-Baryon phase? Although nothing can be rigorously proved, the interactions of pions indicates that

- Lorentz invariance is unbroken.
- Color invariance stays unbroken.
- The chiral symmetries of massless QCD are spontaneously broken.
- The vector symmetries of massless QCD stay unbroken.

Lorentz invariance is obvious since the strong interactions are themselves Lorentz-invariant. The belief in color invariance is substantiated by the inability of experimentalists to isolate quarks.

The consequences of the last two assumptions are much more subtle to identify. If they are true, then the chiral symmetry $SU_L(3) \times SU_R(3)$ is broken to its maximal vectorial subgroup, that is

$$SU_L(3) \times SU_R(3) \to SU_{L+R}(3) \ ,$$

leaving the vectorial $U(1)$ (baryon number) unaffected.

Through what mechanism does this spontaneous breakdown take effect? It is believed to take place through the condensation of quark pairs in the vacuum; although this idea stemmed from analogy with BCS theory, the dynamical mechanisms seem to be vastly different: in the BCS theory the attraction between electrons is generated by interactions with the lattice, described in terms of phonons, the longitudinal quanta associated with the lattice vibrations. In QCD the attraction is believed to be provided by the exchange of transverse virtual gluons. They are different unless one thinks of gluons as the normal modes of some lattice – (if so what kind of a lattice?)

It is remarkable that, independent of the dynamics which brings this about, we can infer a number of detailed consequences, based solely on symmetry considerations. Consider the fermion bilinear made of a quark and an antiquark

$$\mathbf{q}^{\dagger ia}_{L\alpha}\mathbf{q}^{\beta}_{Rjb} \,, \tag{5.28}$$

where i and j are the $SU_L(3)$, and $SU_R(3)$ indices, respectively, α, β the Lorentz indices, and a, b the color indices. Under the Lorentz group it transforms according to

$$(\mathbf{2},\mathbf{1})\otimes(\mathbf{2},\mathbf{1})\sim(\mathbf{1},\mathbf{1})\oplus(\mathbf{3},\mathbf{1}) \;. \tag{5.29}$$

The presence of the singlet indicates that one combination preserves Lorentz invariance. Similarly, under the $SU(3)$ color group, this combination of fields transforms as

$$\bar{\mathbf{3}}^c\otimes\mathbf{3}^c=\mathbf{1}^c\oplus\mathbf{8}^c \;. \tag{5.30}$$

The presence of a singlet means that there is a particular combination that does not break color.

On the other hand, under $SU_L(3)\times SU_R(3)$ the bilinear transforms as $(\mathbf{3},\bar{\mathbf{3}})$. This representation breaks up in terms of the diagonal vector-like subgroup $SU_{L+R}(3)$ as

$$(\mathbf{3},\bar{\mathbf{3}})\rightarrow\mathbf{1}\oplus\mathbf{8} \;; \tag{5.31}$$

which shows one component that breaks the chiral symmetry while preserving its vectorial subgroup.

It follows from this discussion that this quark bilinear has the right quantum numbers to achieve the desired result: breakdown of the chiral symmetry to a vectorial symmetry while leaving Lorentz and color symmetries unbroken.

The dynamical *assumption* is simply that the effect of the QCD force between quarks and antiquarks is to force quark-antiquark condensates to form in the vacuum

$$<\mathbf{q}^{\dagger i}_{L}\mathbf{q}_{Rj}>\neq 0 \;. \tag{5.32}$$

This leaves the Lorentz and color symmetries untouched, but breaks spontaneously the chiral symmetries down to their diagonal vectorial subgroup.

It is worth emphasizing that no one has proven that QCD dynamics actually produces these condensates, aligned so as to produce this desired breaking pattern. In the following, we assume that it does, and compare its consequences with experiment.

In the absence of any other forces to help us distinguish one type of quark from another, condensation can take place with any quark flavor. This means that there are many different inequivalent ground states, and until we introduce extra probes, we cannot hope to distinguish between them. This situation is analogous to the ground state of a ferromagnet which is characterized by some non-zero magnetization $\mathbf{M_0}$. For each direction there is an inequivalent ground state, but we must introduce a directional probe (in this case a magnetic field) to distinguish among the possibilities. Actually, the vacuum aligns with the external probe. In the real world QCD, the external probe is provided by the quark masses and by the electromagnetic interactions.

In the spontaneous breaking of $SU_L(3) \times SU_R(3)$ to $SU_{L+R}(3)$, eight generators are "broken". To each of these generators corresponds a massless Nambu-Goldstone boson with exactly the same quantum numbers, which transform as an octet under the vectorial $SU_{L+R}(3)$.

We caution the reader that it is not quite true to say that the chiral symmetry has been "broken". In fact it is still there, but rather than being realized linearly on the quarks, it is realized on the Nambu-Goldstone bosons, in a special nonlinear, but less recognizable way (unless you know what you are looking for). Its identity card is the way the massless Nambu-Goldstone bosons couple to matter.

In this non-linear realization, the fundamental construct which contains the Nambu-Goldstone bosons is the group element which labels the "broken" symmetry operations. In our notation we take it to be the $n \times n$ complex matrix

$$\mathbf{\Sigma}(x) \equiv e^{\frac{2i}{F_\pi}\boldsymbol{\pi}(x)} , \tag{5.33}$$

where F_π is a constant, and

$$\boldsymbol{\pi}(x) = \boldsymbol{\lambda}^a \pi^a(x) , \tag{5.34}$$

where $\pi^a(x)$ are the eight Nambu-Goldstone fields, and $\boldsymbol{\lambda}^a$ are the generators of $SU(3)$, normalized to

$$\mathrm{Tr}\,(\boldsymbol{\lambda}^a \boldsymbol{\lambda}^b) = 2\delta^{ab} . \tag{5.35}$$

Its transformation properties under $SU_L(3) \times SU_R(3)$ are exactly the same as those of the condensate, that is:

$$\boldsymbol{\Sigma} \to \boldsymbol{\Sigma}' = U_L \boldsymbol{\Sigma} U_R^{-1} , \tag{5.36}$$

where $U_{L,R}$ are unitary matrices generating $SU_{L,R}(3)$

$$U_L = e^{\frac{i}{2}\omega_L} ; \qquad U_R = e^{\frac{i}{2}\omega_R} , \tag{5.37}$$

where

$$\boldsymbol{\omega}_{L,R} = \omega^a_{L,R} \boldsymbol{\lambda}_a . \tag{5.38}$$

Under the vectorial transformations, $\boldsymbol{\omega}_L = \boldsymbol{\omega}_R \equiv \boldsymbol{\omega}_V$, $U_L = U_R$ and the Nambu-Goldstone fields transform linearly as

$$\delta \boldsymbol{\pi}(x) = \frac{i}{2}[\boldsymbol{\omega}_V, \boldsymbol{\pi}(x)] , \tag{5.39}$$

shown here for infitesimal transformations. On the other hand, the Nambu-Goldstone fields transform inhomogeneously under the chiral transformations, for which $\boldsymbol{\omega}_L = -\boldsymbol{\omega}_R \equiv \boldsymbol{\omega}_A$,

$$\delta \boldsymbol{\pi}(x) = \frac{F_\pi}{2}\boldsymbol{\omega}_A - \frac{1}{F_\pi}[\boldsymbol{\pi}, [\boldsymbol{\pi}, \boldsymbol{\omega}_A]] + \cdots . \tag{5.40}$$

This characteristic transformation law is the signature of Nambu-Goldstone bosons. The ω^a_A label the degeneracy of the vacuum. We can immediately infer two consequences from invariance under this transformation law: mass terms which are quadratic in the pion fields are not invariant, and thus these pions must be massless; second pions must couple to matter in such a way that the action is left invariant when the pion field is shifted. Hence the Lagrangian density may change by at most a total derivative: pions must couple to terms which are themselves divergences, and thus decouple at zero momentum.

Pion dynamics is to be described in terms of an effective chiral Lagrangian, invariant under $SU_L(3) \times SU_R(3)$, expressed solely in terms of the unitary matrix $\boldsymbol{\Sigma}(x)$. There is a unique term that involves the least number of derivatives

$$\mathcal{L}_0 = \frac{F_\pi^2}{16} \operatorname{Tr} \left(\partial_\mu \boldsymbol{\Sigma} \partial^\mu \boldsymbol{\Sigma}^{-1} \right) . \tag{5.41}$$

The dimensionful constant F_π is added to make the dimensions come out right. By expanding in powers of F_π^{-1},

$$\boldsymbol{\Sigma} = 1 + \frac{2i}{F_\pi} \boldsymbol{\lambda}^a \pi^a + \dots , \tag{5.42}$$

we obtain the conventional kinetic term

$$\mathcal{L}_0 = \frac{1}{2}\partial_\mu \pi^a \partial_\mu \pi^a \quad + \cdots , \tag{5.43}$$

together with many other interesting terms, but all suppressed by orders of $\frac{1}{F_\pi}$. This suppression is effective only for energies low compared to F_π, indicating the limited validity of this Lagrangian. Note that the eight Nambu-Goldstone fields are massless. In terms of the physical fields,

$$\boldsymbol{\pi} = \sqrt{2}\begin{pmatrix} \frac{1}{\sqrt{2}}\pi_3 + \frac{1}{\sqrt{6}}\pi_8 & \pi^+ & K^+ \\ \pi^- & -\frac{1}{\sqrt{2}}\pi_3 + \frac{1}{\sqrt{6}}\pi_8 & K^0 \\ K^- & \overline{K}^0 & -\sqrt{\frac{2}{3}}\pi_8 \end{pmatrix} . \tag{5.44}$$

The superscripts refer to the electric charge of these fields, anticipating the inclusion of electromagnetic interactions which we have yet to do. The higher order terms in this Lagrangian describe multi Nambu-Goldstone processes such as $\pi - \pi$, $\pi - K$, $K - K$,.. scatterings.

For the moment we restrict ourselves only to this Lagrangian, although it is possible to include other terms with larger number of derivatives. For a detailed comparison with data, many more terms with higher powers of derivatives need to be added, but the coefficients in front of them are quite small. Hence, for our modest purposes, this Lagrangian will suffice.

PROBLEMS

A. Determine the P and CP properties of the quark bilinear (5.28).

B. Apply the discussion of this section to the hypothetical $SU(4)$ interaction introduced in problem A of the last section. In particular, analyze the consequences to the breaking of the electroweak symmetry. What value must its coupling have at Planck mass for this interaction to provide the correct value of the Fermi constant?

C. Using just $\boldsymbol{\Sigma}$ and its derivatives, construct new invariants under the chiral symmetry. In what form would these terms enter the chiral Lagrangian? Discuss their significance.

5.3 EXPLICIT CHIRAL SYMMETRY BREAKING

In the real world, quarks have masses, and their effect must be incorporated in the Lagrangian. The inclusion of the light quark masses explicitly breaks the chiral symmetry, and generates masses for the hitherto massless Nambu-Goldstone bosons. In the QCD Lagrangian this symmetry breaking effect is of course *linear* in the quark masses. In order to see how to incorporate their effect in the chiral Lagrangian, it is necessary to carefully analyze how these masses transform under $SU_L(3) \times SU_R(3)$.

Our first observation is that the simplest potential mass term for the Nambu-Goldstone bosons is simply $\mathrm{Tr}\,(\mathbf{\Sigma})$, which is invariant under the vectorial subgroup. A common mass for all the light quarks in the QCD Lagrangian would produce the same effect: break the chiral symmetry completely, while keeping the vectorial subgroup $SU_{L+R}(3)$ invariant. A common light quark mass corresponds to adding to the chiral Lagrangian a term with the same invariances, namely proportional to $\mathrm{Tr}\,(\mathbf{\Sigma})$.

In the realistic case where all the light quark masses are different, the quark mass differences do not preserve the vectorial subgroup; rather they transform as members of an octet with respect to $SU_{L+R}(3)$. In this basis, we are led to introduce the light quark mass matrix,

$$\mathbf{M} \equiv \begin{pmatrix} m_u & 0 & 0 \\ 0 & m_d & 0 \\ 0 & 0 & m_s \end{pmatrix} , \tag{5.45}$$

in terms of which we can write the symmetry breaking term as

$$\mathcal{L}_{SB} = \frac{F_\pi^2}{8} m_0 \; \mathrm{Tr}\,(\mathbf{\Sigma}^\dagger \mathbf{M}) + \mathrm{c.c}\ . \tag{5.46}$$

where m_0 is an unknown parameter with dimension of mass. The masses are identified as the coefficients in front of the pseudoscalar fields. A straightforward calculation directly yields the mass squared of the off-diagonal fields

$$\begin{aligned} m_{\pi^\pm}^2 &= m_0(m_u + m_d)\ , \\ m_{K^\pm}^2 &= m_0(m_u + m_s)\ , \\ m_{K^0}^2 &= m_0(m_d + m_s)\ . \end{aligned} \tag{5.47}$$

It follows that $m_u < m_d$, since $m_{K^0}(497.6\ \mathrm{MeV}) > m_{K^\pm}(493.6\ \mathrm{MeV})$. The diagonal fields appear in the terms

$$m_0 \left((m_u + m_d)\pi^3\pi^3 + \frac{2}{\sqrt{3}}(m_u - m_d)\pi^3\pi^8 + \frac{1}{3}(m_u + m_d + 4m_s)\pi^8\pi^8 \right) .$$

A canonical transformation, from π^3, π^8 to the physical π^0, η, diagonalizes this expression, yielding their masses

$$\begin{aligned} m_{\pi^0}^2 =& m_0 \left(m_u + m_d - \frac{1}{2} \frac{(m_u - m_d)^2}{2m_s - m_u - m_d} \right) , \\ m_\eta^2 =& m_0 \left(\frac{4m_s + m_u + m_d}{3} + \frac{1}{2} \frac{(m_u - m_d)^2}{2m_s - m_u - m_d} \right) . \end{aligned} \tag{5.48}$$

Although of the correct sign, the second term is too small to account for the mass difference between π^0 (134.96 MeV) and π^+ (139.57 MeV): electromagnetic effects must be included. Still, from these expressions, we can deduce the famous Gell-Mann-Okubo sum rule,

$$m_\eta^2 + m_{\pi^0}^2 = \frac{2}{3}(m_{K^0}^2 + m_{K^+}^2 + m_{\pi^+}^2) , \tag{5.49}$$

which is in agreement with experiment. In this form, it is yet to be corrected for electromagnetic contributions to the masses. This is done in the next section.

Besides the mass terms, $\mathcal{L}_{SB}$ also provides (in higher order) corrections to scattering of the Nambu-Goldstone bosons. This simple effective Lagrangian provides a model for the interactions of the pseudoscalar mesons π, K, η. We leave it as an exercise to compute the corrections it gives to these processes.

PROBLEMS

A. Show that in the QCD Lagrangian the light quark masses transform as an octet under $SU_{L+R}(3)$.

B. Repeat problem C of the last section, using this time the matrices $m_0\mathbf{M}$, $\mathbf{\Sigma}$ and its derivatives. List such terms with the lowest dimensions which do not appear in the Lagrangian (see D. B. Kaplan and A. V. Manohar, *Phys. Rev. Lett.* **56**, 2004(1986)).

5.4 ELECTROWEAK INTERACTIONS

So far we have only examined strong interactions among the Nambu-Goldstone bosons. Their electroweak interactions are determined by their couplings to the gauge bosons of the standard model. In the QCD Lagrangian, the W-boson, the Z-boson and the photon couple to currents bilinear in quark and lepton fields, corresponding to symmetries which appear in the chiral Lagrangian as well, albeit realized differently. We therefore proceed as before: express these currents in terms of $\mathbf{\Sigma}$, and then couple them to the gauge bosons, which appear as fundamental fields.

The canonical $SU_L(3) \times SU_R(3)$ currents of the chiral Lagrangian can be obtained through the Noether process. They are given by

$$J^{\mu a}_{L,R} = \frac{\delta \mathcal{L}_0}{\delta \partial_\mu \pi^b} \frac{\delta \pi^b}{\delta \omega^a_{L,R}} \; . \tag{5.50}$$

We expand this complicated expression in powers of F_π^{-1}, using the expansion Eq. (5.42), and use the first term in the transformation of the fields. The result is

$$J^{\mu a}_{L,R} = \pm \frac{F_\pi}{4} \partial_\mu \pi^a(x) + \cdots \; . \tag{5.51}$$

By comparing this form with the expansion of the Lagrangian Eq. (5.43), we note that, after dropping a surface term, the chiral Lagrangian can be rewritten as

$$\mathcal{L}_0 = -\frac{1}{F_\pi} \pi^a \partial^\mu (J^a_{\mu L} - J^a_{\mu R}) \; . \tag{5.52}$$

This is exactly the expected coupling of Nambu-Goldstone bosons, with F_π playing the role of the vacuum value of the order parameter.

It is easiest to find the full currents by using the following covariance arguments. The left handed current must be built out of $\mathbf{\Sigma}$, transform as a Lorentz four-vector, as an octet under $SU_L(3)$, and as a singlet under $SU_R(3)$. It must therefore contain $\partial_\mu \mathbf{\Sigma}$ or $\partial_\mu \mathbf{\Sigma}^{-1}$. The octet index on the currents signifies that they must contain the matrix $\boldsymbol{\lambda}^a$ as well. Using their transformation properties

$$\boldsymbol{\lambda}^a \to U^{-1}_{L,R} \boldsymbol{\lambda}^a U_{L,R} \; , \tag{5.53}$$

under the chiral group, we check that the expression

$$\mathrm{Tr}\,(\mathbf{\Sigma}^{-1} \boldsymbol{\lambda}^a \partial_\mu \mathbf{\Sigma}) \; , \tag{5.54}$$

transforms the same way as the $SU_L(3)$ current. In order to fix the normalization, we expand it in powers of F_π^{-1}, and compare with Eq. (5.51), with the result

$$J^a_{\mu L} = -i\frac{F_\pi^2}{16}\,\mathrm{Tr}\,(\boldsymbol{\Sigma}^{-1}\boldsymbol{\lambda}^a\partial_\mu\boldsymbol{\Sigma}). \tag{5.55}$$

A similar procedure yields the $SU_R(3)$ current

$$J^a_{\mu R} = -i\frac{F_\pi^2}{16}\,\mathrm{Tr}\,(\boldsymbol{\Sigma}\boldsymbol{\lambda}^a\partial_\mu\boldsymbol{\Sigma}^{-1})\ . \tag{5.56}$$

Expanding in inverse powers of F_π, we obtain

$$J^a_{\mu L,R} = \pm\frac{F_\pi}{8}\,\mathrm{Tr}\,(\boldsymbol{\lambda}^a\partial_\mu\boldsymbol{\pi}) - \frac{i}{8}\,\mathrm{Tr}\,([\boldsymbol{\pi},\boldsymbol{\lambda}^a]\partial_\mu\boldsymbol{\pi}) \mp \frac{1}{12F_\pi}\,\mathrm{Tr}\,([\boldsymbol{\pi},[\boldsymbol{\pi},\boldsymbol{\lambda}^a]]\partial_\mu\boldsymbol{\pi}) + \cdots \tag{5.57}$$

The chiral current contains only those terms that are odd under $F_\pi \to -F_\pi$.

There are other ways to check the form of these currents, for instance by adding the electromagnetic interactions to the chiral system. The photon couples vectorially to the three light quarks with strength equal to their electric charge. Thus, the photon coupling is determined by demanding invariance under local vector-like transformations of the type

$$\boldsymbol{\Sigma} \to e^{i\alpha(x)\mathbf{Q}}\boldsymbol{\Sigma}e^{-i\alpha(x)\mathbf{Q}}\ , \qquad \delta\boldsymbol{\Sigma} = i\alpha(x)[\mathbf{Q},\boldsymbol{\Sigma}]\ , \tag{5.58}$$

where $\mathbf{Q}$ is the diagonal charge matrix

$$\mathbf{Q} = \begin{pmatrix} 2/3 & 0 & 0 \\ 0 & -1/3 & 0 \\ 0 & 0 & -1/3 \end{pmatrix} = \frac{1}{2}\boldsymbol{\lambda}_3 + \frac{1}{2\sqrt{3}}\boldsymbol{\lambda}_8\ . \tag{5.59}$$

Since the Lagrangian density is already globally invariant under this transformation, we simply replace ∂_μ by the covariant derivative

$$\partial_\mu\boldsymbol{\Sigma} \to \partial_\mu\boldsymbol{\Sigma} + ieA_\mu[\mathbf{Q},\boldsymbol{\Sigma}]\ . \tag{5.60}$$

It is easy to verify that we obtain the electromagnetic current of the required form, namely

$$J^{em}_\mu = J^3_{\mu L} + \frac{1}{\sqrt{3}}J^8_{\mu L} + (L \to R)\ . \tag{5.61}$$

Note that the electromagnetic interactions are not invariant under all the chiral symmetries. The broken symmetries in the directions associated with the charged NG bosons, that is $\pi^\pm$ and $K^\pm$, are broken equally since they have the same charge. One can therefore expect the same contribution to the mass squared of these two particles, coming from the electromagnetic interactions. These affect the Gell-Mann-Okubo mass relation, by adding the same correction to the charged NG boson squared masses.

We now see how to include electromagnetic effects in the meson mass analysis. We simply add to both $m^2_{\pi^+}$ and $m^2_{K^+}$ an unknown equal contribution Δ_{em},

$$m^2_{\pi^\pm} = m_0(m_u + m_d) + \Delta_{em} \ , \qquad m^2_{K^\pm} = m_0(m_u + m_s) + \Delta_{em} \ , \quad (5.62)$$

and none to the masses of the neutral mesons. Fitting to the five meson masses, we find (in MeV^2)

$$m_0 m_u = 6528 \ , \qquad m_0 m_d = 11720 \ , \qquad m_0 m_s = 11720 \ , \qquad (5.63)$$

and $\Delta_{em} = 1232$ MeV^2, yielding an accurate determination of the quark mass ratios

$$\frac{m_u}{m_d} \approx \frac{1}{2} \ , \qquad \frac{m_d}{m_s} \approx \frac{1}{20} \ , \qquad (5.64)$$

as well as a determination of the η mass.

In order to couple the weak gauge bosons, we first identify the quantum numbers of the current to which they couple in the QCD Lagrangian. Their couplings in the chiral Lagrangian are then simply taken to be to the same current, but the light quark contributions are replaced by the expression Eqs. (5.55) and (5.56). Thus, the Z-boson couples according to

$$J^Z_\mu = J^3_{\mu L} - \sin^2\theta_{\rm W} J^{\rm em}_\mu + \cdots \ , \qquad (5.65)$$

and the charged W-boson couples to the current

$$J^-_\mu = V_{ud} J^{1+i2}_{\mu L} + V_{us} J^{4+i5}_{\mu L} + \cdots \ , \qquad (5.66)$$

where we have inserted the relevant CKM matrix elements, and where the dots refer to the leptonic weak currents as well as to the quark weak currents that contain at least one heavier quark. For the charged current they are of two types, one containing one heavy and one light quark, and one containing only heavy quarks. For the neutral current, they are only of the heavy-heavy type. The mixed quark charged current requires special handling, which we do not describe here.

These couplings enable us to compute the weak decays of the Nambu-Goldstone bosons, starting with

$$\pi^+ \to \mu^+ + \nu_\mu \ , \tag{5.67}$$

which, upon comparison with experiment, will determine the value of F_π. This decay occurs through the coupling to the W-boson which then decays into an antimuon and a muon neutrino, as shown below

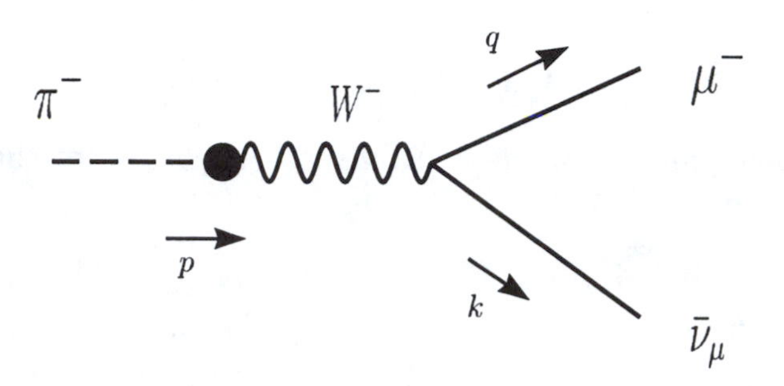

It is caused by the interactions

$$\mathcal{L}_{int} = \frac{i}{\sqrt{2}} g_2 W^{+\rho}(J^-_\rho + \nu^\dagger_\mu \sigma_\rho \mu_L) \ . \tag{5.68}$$

The matrix element for the transition can be calculated to be

$$\hat{T}_{fi} = i\sqrt{2} G_F V_{ud} F_\pi \sqrt{\frac{m_\mu m_\nu}{2E_\pi E_\mu E_\nu}} \bar{v}^{(a)}(k) \not{p} L u^{(b)}(q) \ , \tag{5.69}$$

where we have used

$$\langle 0|\pi^-(x)|\pi^+\rangle = \frac{e^{ip\cdot x}}{\sqrt{2E_\pi}} \ . \tag{5.70}$$

By using momentum conservation, $p = q + k$, and the Dirac equation for both muon and neutrino fields, we can simplify this expression to

$$\hat{T}_{fi} = i m_\mu \sqrt{2} G_F V_{ud} F_\pi \sqrt{\frac{m_\mu m_\nu}{2E_\pi E_\mu E_\nu}} \bar{v}^{(a)}(k) R u^{(b)}(q) \ , \tag{5.71}$$

We apply the techniques of chapter 4 to find

$$\sum_{a,b} |\bar{v}^{(a)}(k) R u^{(b)}(q)|^2 = \frac{(m_\pi^2 - m_\mu^2)}{4 m_\mu m_\nu} \ , \tag{5.72}$$

after using the kinematical relation

$$2p \cdot k = m_\pi^2 - m_\mu^2 \ . \tag{5.73}$$

The rest frame differential decay rate is given by

$$d\Gamma = \frac{(G_F F_\pi m_\mu V_{ud})^2}{4m_\pi (2\pi)^2}(m_\pi^2 - m_\mu^2)\frac{d^3q}{E_\mu}\frac{d^3k}{E_\nu}\delta^{(4)}(p - q - k) \ . \tag{5.74}$$

Performing the phase space integrals yields the decay rate

$$\Gamma(\pi^+ \to \mu^+ + \nu_\mu) = \frac{m_\pi}{16\pi}(G_F F_\pi m_\mu V_{ud})^2(1 - \frac{m_\mu^2}{m_\pi^2})^2 \ . \tag{5.75}$$

Comparison with experiment yields

$$F_\pi = 186 \text{ MeV} \ . \tag{5.76}$$

It is reassuring that this value is of the same order of magnitude as the scale at which the QCD coupling constant becomes strong, Λ_{QCD}. Since F_π is the value of the order parameter, it is consistent with the interpretation that the QCD force spontaneously breaks chiral symmetry.

A host of other processes can be predicted. For instance, the semileptonic decays of the τ lepton

$$\tau^- \to \pi^- + \nu_\tau \ , \tag{5.77}$$

and the strangeness-changing decay of the charged Kaons

$$K^- \to \mu^- + \bar{\nu}_\mu \ , \tag{5.78}$$

which fixes the ratio of the CKM matrix elements,

$$\frac{V_{us}^2}{V_{ud}^2} \approx .05 \ . \tag{5.79}$$

Consistency can then be tested by predicting the decay rate of

$$\tau^- \to K^- + \nu_\tau \ . \tag{5.80}$$

We can test other parts of the current which are not proportional to F_π. Such is the semi-leptonic pion β-decay

$$\pi^+ \to \pi^0 + e^+ + \nu_e \ . \tag{5.81}$$

We relegate their evaluation to the problems.

Our Lagrangian also predicts all types of non-leptonic weak decays, such as $K^0 \to \pi^+\pi^-$, $K^0 \to \pi^0\pi^0$, $K^0 \to \pi^+\pi^-\pi^0$ etc... Naively, these processes come about by W-boson exchange between two chiral currents. However the chiral Lagrangian prediction does not agree with experiment for this type of decays. In order to understand the reason, we must go back to the QCD Lagrangian, where there are strong gluon corrections to the simple tree level quark current-quark current interaction. We do not know how to calculate them because the QCD coupling is too strong. In other words, since both currents contain quarks, gluons are exchanged between the quarks in both currents, and correct the simple W-exchange picture by an amount we are unable to calculate. This is to be contrasted with semi-leptonic decays, where one current is purely leptonic, so that the only corrections to the current-current form are the calculable small weak corrections.

There are also semileptonic interactions which come about by Z-exchange. One such process is the decay

$$\pi^0 \to e^+ + e^- \ . \tag{5.82}$$

A straightforward computation yields the rate

$$\Gamma(\pi^0 \to e^+ + e^-) = \frac{m_\pi}{16\pi}(G_F F_\pi m_e)^2(\frac{1}{4} - \sin^2\theta_w + 2\sin^4\theta_w) \ , \tag{5.83}$$

where we have neglected the electron's mass in the kinematics. This rate is comparable to that for the charged decay, yet a quick glance at the Blue Book shows that this decay has not yet been observed. The reason is that the neutral pion decays much faster into two photons. Our Lagrangian does not contain such an interaction; it is incomplete.

PROBLEMS

A. Calculate $\Gamma(\tau^- \to \pi^- + \nu_\tau)$ and check that the strength of the strangeness-changing decay, $\tau^- \to K^- + \nu_\tau$, agrees with experiment.

B. Calculate $\Gamma(\pi^+ \to \pi^0 + e^+ + \nu_e)$.

C. Derive the couplings of the chiral Lagrangian by making it locally invariant under the Cabibbo-rotated $SU(2)_L$, generated by the matrices

$$I^\pm = \frac{1}{2}(\cos\theta_c \boldsymbol{\lambda}^{1\mp i2} + \sin\theta_c \boldsymbol{\lambda}^{4\mp i5}) \ ; \qquad I^3 = \frac{1}{2}\boldsymbol{\lambda}^3 \ .$$

Deduce to the lowest order in F_π^{-1} the couplings to the chiral system to two W-bosons ("seagull").

5.5 ANOMALIES AND THE CHIRAL LAGRANGIAN

The above discussion makes it clear that the effective Lagrangian of the previous section is not complete. We used symmetry requirements based on the Lorentz group and on $SU_L(3) \times SU_R(3)$ as a guide to infer the chiral Lagrangian from its QCD counterpart, but we did not include discrete symmetries. We have already remarked that $\mathcal{L}_{QCD}$ is invariant under both C(harge conjugation) and P(arity). It behooves us to see how these translate in the language of the chiral Lagrangian.

Charge conjugation is easily identified since it changes particles into antiparticles

$$\pi^0, \eta \to \pi^0, \eta \ ; \qquad \pi^\pm, K^\pm \to \pi^\mp, K^\mp \ ; \qquad K^0 \to \overline{K}^0 \ .$$

In terms of $\mathbf{\Sigma}$, this is equivalent to taking its transpose

$$\mathrm{C}: \qquad \mathbf{\Sigma} \to \mathbf{\Sigma}^T \ , \tag{5.84}$$

and because of the cyclic property of the trace, it is clear that $\mathcal{L}$ is invariant under charge conjugation.

The Nambu-Goldstone particles are thought to be bound states of quarks and antiquarks. Since the parity of an antifermion is opposite that of a fermion, they must be pseudoscalars. Thus under P(arity), we must have $\pi^a \to -\pi^a$, which translates in the combined operations

$$\mathbf{\Sigma} \to \mathbf{\Sigma}^{-1} \ ; \qquad \vec{x} \to -\vec{x} \ , \qquad t \to t \ . \tag{5.85}$$

These obviously leave the chiral Lagrangian, $\mathcal{L}_0$ invariant. Thus we have, as in QCD, invariance under both C and P.

However, $\mathcal{L}_0$ is also invariant under a discrete symmetry that is not present in QCD. Indeed the simpler transformation

$$(-1)^{N_B}: \qquad \mathbf{\Sigma} \to \mathbf{\Sigma}^{-1} \ , \tag{5.86}$$

leaves $\mathcal{L}_0$ invariant, where N_B is the number of bosons. As remarked by Witten (*Nucl. Phys.* **B223**, 422(1983)), there exist physical processes allowed by P but forbidden by $(-1)^{N_B}$. For instance, $K^+K^- \to \pi^+\pi^-\pi^0$. Although difficult to realize experimentally, the ϕ meson does decay strongly to both K^+K^- and $\pi^+\pi^-\pi^0$. Thus we must conclude that the chiral model is not complete – one must add to it a term which violates $(-1)^{N_B}$ but preserves P. It amounts to finding a term that is odd under the interchange of the spatial coordinates, all other things being equal.

In three space dimensions it is easy to manufacture such a term, since any such Poincaré invariant term must contain $\epsilon_{\mu\nu\rho\sigma}$, the Levi-Cività symbol. The quest begins with the equations of motion

$$\frac{F_\pi^2}{8}(\partial_\mu\partial^\mu\mathbf{\Sigma} + \mathbf{\Sigma}\partial_\mu\mathbf{\Sigma}^{-1}\partial_\mu\mathbf{\Sigma}) = \frac{F_\pi^2}{8}\mathbf{\Sigma}\partial^\mu(\mathbf{\Sigma}^{-1}\partial_\mu\mathbf{\Sigma}) = 0 \ , \tag{5.87}$$

which can be rewritten in the simple form

$$\frac{F_\pi^2}{8}\partial^\mu\mathbf{X}_\mu = 0 \ , \tag{5.88}$$

where

$$\mathbf{X}_\mu \equiv \mathbf{\Sigma}^{-1}\partial_\mu\mathbf{\Sigma} \ , \tag{5.89}$$

transforms as

$$\mathbf{X}_\mu \to U_R\mathbf{X}_\mu U_R^{-1} \ . \tag{5.90}$$

The Lorentz invariant combination

$$\epsilon^{\mu\nu\rho\sigma}\mathbf{X}_\mu\mathbf{X}_\nu\mathbf{X}_\rho\mathbf{X}_\sigma \ , \tag{5.91}$$

contains three space derivatives, and transforms the same way as the equations of motion derived from $\mathcal{L}_0$. It is also invariant under P. Thus, by adding it as a force term to the equations of motion, we produce a theory that is invariant only under P

$$\frac{F_\pi^2}{8}\partial^\mu\mathbf{X}_\mu = c\epsilon^{\mu\nu\rho\sigma}\mathbf{X}_\mu\mathbf{X}_\nu\mathbf{X}_\rho\mathbf{X}_\sigma \ , \tag{5.92}$$

with some strength c. Alas, it is not so easy to write a Lagrange density which involves $\epsilon_{\mu\nu\rho\sigma}$ and is invariant under $SU_L(3) \times SU_R(3)$, since the obvious term,

$$\epsilon^{\mu\nu\rho\sigma}\mathrm{Tr}(\mathbf{X}_\mu\mathbf{X}_\nu\mathbf{X}_\rho\mathbf{X}_\sigma) \ , \tag{5.93}$$

is identically zero due to the cyclic property of the trace. Thus we do not seem to be able to derive this equation of motion from a Lagrangian

At this stage the author cannot help but digress to present the beautiful reasoning by analogy of Witten, who finds a similar paradox in a much simpler situation, a charged point particle on a sphere with a magnetic monopole at its center. The Lagrangian that describes a a point particle on a sphere is given by

$$L = \frac{1}{2} m \dot{\mathrm{x}}_i^2(t) + \lambda[\mathrm{x}_i(t)\mathrm{x}_i(t) - R^2] \ , \tag{5.94}$$

where λ is a Lagrange multiplier. The equations of motion are

$$m\ddot{\mathrm{x}}_i(t) - 2\lambda \mathrm{x}_i(t) = 0 \ . \tag{5.95}$$

We can eliminate the Lagrange multiplier by dotting them with x_i, and using the constraint; this yields

$$\begin{aligned} 0 &= 2\lambda \mathrm{x}_i \mathrm{x}_i + m \mathrm{x}_i \ddot{\mathrm{x}}_i = 2\lambda R^2 + m\frac{d}{dt}(\mathrm{x}_i \dot{\mathrm{x}}_i) - m\dot{\mathrm{x}}_i \dot{\mathrm{x}}_i \ , \\ &= 2\lambda R^2 - m\dot{\mathrm{x}}_i \dot{\mathrm{x}}_i \ . \end{aligned} \tag{5.96}$$

The equations of motion are then

$$m\ddot{\mathrm{x}}_i + \frac{m\dot{\mathrm{x}}_j \dot{\mathrm{x}}_j}{R^2} \mathrm{x}_i = 0 \ . \tag{5.97}$$

They are invariant under two separate discrete symmetries, $t \to -t$ and $\mathrm{x}_i \to -\mathrm{x}_i$. Suppose we wanted invariance under the *combined* $t \to -t$, $\mathrm{x}_i \to -\mathrm{x}_i$. This can easily be achieved in the equations of motion, by adding a force term

$$m\ddot{\mathrm{x}}_i + \frac{m}{R^2} \dot{\mathrm{x}}_l \dot{\mathrm{x}}_l \mathrm{x}_i = \alpha \ \epsilon_{ijk} \mathrm{x}_j \dot{\mathrm{x}}_k \ . \tag{5.98}$$

However this extra term cannot be derived from a Lagrangian since the obvious candidate $\epsilon_{ijk}\mathrm{x}_i\mathrm{x}_j\dot{\mathrm{x}}_k$ vanishes identically.

The solution of the puzzle lies in recognizing that the right hand side can be interpreted as a Lorentz force due to the presence of a magnetic field

$$B_j = g\frac{\mathrm{x}_j}{R^3} = \frac{\alpha}{e}\mathrm{x}_j \ , \tag{5.99}$$

generated by a magnetic monopole of strength $g = \alpha R^3/e$, located at the center of the sphere. The Lagrangian that describes the interaction of the particle with the magnetic field involves the vector potential, $\vec{A}$. However, it is well know that for a magnetic monopole, $\vec{A}$ is not everywhere well defined. Gauss's law applied to a sphere with a magnetic pole at its center yields a net magnetic flux, while it should vanish for a potential description of the magnetic field. One way out is to imagine that there is a hole in the Gaussian surface out of which flows the magnetic flux. By considering smaller and smaller Gaussian surfaces one engenders a Dirac string which emanates out of the monopole. The potential is well-defined everywhere except on the string. Another way to look at it is to note that the presence of a magnetic charge means that somewhere you cannot interchange ∂_i and ∂_j when operating on $\vec{A}$, and that somewhere is the string. Having said this, we are free to consider the Lagrangian

$$L = \frac{1}{2} m \dot{\mathrm{x}}_i^2 + e\ A_i \dot{\mathrm{x}}_i - \lambda [\mathrm{x}_i \mathrm{x}_i - R^2] \ , \tag{5.100}$$

with A_i defined in terms of two coordinate systems, and e is the charge of the particle. However this Lagrangian is no longer-gauge invariant because the vector potential has singularities on the Dirac string.

A correct quantum mechanical treatment requires that e be quantized (Dirac quantization). Quantum amplitudes are obtained from the path integral of the exponential of this Lagrangian over all types of paths of the particle. Consider the part of the path integral integrand

$$e^{ie \oint_C A_i d\mathrm{x}_i} \tag{5.101}$$

over a closed path C on the sphere. By Stokes' theorem, it can be rewritten as

$$e^{ie \oint_C A_i d\mathrm{x}_i} = e^{ie \int_D F_{ij} d\sigma^{ij}} \ , \tag{5.102}$$

where D is a disc bounded by the closed orbit C. Since F_{ij} is well-defined everywhere, this serves as a perfectly good definition of the phase. Thus we should have started directly with this term in the action since it is obviously gauge invariant. There is, however, an ambiguity, since for a given orbit C there are two types of disks that abut it, the disk inside C, D, and the disk outside C, $\overline{D}$. It is easy to see that

$$e^{ie \oint_C A_i d\mathrm{x}_i} = e^{ie \int_D F_{ij} d\sigma^{ij}} = e^{-ie \int_{\overline{D}} F_{ij} d\sigma^{ij}} \ , \tag{5.103}$$

where the minus sign comes up because of the sense of the circulation around C. Consistency between the two choices can be achieved only if they are equal, that is when

$$\exp \left\{ ie \int_{D+\overline{D}} F_{ij} d\sigma^{ij} \right\} = 1 \ ; \tag{5.104}$$

but $D + \overline{D}$ is nothing but the whole sphere that surrounds the monopole. It follows from Gauss's law that the integral is exactly equal to $4g\pi$, so that $4ge\pi$ has to be a multiple of 2π, which is exactly the Dirac quantization condition

$$eg = \frac{n}{2} ; \qquad n = \pm 1, \pm 2, \cdots . \tag{5.105}$$

We draw two lessons from this analogy. One is that we should look for the extra term in the chiral Action in the form of a 5-dimensional integral of a total five-dimensional divergence. While it reduces to a four-dimensional integral, its integrand is no longer invariant under $SU_L(3) \times SU_R(3)$. Its four dimensional integral, the action, is of course invariant. This is why we have failed in finding an invariant Lagrangian to account for the extra symmetry– it does not exist. The second lesson is that we should expect that the coefficient in front of the extra term in the Action should be quantized, in analogy to the Dirac quantization condition. Remarkably, this is exactly the way it happens for the chiral model.

We now follow Witten's construction which applies only in the case of three massless quarks. Space-time is viewed as a four dimensional sphere M. The coset field $\mathbf{\Sigma}$ can be viewed as the mapping of M into the $SU(3)$ group manifold. There are no topological subtelties in this mapping because the group manifold of $SU(3)$ is eight-dimensional. In this manifold, the four-sphere M is the boundary of a five-dimensional disk, D.

We are therefore looking for a term of the form

$$\Gamma = \int_D d\sigma^{ijklm} \Omega_{ijklm} , \tag{5.106}$$

where $d\sigma$ is the surface element, and Ω is a fifth-rank antisymmetric tensor, invariant under the chiral group $SU_L(3) \times SU_R(3)$. To see if such a tensor exists, we note that it must exist around the point $\mathbf{\Sigma} = 1$, where it must be invariant under the group that leaves this point invariant, the vectorial subgroup $SU_{L+R}(3)$. It can thus be expressed on the tangent space at that point, which is spanned by the Lie algebra of $SU(3)$, and the problem has been reduced to one in Lie algebra theory. Does the antisymmetrized product of five octets contain an $SU(3)$ singlet? We first note that the quintuple product is dual to the three-times antisymmetrized product of octet, which necessarily contains a singlet, due to the totally antisymmetric structure constant of the Lie algebra. It follows that the five-product also has a singlet; indeed it can be verified that

$$(\mathbf{8} \times \mathbf{8} \times \mathbf{8} \times \mathbf{8} \times \mathbf{8})_A = \mathbf{1} \oplus \mathbf{8} \oplus \mathbf{10} \oplus \overline{\mathbf{10}} \oplus \mathbf{27} , \tag{5.107}$$

the same as the three times antisymmetrized product. Having constructed the tensor at $\mathbf{\Sigma} = 1$, we can then transport it to any other point by chiral transformations.

We do not expect Γ to be unique. As in the magnetic monopole case, the four-sphere M serves as the boundary of two disks, one "inside" and one "outside", $\overline{D}$. The sum of these two disks is a closed five-dimensional sphere, S. The ambiguity on which to use, D or $\overline{D}$ is resolved if

$$\int_S d\sigma^{ijklm}\Omega_{ijklm} = 2\pi N \; ; \qquad N = \pm 1, \pm 2, \cdots \; . \tag{5.108}$$

The mapping of a five-sphere into the $SU(3)$ manifold is not topologically trivial; it has a winding number, corresponding to the integer N above. It means that not all spheres are equal, but there is one basic sphere S_0 over which the flux is normalized

$$\int_{S_0} d\sigma^{ijklm}\Omega_{ijklm} = 2\pi \; . \tag{5.109}$$

The total chiral action is now given by

$$\mathcal{L} = \mathcal{L}_0 + N\Gamma_0 \; . \tag{5.110}$$

It is but a mathematical problem to find the invariant tensor Ω which satisfies this normalization condition. An inspired glimpse at the mathematical literature yields

$$\Omega_{ijklm} = -\frac{1}{240\pi^2}\mathrm{Tr}(\mathbf{X}_{[i}\mathbf{X}_j\mathbf{X}_k\mathbf{X}_\ell\mathbf{X}_{m]}) \; , \tag{5.111}$$

where the square brackets denote antisymmetrization, and

$$\mathbf{X}_i = \mathbf{\Sigma}^{-1}\frac{\partial \mathbf{\Sigma}}{\partial y^i}, \tag{5.112}$$

and the y^i are the five coordinates on the disk. The expansion in F_π^{-1},

$$\mathbf{X}_i = 1 + \frac{2i}{F_\pi}\boldsymbol{\lambda}^A\partial_i\pi^A + \cdots \; , \tag{5.113}$$

yields

$$\Omega_{ijklm} = \frac{2}{15\pi^2 F_\pi^5}\partial_{[i}\left\{\mathrm{Tr}(\boldsymbol{\pi}\partial_j\boldsymbol{\pi}\partial_k\boldsymbol{\pi}\partial_l\boldsymbol{\pi}\partial_{m]}\boldsymbol{\pi})\right\} + \cdots \; . \tag{5.114}$$

Since it is a total divergence, its contribution leads to the extra action, written as a four dimensional integral,

$$S_{\text{extra}} \equiv S_{WZ} = \frac{2N}{15\pi^2 F_\pi^5} \int d^4x \epsilon^{\mu\nu\rho\sigma} \text{Tr}(\boldsymbol{\pi}\partial_\mu\boldsymbol{\pi}\partial_\nu\boldsymbol{\pi}\partial_\rho\boldsymbol{\pi}\partial_\sigma\boldsymbol{\pi}) + \cdots . \tag{5.115}$$

This expression was first derived by Wess and Zumino, in an elegant analysis of the gauging of anomalous symmetries in the chiral Lagrangian.

Clearly this term accounts for the hitherto forbidden process $K^+K^- \rightarrow \pi^+\pi^-\pi^0$, which occurs with a strength fixed from π^+ decay, up to the mysterious integer. The meaning of the Witten quantization condition, as well as the value of the integer N can be understood when electromagnetic interactions are added to the Wess-Zumino-Witten Lagrangian.

The gauging of the Wess-Zumino term is not straightforward. The reason is that its integrand is not invariant under the chiral group. Hence we will have to rely on the trickle down technique: first gauge naively, then see what is left over; invent a new term to take care of the leak, and repeat until there is no more gauge leakage.

The Wess-Zumino term is invariant under a global charge transformation, but a local infinitesimal transformation

$$\delta\boldsymbol{\Sigma} = i\alpha(x)[\mathbf{Q}, \boldsymbol{\Sigma}] , \tag{5.116}$$

does not leave it invariant. Rather, by varying the five-dimensional integral, we find (notice the extra W)

$$\delta S_{WZW} = -N \int d^4x J^\mu \partial_\mu \alpha , \tag{5.117}$$

where the current is given by

$$J^\mu = \frac{1}{48\pi^2} \epsilon^{\mu\nu\rho\sigma} \text{Tr} \left\{ \boldsymbol{\Sigma}^{-1}\mathbf{Q}(\partial_\nu\boldsymbol{\Sigma})\mathbf{X}_\rho\mathbf{X}_\sigma + \mathbf{Q}\mathbf{X}_\nu\mathbf{X}_\rho\mathbf{X}_\sigma \right\} \tag{5.118}$$

$$= \frac{1}{3\pi^2 F_\pi^3} \epsilon^{\mu\nu\rho\sigma} \text{Tr} \left\{ \mathbf{Q}\partial_\nu\boldsymbol{\pi}\partial_\rho\boldsymbol{\pi}\partial_\sigma\boldsymbol{\pi} \right\} + \cdots \tag{5.119}$$

This current starts at order F_π^{-3}. Had we started by directly varying expression (5.115), we would have missed the lowest order terms. The change of the potential under a gauge transformation

$$\delta A_\mu = -\frac{1}{e}\partial_\mu\alpha , \tag{5.120}$$

leads us to consider the new action

$$S'_{WZW} = S_{WZW} + eN \int d^4x A_\mu J^\mu \ . \qquad (5.121)$$

Upon variation of the photon field, it would seem that the new term takes care of the extra piece Eq. (5.117). However things are not what they seem to be: the current J^μ is itself gauge dependent. A "little" calculation shows that its variation can be written in the form

$$\delta J^\mu = i\epsilon^{\mu\nu\rho\sigma}\partial_\rho K_\sigma \partial_\nu \alpha \ , \qquad (5.122)$$

where

$$\begin{aligned} K_\sigma &= \frac{1}{24\pi^2}\,\mathrm{Tr}\,\left\{\boldsymbol{\Sigma}^{-1}\mathbf{Q}^2\partial_\sigma\boldsymbol{\Sigma} + \mathbf{Q}^2\mathbf{X}_\sigma + \boldsymbol{\Sigma}^{-1}\mathbf{Q}\boldsymbol{\Sigma}\mathbf{Q}\mathbf{X}_\sigma\right\}, \qquad (5.123)\\ &= \frac{-i}{4\pi^2 F_\pi}(\partial_\sigma\pi^3 + \frac{1}{\sqrt{3}}\partial_\sigma\pi^8) + \cdots \ . \end{aligned}$$

This allows us to rewrite (after integration by parts) the variation of S'_{WZW} in the suggestive form

$$\delta S'_{WZW} = -\int d^4x \epsilon^{\mu\nu\rho\sigma}\partial_\rho A_\mu K_\sigma \partial_\nu \alpha \ . \qquad (5.124)$$

Thus in order to plug this leak, we have to add yet another term of the form

$$\epsilon^{\mu\nu\rho\sigma}\int d^4x K_\sigma A_\mu \partial_\rho A_\nu \ . \qquad (5.125)$$

A further calculation shows that K_σ is gauge invariant. The process ends and we finally arrive at the gauge invariant Action

$$S_{WZW} + Ne\int d^4x A_\mu J^\mu - iNe^2\epsilon^{\mu\nu\rho\sigma}\int d^4x \partial_\rho A_\mu A_\nu K_\sigma \ . \qquad (5.126)$$

The first piece of this action describes improbable processes such as $\pi^+\pi^- \to \gamma + \pi^0$, which are of order F_π^{-3}; on the other hand the second term describes very familiar processes such as the $\pi^o \to 2\gamma$ anomalous decay, of order $\frac{1}{F_\pi}$, viz

$$\frac{e^2}{48\pi^2}\frac{N}{F_\pi}\pi^0\epsilon^{\mu\nu\rho\sigma}F_{\mu\nu}F_{\rho\sigma} \ . \qquad (5.127)$$

A similar expression explains η decay into two photons. In the language of quarks, these processes take place through the famous triangle diagram. The neutral pion is associated with the operator $\overline{\mathbf{u}}\gamma_5\mathbf{u} - \overline{\mathbf{d}}\gamma_5\mathbf{d}$, the divergence of the relevant axial current

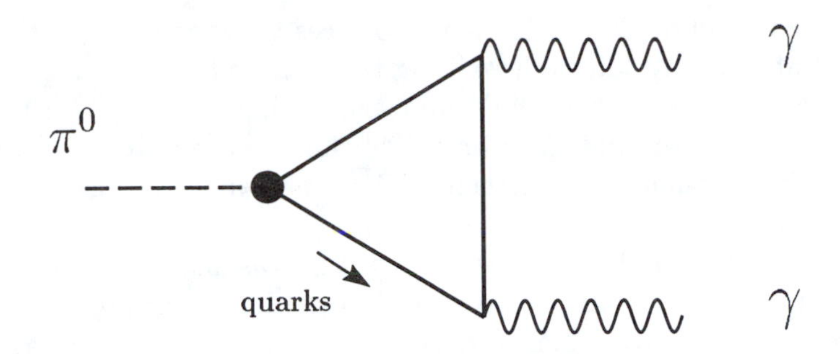

The diagram is proportional to the number of quarks circulating in the triangle, and therefore to the number of colors. Comparison with the expression we have just derived compels us to identify N with the number of colors $N_c = 3$! This remarkable result sheds much light on Witten's quantization condition. However, at the chiral level, N appears only as a parameter, to be fixed by experiment.

It is plausible, based on symmetry considerations, that this effective chiral Lagrangian does indeed describe the static limit of QCD. It is clearly not complete since it does not include the baryons. However, 't Hooft (*Nucl. Phys.* B72, 461(1974)) has shown that the large N_c limit of QCD at low energy is described only by the pseudoscalar mesons. Baryons do not survive in the large N_c limit, as their masses are proportional to N_c. It would seem that only in that limit our chiral Lagrangian describes the whole static limit of QCD.

Yet, things are even better than that. Baryons can be shown to be solitons of the chiral model! This idea due to Skyrme (*Proc. R. Soc.* **A260**, 127(1961)) at the inception of the chiral model, revived in recent years by A. P. Balachandran, V. P. Nair, S. G. Rajeev and A. Stern *Phys. Lett.* **49**, 1124(1982), and by G. Adkins, C. Nappi and E. Witten *Nucl. Phys.* **B228**, 552(1983). This is a beautiful subject of its own, but it is not our main topic of interest, so we only mention it in passing. For a complete description of the low energy world, we would also need to incorporate the vector particles $\rho, \omega, ...$ as well as the glueballs. Since they are related to the pseudoscalars by SU(6), it would be interesting to study in detail how SU(6) emerges from the skyrmion picture.

In this brief analysis of the chiral model of QCD, we have not concerned ourselves with the limits of the model. We have seen how the model is an expansion in inverse powers of F_π, which makes us think that it is valid as long as it describes processes with energies small on the scale of F_π. It is however very difficult to estimate the errors when the chiral model is used beyond this point. For that reason, nobody is amazed that the Skyrme model of the nucleons reproduces the static properties of the baryons, but starts to fail in describing their extended properties. Yet the failure is not too large ($\sim 30\%$), and estimating this error is a subject of much research. In that gray area lies the ρ meson. It appears in the real world in pion-nucleon scattering. Can one expect its presence in the Skyrme picture?

We would like to close this section with a discussion of the radiative corrections to the chiral model. We can in principle compute the radiative loop corections, but we are faced with one problem: the chiral Lagrangian is not a renormalizable theory and its ultraviolet divergences cannot be buried in a redefinition of its input parameters and fields. But wait! we have said all along that the chiral Lagrangian provides only a low energy description of QCD, a renormalizable Lagrangian. Hence we cannot expect its description to be valid at an arbitrary high energy. Thus we are fully justified in just cutting-off the divergent integrals at some scale Λ_0. For details, see H. Georgi, *Weak Interactions and Modern Particle Physics*, (Benjamin/Cumming, 1984).

We expect this scale to be commensurate with F_π. A beautiful argument due to Weinberg (*Acta Physica* **96A**,327(1979)), makes their relation more explicit. In chiral perturbation theory, the cut-off will appear either as a polynomial or as the argument of a logarithm. The polynomial dependence on the cut-off can always be absorbed in a redefinition of the parameters $\mathcal{L}_0$. Otherwise, it would leave us with extra dimensions, robbing the fields of their momentum, in contradiction with the symmetry construction. This leaves only the logarithmic terms. They will be of the form

$$\frac{1}{(4\pi)^2} \ln \left(\frac{\Lambda_0}{\mu} \right) , \tag{5.128}$$

where μ is an arbitrary renormalization scale. These terms give rise to new interactions. Consider for instance the one loop correction to $\pi - \pi$ scattering

The tree-level amplitude is quadratic in the pion's momenta. However, the one-loop correction is quartic in the same momenta. There is no such interaction in $\mathcal{L}_0$. Thus the radiative corrections create new interactions. In this case, the correction is like

$$\mathcal{L}_{corr.} = \frac{F_\pi^2}{16} \frac{1}{\Lambda^2} \operatorname{Tr} \left(\partial_\mu \partial_\nu \mathbf{\Sigma} \partial^\mu \partial^\nu \mathbf{\Sigma}^{-1} \right) , \tag{5.129}$$

where Λ is a parameter with dimensions of mass. This term can be viewed as a nonlocal correction, using $\frac{\partial_\mu}{\Lambda}$ as an expansion parameter.

A reasonable variation in the renormalization scale is seen to be equivalent to the shift

$$\frac{F_\pi^2}{\Lambda^2} = \frac{1}{(4\pi)^2} \, . \tag{5.130}$$

Thus the effect of the ultraviolet divergences is to allow for terms of higher dimensions in the chiral Lagrangian, with the dimensionful parameter given by

$$\Lambda = 4\pi F_\pi \, . \tag{5.131}$$

The one-loop corrections have other more dramatic effects. At the lower end of integration of the loop integral, one finds infrared divergences in the strict chiral limit. In the realistic case where the pion has a small mass, the infrared divergent terms will dominate. Such an effect appears in the computation of the electric dipole moment of the neutron.

Finally we have not discussed how to include the heavier quarks in this picture. These are such that their masses are large compared to the QCD scale. Thus one can hope to describe their interactions, using the inverse as a perturbation parameter. This beautiful subject has been extensively discussed, following the work of N. Isgur and M. Wise, *Phys. Letters* **B232** 113(1989).

PROBLEMS

A. There are other possible condensates which preserve both color and Lorentz invariance; one is the six-quark operator $\langle \mathbf{qqqqqq} \rangle$. List all the symmetries of the standard model it would break if it were to occur. Can you offer a good reason why it does not occur?

B. Verify that the variation of Γ_0 does lead to the current (5.118). Use this result to find the matrix element responsible for the annihilation of three pions into one photon. Proceed to show that the variation of the current is indeed given by (5.122), with K_σ given by (5.123).

C. Show that K_σ is gauge invariant.

5.6 THE QCD VACUUM

The classical QCD energy density is the sum of the squares of the chromo-electric and chromo-magnetic fields. It is zero only when both fields vanish. This happens whenever the potentials are in a pure gauge configuration. In Non-Abelian gauge theories, the structure of these configurations is rather complicated and leads to striking physical predictions.

All connected Green's functions of the quantum theory are generated by the path integral of the exponential of the classical action over all configurations of the gauge potentials. Because of their gauge dependence, not all configurations are included in the sum, only those which are gauge inequivalent. Possible double counting is avoided by redefining the measure (Haar measure) that sums only once over gauge equivalent configurations. This measure considers two potentials to be gauge equivalent if they are connected by a *continuous* gauge transformation. In perturbation theory, the Haar measure leads to the inclusion of Fadeev-Popov ghosts in some gauges.

We must also specify the boundary configurations in the path integral. For the vacuum to vacuum amplitude, they are all pure gauges. In these configurations, the potentials are given by

$$\mathbf{A}_\mu = i\mathbf{U}^\dagger(x)\partial_\mu \mathbf{U}(x) \ , \tag{5.132}$$

where $\mathbf{U}(x)$ is any SU(3) group operation

$$\mathbf{U}(x) = \exp\left\{ i\omega^A(x)\frac{\lambda^A}{2} \right\} \ . \tag{5.133}$$

Each configuration corresponds to a particular map $\omega^A(x)$ of spacetime into the eight-dimensional group manifold of $SU(3)$. These mappings have non-trivial topological properties with observable consequences in the quantum theory.

Consider the possible vacuum configurations in the "gauge" $\mathbf{A}_0 = 0$. This restriction entails no loss of generality and means the unitary maps $\mathbf{U}(\vec{x})$ which characterize the vacuum gauge functions depend only on space.

These maps, from one vacuum configuration to the next, are local in space, and thus must satisfy the boundary conditions

$$\lim_{|\vec{x}|\to\infty} \mathbf{U}(\vec{x}) = \mathbf{I} \ , \tag{5.134}$$

which do not depend on the direction at spatial infinity. Hence, the maps are defined in a three dimensional space, with all of its directions at ∞

identified. This is a compact manifold, the 3-sphere. The mappings $\omega^A(\vec{x})$ can be viewed as mapping of the 3-sphere over the group manifold.

Since any non-Abelian group contains the group $SU(2)$, consider for a moment its group manifold, parametrized by three angles; these label points on its group manifold, the 3-sphere. To see this in another way, note that we can write any $SU(2)$ unitary transformation in the form

$$\mathbf{U} = a_0 + i\vec{\sigma} \cdot \vec{a} \ , \tag{5.135}$$

where unitarity requires

$$a_0^2 + \vec{a} \cdot \vec{a} = 1 \ , \tag{5.136}$$

the equation for a 3-sphere. The maps which relate the different vacuum configurations can therefore be viewed as mapping the 3-sphere of space (with directions at infinity identified) onto the 3-sphere of the group manifold. These maps are characterized by a topological winding number which counts the number of times each sphere is mapped into the other. This is the three-dimensional generalization of the mapping of a circle onto a circle.

An example of a map with unit winding number is

$$a_0 = \frac{\vec{x}^2 - \lambda^2}{\vec{x}^2 + \lambda^2} \ ; \qquad \vec{a} = \frac{2i\vec{x}}{\vec{x}^2 + \lambda^2} \ , \tag{5.137}$$

where λ is a parameter.

Clearly, maps with different winding number cannot be smoothly deformed into one another. Hence, as far as the path integral is concerned, we need to specify the winding number of the gauge configurations we use at the boundaries. There is an infinite number of such configurations, each labelled by the winding number which tells how many times one sphere is wrapped around the other. The winding number can be written in terms of the group element as

$$n = \frac{\epsilon^{\mu\nu\rho\tau}}{16\pi^2} \int d\sigma_\tau \, \mathrm{Tr} \left(\mathbf{U}^\dagger \partial_\mu \mathbf{U} \mathbf{U}^\dagger \partial_\nu \mathbf{U} \mathbf{U}^\dagger \partial_\rho \mathbf{U} \right) \ , \tag{5.138}$$

where the integration extends over the three-sphere at infinity. Maps with the same winding number are said to be *homotopically* equivalent; they can be deformed continuously into one another. Although we have used SU_2 for a gauge group, the result is generalizable since all other groups contain SU_2 and the relevant thing is the mapping of S_3 into some S_3 within the group manifold. In fact there is even a theorem to that effect. Thus our discussion applies to $SU(3)$, which has $S_3 \times S_5$ as group manifold.

The usual starting configuration in perturbation theory is $\mathbf{A}_\mu = 0$, with zero winding number. Suppose we start perturbation theory with a pure gauge configuration of winding number one? Are these two equivalent?

It would seem that we have many inequivalent perturbation theories, each starting from a different winding number, living in a different Hilbert space with vacuum state $|\Omega_n\rangle$, labelled by an integer n. We also know that there exist gauge transformations which can change the winding number. Does that mean that there are many possible disconnected Hilbert spaces, one for each winding number, and that the theory is only invariant under gauge transformations that do not change the winding number?

A naïf might think that any pure gauge configuration can serve equally well as a starting point for perturbation theory. However that is not so, because there is more to it than perturbation theory.

The answer depends on the existence of transition amplitudes between these different vacuum states. If no such transitions exist, unitarity is realized in each Hilbert space, and we have many equivalent theories. On the other hand, if transitions between Hilbert spaces do exist, and all the (sub)Hilbert spaces must be included in order to have a unitary theory.

Translated in the path integral language, the same question reads: does the path integral vanish when it starts from a configuration of winding number n at the lower end of integration, to one of winding number m at the upper end?

Actually, winding number is not conserved by QCD; there is quantum tunnelling between vacua of different winding number, which are analyzed in the saddle-point approximation of the path integral, by expanding away from configurations of minimum action in Euclidean space. These configurations are called "instantons". They mediate the tunnelling. Their contributions the tunnelling amplitude is of the order of

$$e^{-\frac{8\pi^2}{g_3^2}} , \tag{5.139}$$

where g_3^2 is the strong (QCD) coupling constant; it is certainly not perturbative! Normally, such an amplitude would lead to a negligible rate, but the QCD coupling constant can be very large in the infrared, effectively negating this typical exponential suppression of tunnelling processes.

Since all vacua of different winding numbers can be reached this way, *the* physical vacuum state must be a linear superposition of all states of zero energy and different winding numbers, $|\Omega_n\rangle$. Its precise form is determined by gauge invariance. Under a gauge transformation T which changes winding number by one unit, we have

$$T|\Omega_n\rangle = |\Omega_{n+1}\rangle \ . \tag{5.140}$$

None of these states are physical since they can be transformed into one another by a gauge transformation. *The* vacuum state is that on which a gauge transformation can at most result in a phase transformation. This fixes the physical vacuum to be given by the Bloch superposition

$$|\Omega_{QCD}\rangle_\theta = \sum_n e^{-in\theta}|\Omega_n\rangle \ , \tag{5.141}$$

which is designed such that

$$T|\Omega_{QCD}\rangle_\theta = e^{i\theta}|\Omega_{QCD}\rangle_\theta \ . \tag{5.142}$$

The true vacuum state of gauge theories depends on an angle θ, defined modulo 2π.

Standard arguments applied to the QCD path integral with this vacuum as boundary has for effect to alter the Lagrangian by a "surface term" which does not appear in the equations of motion, yielding the QCD effective Lagrangian

$$\mathcal{L}_{eff} = \mathcal{L}_{QCD} + \frac{\theta}{16\pi^2}\mathrm{Tr}(\mathbf{G}_{\mu\nu}\widetilde{\mathbf{G}}^{\mu\nu}) \ . \tag{5.143}$$

To arrive at this result one uses the fact that the winding number can be written in terms of the field strengths as

$$n = \frac{1}{16\pi^2}\int d^4x \mathrm{Tr}(\mathbf{G}_{\mu\nu}\widetilde{\mathbf{G}}^{\mu\nu}) \ , \tag{5.144}$$

where the tilde indicates the dual

$$\widetilde{\mathbf{G}}_{\mu\nu} = \frac{1}{2}\epsilon_{\mu\nu\rho\sigma}\mathbf{G}^{\rho\sigma} \ . \tag{5.145}$$

This new term is a pure divergence, the divergence of the Chern-Simons current, and it causes no change in the equations of motion. However, it violates the discrete symmetries P and CP, both of which are symmetries of $\mathcal{L}_{YM}$. This term can also be written as $\mathbf{E}\cdot\mathbf{B}$, which is manifestly odd under both P and T, which is the same as CP by the CPT theorem.

In order to extract physical consequences of this extra term, it is necessary to consider the coupling to fermions, where it is identified with the divergence of the fermion chiral current.

Let us first examine this coupling in a much simpler theory, QED. There the presence of the $\vec{E}\cdot\vec{B}$ term cannot be inferred from the previous analysis because the gauge transformations which label the pure gauge field configurations of the vacuum state are too simple to have a winding number (maps of a circle onto a two-sphere have no winding number: one cannot lasso a tennis ball!). Yet, this term is present through the Adler-Bell-Jackiw anomaly as the anomalous divergence of the chiral current

$$\partial_\mu J^{\mu 5} = \frac{iNe^2}{32\pi^2}\epsilon^{\mu\nu\rho\sigma}F_{\mu\nu}F_{\rho\sigma}, \tag{5.146}$$

which is obtained from the triangle graph; N is the number of fermions of charge e, and the chiral current is

$$J_5^\mu = \sum_i^N i\ \overline{\Psi}_i\gamma^\mu\gamma_5\Psi_i. \tag{5.147}$$

Since the right-hand side of Eq. (5.146) is itself a surface term, we can define a new conserved current. This new current leads to a gauge invariant conserved charge in the case of QED. In QCD, it is does not lead to a gauge invariant conserved charge (see problem). This is the difference between Abelian and Non-Abelian theories. There is no θ parameter in QED because the U(1) gauge transformations have no winding number in four space-time dimensions (do they in any other number of dimensions?).

This current is generated by the chiral transformation,

$$\Psi \to e^{i\alpha\gamma_5}\Psi\ . \tag{5.148}$$

Noether's theorem tells us that the divergence of the current is related to the explicit change in the Lagrangian density through the equation

$$\partial_\mu J_5^\mu = -\frac{\delta\mathcal{L}}{\delta\alpha}\ . \tag{5.149}$$

Comparison with the anomaly condition implies that we add to the QED Lagrangian the surface term

$$\theta\frac{ie^2}{32\pi^2}\epsilon^{\mu\nu\rho\sigma}F_{\mu\nu}F_{\rho\sigma}\ , \tag{5.150}$$

and that under the chiral transformation the angle θ gets shifted by

$$\theta \to \theta + 2\alpha N\ . \tag{5.151}$$

We interpret this equation to mean that θ can be shifted to zero by a chiral transformation, which is a manifest symmetry of the massless QED action. In addition, we should not forget that θ is an angle, and that all of this discussion is valid modulo 2π.

Thus, if there is as much as one massless fermion in the theory, the θ angle can always be shifted to zero by performing a chiral transformation on that fermion field. In this case it has no physical consequences.

When all the fermions are massive, the angle θ can no longer be absorbed in the fermion fields. The Lagrangian has now the additional (mass) term

$$\mathcal{L}_m = \overline{\Psi} M \Psi \ . \tag{5.152}$$

As a result, chiral invariance is explicitly violated, yielding the extra contribution

$$\frac{\delta \mathcal{L}}{\delta \alpha} = \frac{\delta}{\delta \alpha} \overline{\Psi} M \Psi = 2i \overline{\Psi} M \gamma_5 \Psi \ . \tag{5.153}$$

Thus we can shift the angle θ to zero in the surface term, only at the cost of reintroducing it in the following CP-violating term

$$\delta \mathcal{L}_{CP} = 2i \bar{\Psi} M \gamma_5 \Psi \ . \tag{5.154}$$

Note that if any one fermion is massless, we can still perform a chiral transformation on that fermion alone to absorb the full θ phase, but if none are massless, this analysis tells us that the only relevant quantity is the difference

$$\bar{\theta} = \theta - \arg \det M \ , \tag{5.155}$$

which is left invariant by such manipulations.

In real life QCD, none of the quarks appear to be massless, although the u quark is the lightest, so that $\bar{\theta}$ cannot be redefined away; it is a parameter of QCD, to be added to the list of parameters of the standard model. It can lead to real observable CP-violating effects in the strong interactions, such as an electric dipole moment for the neutron. It can be shown to lead to the CP-violating interaction (see V. Baluni, *Phys. Rev.* **D**, 2227(1979))

$$\delta \mathcal{L}_{CP} = i \bar{\theta} \frac{m_u m_d m_s}{m_u m_d + m_d m_s + m_s m_u} (\overline{\mathbf{u}} \gamma_5 \mathbf{u} + \overline{\mathbf{d}} \gamma_5 \mathbf{d} + \overline{\mathbf{s}} \gamma_5 \mathbf{s}) \ , \tag{5.156}$$

when $\bar{\theta} \ll 1$, which is justified since there is no apparent experimental CP violation by the strong interactions.

PROBLEMS

A. Show that for $SU(2)$, the gauge transformation

$$\mathbf{U}(x) = \exp\left[i\pi\vec{\tau}\cdot\vec{\mathbf{x}}F(x)\right] ,$$

where $F(x) = (t^2 + \vec{\mathbf{x}}\cdot\vec{\mathbf{x}} + \lambda^2)^{-1/2}$, changes the winding number by one unit.

B. Show that one can always define a new conserved chiral current, since the anomalous divergence is itself a surface term. Find the form of this current both in the Abelian and non-Abelian cases.

C. For $SU(2)$, show that the conserved charge obtained from this conserved current is not invariant under the gauge transformations of problem A. It is therefore a fake conservation law.

D. Verify the form of CP violation given by (5.156).

5.7 U(1) ANOMALY AND THE CHIRAL LAGRANGIAN

We have shown how the properly gauged Wess-Zumino-Witten term yields the correct π^o decay rate into two photons, which in QCD comes from the flavor anomaly. We now turn our attention to incorporating the *flavor-independent* $U(1)$ chiral anomaly in the low energy Lagrangian. In QCD, the flavor-blind chiral current is anomalous, with divergence

$$\partial^\mu J^5_\mu = i n_f \frac{g_3^2}{16\pi^2}\,\mathrm{Tr}\left(\mathbf{G}_{\mu\nu}\tilde{\mathbf{G}}^{\mu\nu}\right) , \tag{5.157}$$

where n_f is the number of quark flavors. Hence it is only a clasical symmetry, explicitly broken by quantum effects. That is the reason we did not include it in our analysis of the previous section. However, in the limit of large N_c (number of colors) advocated by 't Hooft, this particular anomaly disappears (see problem), and the chiral $U(1)$ symmetry is restored. Since the large N_c theory is a real quantum theory, it makes sense to reconsider the chiral Lagrangian approach, taking this symmetry into account.

In the large N_c limit, this chiral symmetry is broken spontaneously by the condensation of quark pairs in the vacuum, the same mechanism that breaks the flavor chiral symmetries. Thus we can expect an extra massless Nambu-Goldstone boson. To incorporate such a particle in the chiral Lagrangian, we simply replace $\mathbf{\Sigma}(x)$ by

$$\mathbf{\Sigma}'(x) = e^{\frac{2i}{F_\pi}(\boldsymbol{\lambda}^0\eta' + \boldsymbol{\lambda}^A\pi^A)} , \tag{5.158}$$

where $\boldsymbol{\lambda}^0$ is proportional to the unit matrix, normalized like the $\boldsymbol{\lambda}^A$ matrices:

$$\mathrm{Tr}\,(\boldsymbol{\lambda}^0\boldsymbol{\lambda}^0) = 2 \; . \tag{5.159}$$

This introduces an extra massless pseudoscalar field $\eta'(x)$ in the theory. Its mass happens to be slighly larger than the proton's, and thus can hardly be considered light in the same sense as the other pseudoscalars. This just means that for $N_c = 3$, the large N_c approximation is not that good, since the proton mass is expected to be proportional to N_c. The η' should be viewed as light, in the sense of the large N_c theory, where the first correction to this theory gives the η' a mass which is of the order of $1/N_c$. Other more complicated interactions come in higher orders of the expansion. The new chiral Lagrangian is given by

$$\mathcal{L}_0 = \frac{F_\pi^2}{16}\,\mathrm{Tr}\,(\partial^\mu\mathbf{\Sigma}'\partial_\mu\mathbf{\Sigma}'^{-1}) \; , \tag{5.160}$$

which has, besides the old $SU_L(3)\times SU_R(3)$ invariance, the flavor-independent chiral invariance

$$\mathbf{\Sigma}'(x) \to e^{i\beta}\mathbf{\Sigma}'(x)e^{i\beta} = e^{2i\beta}\mathbf{\Sigma}'(x) \; . \tag{5.161}$$

Under this invariance, the $\eta'(x)$ field does behave as a Nambu-Goldstone boson

$$\eta'(x) \to \eta'(x) + F_\pi\beta \; . \tag{5.162}$$

The Noether current which generates this transformation is (dropping the primes)

$$\begin{aligned} J_\mu^5 &= \mathrm{Tr}\,(\frac{\delta\mathcal{L}}{\delta\partial_\mu\mathbf{\Sigma}}\delta\mathbf{\Sigma} + \frac{\delta\mathcal{L}}{\delta\partial_\mu\mathbf{\Sigma}^{-1}}\delta\mathbf{\Sigma}^{-1}) \; , \\ &= \frac{iF_\pi^2}{8}\,\mathrm{Tr}\,(\partial_\mu\mathbf{\Sigma}^{-1}\mathbf{\Sigma} - \partial_\mu\mathbf{\Sigma}\mathbf{\Sigma}^{-1}) \; . \end{aligned} \tag{5.163}$$

This current is conserved in $\mathcal{L}_0$. In order to account for its breaking by the anomaly in QCD, we must add to $\mathcal{L}_0$ an explicit breaking term. We must require that this extra term, call it $\mathcal{L}_A$, be invariant under

$SU)_L(n_f) \times SU_R(n_f)$, and be constructed out of $\mathbf{\Sigma}'(x)$. Since under a chiral transformation,

$$\det \mathbf{\Sigma} \to \det \mathbf{U}_L \det \mathbf{\Sigma} \det \mathbf{U}_R^\dagger , \tag{5.164}$$

and the determinants of the unitary transformations are unity, the determinants of $\mathbf{\Sigma}$ and $\mathbf{\Sigma}^\dagger$, have the requisite invariances. It is convenient at this point to introduce the variables

$$z = \det \mathbf{\Sigma} , \qquad \overline{z} = \det \mathbf{\Sigma}^{-1} ; \tag{5.165}$$

they transform under the chiral $U(1)$ as

$$\delta z = 2i\beta n_f z , \qquad \delta\overline{z} = -2i\beta n_f \overline{z} . \tag{5.166}$$

Hence we take

$$\mathcal{L}_A = \mathcal{L}_A(z, \overline{z}) , \tag{5.167}$$

which will affect the divergence of the chiral total through

$$\partial^\mu J^5_\mu = -\frac{\delta \mathcal{L}_A}{\delta\beta} ; \tag{5.168}$$

explicit differentiation yields

$$\frac{\delta \mathcal{L}_A}{\delta\beta} = 2i \; n_f \left[\frac{\partial \mathcal{L}_A}{\partial z} z - \frac{\partial \mathcal{L}_A}{\partial \overline{z}} \overline{z} \right] . \tag{5.169}$$

This general kinematic analysis does not yield the functional form of $\mathcal{L}_A$ in terms of z and $\overline{z}$. Notice the emergence of the number of flavors, n_f, in the equation. To proceed, we have to find a way to incorporate the form of the anomaly as given by QCD.

One way to proceed is that advocated by Rosensweig, Schechter and Trahern, *Phys. Rev.* **D21**,3388(1980). They introduce in the action an auxiliary field $A(x)$ which plays the role of the anomaly, in the sense that the equation

$$\partial_\mu J^{\mu 5} = n_f A(x) , \tag{5.170}$$

is satisfied. $A(x)$ can be viewed as the vacuum value of the condensate

$$A(x) \sim i < \mathrm{Tr}\,(\mathbf{G}_{\mu\nu}\widetilde{\mathbf{G}}^{\mu\nu}) >_0 ; \tag{5.171}$$

it has the dimension of a Lagrange density. The auxiliary field A is invariant under the same transformation. Suppose we add to the Lagrangian the extra terms

$$\mathcal{L}_A = \frac{1}{2\mu^4}A^2 - AB(z,\overline{z}) , \qquad (5.172)$$

where μ is an unknown mass parameter, and B is a dimensionless function to be determined. Variation with respect to A yields

$$A = \mu^4 B(z,\overline{z}) . \qquad (5.173)$$

Applying the Noether procedure to this new Lagrangian yields the divergence of the flavor-blind chiral current

$$\partial^\mu J^5_\mu = A(x)\frac{\delta B}{\delta\beta} = 2in_f A(x)(z\frac{\partial B}{\partial z} - \overline{z}\frac{\partial B}{\partial \overline{z}}) . \qquad (5.174)$$

Comparison with Eq. (5.170) yields

$$2i(z\frac{\partial B}{\partial z} - \overline{z}\frac{\partial B}{\partial \overline{z}}) = 1 , \qquad (5.175)$$

which has the solution

$$B = -\frac{i}{4}(\ln z - \ln \overline{z}) , \qquad (5.176)$$

up to an arbitrary function of the constant $z\overline{z}$. Thus we arrive at the Lagrangian

$$\mathcal{L} = \frac{F_\pi^2}{16}\,\mathrm{Tr}\,(\partial_\mu \mathbf{\Sigma}\partial^\mu \mathbf{\Sigma}^{-1}) + \frac{A^2(x)}{2\mu^4} + \frac{iA(x)}{4}(\ln\ \frac{\det \mathbf{\Sigma}}{\det \mathbf{\Sigma}^{-1}}) . \qquad (5.177)$$

The auxiliary field $A(x)$ can then be eliminated to yield

$$\mathcal{L} = \frac{F_\pi^2}{16}\,\mathrm{Tr}\,(\partial_\mu \mathbf{\Sigma}\partial^\mu \mathbf{\Sigma}^{-1}) + \frac{\mu^4}{32}[\ln\ \frac{\det \mathbf{\Sigma}}{\det \mathbf{\Sigma}^{-1}}]^2 . \qquad (5.178)$$

The meaning of the symmetry breaking term becomes transparent when expressed in terms of the η' field. Since

$$\det \mathbf{\Sigma} = \det(e^{\frac{2i}{F_\pi}\lambda_0\eta' + \frac{2i}{F_\pi}\lambda^A\pi^A}) = \exp\left\{\frac{2in_{fl}}{F_\pi}\eta'(x)\right\} , \qquad (5.179)$$

we obtain for the breaking term

$$\mathcal{L}_A = -\frac{\mu^4}{2}\left(\frac{n_{fl}^2}{F_\pi^2}\right)\eta'(x)\eta'(x) , \qquad (5.180)$$

which, after all this work, is nothing but the mass term for η'

$$m_{\eta'}^2 = \left(\frac{\mu^2 n_{fl}}{F_\pi}\right)^2 . \tag{5.181}$$

In this approach, μ and F_π are unrelated. However, comparison with experiment yields $\mu \approx 240$ MeV. The form of this mass is to be contrasted with that of the other pseudoscalars, for instance, $m_{\pi^\pm}^2 = m_0(m_u + m_d)$, which vanishes in the massless quark limit. Consider now the full Lagrangian

$$\mathcal{L}_T = \frac{F_\pi^2}{16}\,\mathrm{Tr}\,(\partial^\mu \mathbf{\Sigma}\partial_\mu \mathbf{\Sigma}^{-1}) + \frac{\mu^4}{32}(\ln\det\mathbf{\Sigma} - \ln\det\mathbf{\Sigma}^{-1})^2 + \\ + \frac{F_\pi^2}{8} m_0\,\mathrm{Tr}\,(\mathbf{\Sigma}^{-1}\mathbf{M} + \mathbf{M}^\dagger\mathbf{\Sigma}) , \tag{5.182}$$

which has the explicit flavor chiral breaking from the quark masses as well as the flavor independent breaking from the anomaly. $\mathbf{M}$ is the diagonal quark mass matrix introduced previously. We have not included the Wess-Zumino term, but leave it as an exercise to the reader to do so, and examine its effect on this analysis. Let us expunge an overall phase from $\mathbf{M}$ by the transformation

$$\mathbf{M} \to e^{i\theta/n_f}\mathbf{M} , \tag{5.183}$$

into $\mathbf{\Sigma}$ by the transformation

$$\mathbf{\Sigma} \to e^{i\theta/n_f}\mathbf{\Sigma} , \tag{5.184}$$

but then it reappears in the determinant part, which becomes

$$\frac{\mu^4}{32}(\ln\ \det\mathbf{\Sigma} - \ln\ \det\mathbf{\Sigma}^{-1} + 2i\theta)^2 . \tag{5.185}$$

The full potential energy for this model is

$$V = -\frac{F_\pi^2}{8} m_0\,\mathrm{Tr}\,(\mathbf{\Sigma}^\dagger\mathbf{M} + \mathbf{M}^\dagger\mathbf{\Sigma}) - \frac{\mu^4}{32}(\ln\ \frac{\det\mathbf{\Sigma}}{\det\mathbf{\Sigma}^{-1}} + 2i\theta)^2 . \tag{5.186}$$

We can now find its vacuum configuration. Using

$$\delta\ \det\mathbf{\Sigma} = \delta(e^{\,\mathrm{Tr}\ \ln\mathbf{\Sigma}}) = (\det\mathbf{\Sigma})\mathbf{\Sigma}^{-1}\delta\mathbf{\Sigma} , \tag{5.187}$$

and remembering that the variations of $\mathbf{\Sigma}$ and its dagger are not independent, we obtain the minimum when

$$-\frac{F_\pi^2}{8}m_0(\mathbf{M}^\dagger - \mathbf{\Sigma}^{-1}\mathbf{M}\mathbf{\Sigma}^{-1}) - \frac{\mu^4}{8}(\ln \frac{\det \mathbf{\Sigma}}{\det \mathbf{\Sigma}^{-1}} + 2i\theta)\mathbf{\Sigma}^{-1} = 0 \ . \quad (5.188)$$

Since $\mathbf{M}$ is diagonal, $\mathbf{\Sigma}$ is also diagonal at the minimum. Thus we set, specializing to the realistic case of three light flavors,

$$\mathbf{\Sigma} = \begin{pmatrix} e^{i\varphi_u} & 0 & 0 \\ 0 & e^{i\varphi_d} & 0 \\ 0 & 0 & e^{i\varphi_s} \end{pmatrix} , \quad (5.189)$$

where the angles $\varphi_a(x)$ are proportional to the pseudoscalar fields at minimum times F_π^{-1}. Substituting in the minimum equation, we obtain the three transcendental equations

$$-\frac{F_\pi^2}{8}m_0 m_a \sin\varphi_a = \frac{\mu^4}{8}(\sum_b \varphi_b + \theta) \ , \quad a = u, d, s \ . \quad (5.190)$$

Evaluating the potential away from its minimum yields the mass matrix for the three neutral pseudoscalars

$$\frac{\mu^4}{8F_\pi^2}\begin{pmatrix} 1 & 1 & 1 \\ 1 & 1 & 1 \\ 1 & 1 & 1 \end{pmatrix} + \frac{m_0}{8}\begin{pmatrix} m_u & 0 & 0 \\ 0 & m_d & 0 \\ 0 & 0 & m_s \end{pmatrix} . \quad (5.191)$$

In the realistic regime where

$$m_u \sim m_d << m_s << \frac{8\mu^4}{F_\pi^2 m_0} \equiv \hat{\mu} \ , \quad (5.192)$$

we can approximate these vacuum equations as follows. First we note that since

$$\frac{\sin\varphi_s}{\sin\varphi_{u,d}} = \frac{m_{u,d}}{m_s} << 1 \ , \quad (5.193)$$

we can set

$$\varphi_s \approx 0 + \mathcal{O}(\frac{m_{u,d}}{m_s}) \ . \quad (5.194)$$

Since at the minimum,

$$-m_s \sin\varphi_s = \frac{\hat{\mu}}{8}(\sum_a \varphi_a + \theta) \approx 0 \ , \quad (5.195)$$

from which we deduce

$$\varphi_u + \varphi_d \approx -\theta \ . \tag{5.196}$$

This enables us to solve for $\varphi_{u,d}$, starting from

$$\frac{m_d}{m_u} \sin \varphi_d \approx -\sin(\theta + \varphi_d) \ , \tag{5.197}$$

with the result

$$\sin \varphi_{u,d} = \frac{m_{d,u} \sin \theta}{(m_u^2 + m_d^2 + 2m_u m_d \cos \theta)^{1/2}} \ . \tag{5.198}$$

The value of the potential at minimum, neglecting the $\hat{\mu}^4$ term (is it justified?), is then given by

$$V = -\frac{F_\pi^2 m_0}{4} \sqrt{m_u^2 + m_d^2 + 2m_u m_d \cos \theta} \ . \tag{5.199}$$

We arrive at the important conclusion that the energy assumes its lowest value when $\cos \theta = 1$, that is for $\theta = 0$. The same result can also be obtained directly from path integral considerations.

We must emphasize that θ is just a parameter, which can have any value. However we should keep in mind that if it were a dynamical variable, it would naturally seek the state of lowest energy and thus relax to zero. This is the basis of the work of Peccei and Quinn, which leads to the axion.

Under CP, $\mathbf{\Sigma} \to \mathbf{\Sigma}^\dagger$ and $\theta \to -\theta$. It follows that CP is a symmetry of the Lagrangian if $\theta = 0, \pi$. At $\theta = 0$, invariance under $\varphi_a \to -\varphi_a$ is obvious, while at $\theta = \pi$, we need to have $\varphi_a \to -\varphi_a + 2\pi n_a \quad \sum n_a = 1$. Why is CP a symmetry at $\theta = \pi$?

The angle θ is a fundamental parameter of QCD. Its value is to be determined by experiment. The best way is to look for effects which would be forbidden if it were zero. Such an effect is the electric dipole of the neutron. Its presence is forbidden in the limit of exact CP symmetry. It does get a contribution from CP violation in the weak interactions, which can be shown to yield a value around 10^{-31} esu, which is far from the present experimental limit according to which it should be smaller than 10^{-26} esu. Thus we can use this value to set a bound on the θ parameter (see problem).

There is another anomaly, associated with the breakdown of scale invariance, which we might be tempted to treat in the same way. We know it stems from the necessity to subtract the divergences of the quantum field theory, thereby introducing a scale. To describe it, we start by defining a dilatation current D_μ, which is classically conserved in a scale invariant theory

$$\partial^\mu D_\mu = 0 \; . \tag{5.200}$$

More generally, in any classical field theory, the divergence of the dilatation current is related to the trace of the energy momentum tensor, or the improved energy momentum tensor if spinless fields are around,

$$\partial^\mu D_\mu = \theta^\mu_\mu = 0 \; . \tag{5.201}$$

However, quantum field theories are not generally dilatation invariant because the renormalization process singles out a scale. Thus the classical dilatation invariance is broken by quantum effects. This can be phrased in terms of an anomaly (in the sense that the breaking is of order $\hbar$), leading to a modification of the right hand side of the above equation. In a Yang-Mills theory, it takes the form

$$\theta^\mu_\mu = -\frac{\beta(g)}{g} \mathrm{Tr}\left(\mathbf{G}_{\mu\nu}\mathbf{G}^{\mu\nu}\right) \; , \tag{5.202}$$

where $\beta(g)/g$ is the usual Gell-Mann-Low function. This "anomaly" is very different from the chiral anomaly. First of all, it goes away only when Quantum Mechanics goes away, *i.e.* in the classical limit. Secondly it is contributed to in all orders of $\hbar$. Constrast this with the chiral anomaly which can vanish quantum mechanically, as in the large N_c limit and which gets only contributions of order $\hbar$, as stated by the Adler-Bardeen theorem.

Hence it is only in the classical limit that this dilatation anomaly disappears; it follows that it is inappropriate to treat the dilatation current as anomalous in the low energy chiral Lagrangian.

Another difference lies in the fact that dilatation is a non compact symmetry, unlike the compact internal symmetries. Thus the mocking up of the effective theory in a scale-invariant way does not have the same justification as the chiral anomaly.

PROBLEMS

A. Show that if the mass of the *up* quark is zero, the symmetries of the chiral Lagrangian are not changed.

B. Show from (5.190) that if any quark mass is zero, θ can be eliminated from the potential energy.

C. Show that the physical consequences do not change under the shift $\theta \to \theta + 2n_{fl}\pi$.

D. Diagonalize the mass matrix (5.191), and find its eigenvalues. Hint: use perturbation theory is warranted since $\mu^4 >> m_q m_0 \frac{F_\pi^2}{8}$. Interpret physically.

E. Using chiral perturbation theory, calculate the electric dipole moment of the neutron in terms of $\overline{\theta}$. For reference, see R. J. Crewther, P. DiVecchia, G. Veneziano, and E. Witten, *Phys. Lett.* **88B**, 123(1979). Erratum-*ibid* **B91**, 487(1980).

CHAPTER 6

STANDARD MODEL

ONE-LOOP STRUCTURE

Although the fundamental laws of Nature obey quantum mechanics, microscopically challenged physicists build and use quantum field theories by starting from a classical Lagrangian. The classical approximation, which describes macroscopic objects from physics professors to dinosaurs, has in itself a physical reality, but since it emerges only at later times of cosmological evolution, it is not fundamental. We should therefore not be too surprised if unforeseen special problems and opportunities emerge in the analysis of quantum perturbations away from the classical Lagrangian.

The classical Lagrangian is used as input to the path integral, whose evaluation produces another Lagrangian, the effective Lagrangian, $\mathcal{L}_{eff}$, which encodes all the consequences of the quantum field theory. It contains an infinite series of polynomials in the fields associated with its degrees of freedom, and their derivatives. The classical Lagrangian is reproduced by this expansion in the lowest power of $\hbar$ and of momentum. With the notable exceptions of scale invariance, and of some (anomalous) chiral symmetries, we think that the symmetries of the classical Lagrangian survive the quantization process. Consequently, not all possible polynomials in the fields and their derivatives appear in $\mathcal{L}_{eff}$, only those which respect the symmetries.

The terms which are of higher order in $\hbar$ yield the quantum corrections to the theory. They are calculated according to a specific, but perilous path, which uses the classical Lagrangian as input. This procedure generates infinities, due to quantum effects at short distances. Fortunately, most fundamental interactions are described by theories where these infinities can be absorbed in a redefinition of the input parameters and fields, *i.e.* swept under the rug. These theories, which yield finite quantum corrections, are said to be *renormalizable*; the standard model is one of them. On the other hand, Quantum Gravity is not renormalizable, and its quantum corrections generate an intractable number of infinities. As a result, many physicists believe that gravity is an infrared approximation to a theory devoid of these ultraviolet infinities; likely candidates are the superstring theories.

In a renormalizable theory, quantum corrections generate *a priori* all possible terms consistent with the invariances of the input (classical) Lagrangian that survive quantization. These appear as terms in the effective Lagrangian, with coefficients to be calculated, either through perturbative or non-perturbative techniques.

By absorbing the ultraviolet divergences through a redefinition of the input fields and parameters, scale invariance is necessarily broken. In the process a new scale is introduced in the theory, and all the parameters become scale-dependent. Their runnings are determined by the first order differential equations of the renormalization group. Each has one integration constant which is roughly identified with the measured numerical value of the parameter at the scale determined by the experiment.

The rules for a renormalizable theory in four dimensions are rather easy to state. Start with *all* possible effective interactions of (mass) dimensions less than or equal to four, consistent with symmetries. All the ultraviolet infinities of the theory are then absorbed in its fields and parameters. Quantum corrections generate different types of terms. Some are of the same form as the input terms; they describe finite renormalizations of the basic interactions, and yield the scale-dependence of the input parameters. Some generate new interaction terms that were not in the classical description. The coefficients in front of effective interactions of mass dimension larger than four are finite and calculable in terms of the input parameters of the theory.

In general, the classical input Lagrangian must contain all terms of dimension four. Should one of the terms be absent, it is generated by the quantum corrections, with infinite strength. Thus it must be included as an input, so that its coupling strength can be used to absorb that infinity. There is one important exception to this rule: suppose that by deleting some terms of dimension four or less, the input Lagrangian acquires a larger symmetry. If that symmetry is of the type respected by quantization, (*i.e.* except scale invariance and anomalous symmetries), quantum corrections will not generate those terms and their associated infinities.

6.1 QUANTUM ELECTRODYNAMICS

We begin with the quantum corrections of the mother of all renormalizable theories, Quantum Electrodynamics (QED). This section assumes prior knowledge of QED; it serves as an introduction to our method for analyzing the radiative structure of the electroweak theory.

QED is described by two fields, the electron field Ψ and the photon gauge field A_μ, one dimensionless gauge coupling constant e, the electric charge of the electron, and one mass parameter, M_e, the mass of the electron. The classical QED Lagrangian is Lorentz-invariant, gauge invariant, parity and charge conjugation invariant. The terms generated in its effective Lagrangian must then also be invariant under these same symmetries.

As a result, the gauge field appears in the covariant derivative combination, $\mathcal{D}_\mu = \partial_\mu + ieA_\mu$, acting on either itself or Ψ. The effective QED Lagrangian is just an infinite sum of polynomials in D_μ and Ψ, each of which is gauge invariant, Lorentz-invariant, even under parity and charge conjugation (Furry's theorem).

The most important tool for calculating the quantum corrections is the loop expansion with Feynman diagrams. In terms of diagrams, the effective Lagrangian is generated by one-particle irreducible or *proper* diagrams that cannot be disconnected by cutting one line. The effective Lagrangian is written as an expansion in $\hbar$,

$$\mathcal{L}_{eff} = \mathcal{L}_{cl.} + \hbar\mathcal{L}_1 + \hbar^2\mathcal{L}_2 + \cdots \ .$$

In the above, $\mathcal{L}_{cl.}$ is the same as the input Lagrangian: it contains only combinations of fields and derivatives of dimensions less than or equal to four. The higher order terms, $\mathcal{L}_n$ denote the terms generated by n-loop corrections; each includes infinite polynomials in the input fields and their derivatives.

Dimensional analysis is a potent tool in organizing the results. Since the action is dimensionless ($\hbar = 1$), the Lagrangian has (mass) dimension 4. The derivative has dimension 1, the electron field has dimension 3/2, and the photon field has dimension 1. Terms of dimension less than or equal to 4 appear in the input Lagrangian. In a renormalizable theory, these are the only ultraviolet-divergent terms, and their divergences can all be absorbed in a redefinition of the input fields and parameters.

Polynomials of higher dimensions either yield new interactions or finite corrections to the basic interactions; their strengths are in principle computable in terms of the input parameters.

We begin by organizing the expansion in terms of Feynman diagrams. Consider all possible one-loop Feynman diagrams of the same structure as the terms in the classical Lagrangian; these are the photon two-point function, the photon-electron vertex, and the electron two-point function. The relevant diagrams are

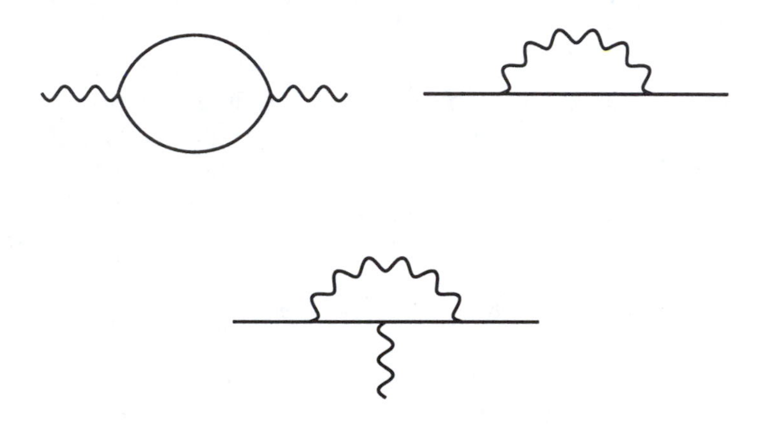

There are other one-loop diagrams which describe interactions absent from the classical Lagrangian; they are the box diagrams, which, in QED, generate four-fermion and multi-photon interactions.

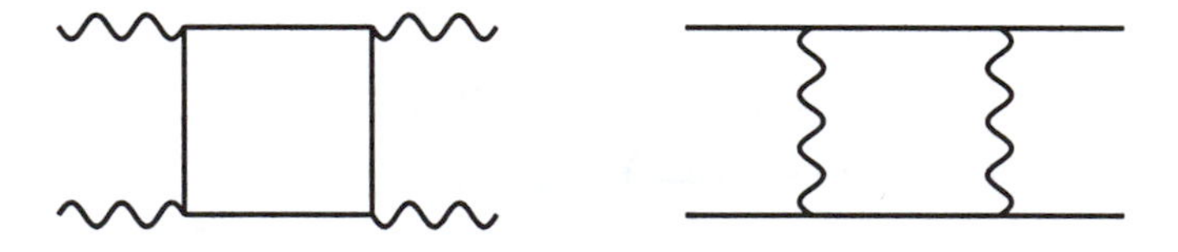

By our rules, both of these must be finite, since QED is renormalizable. For example the four-fermion interaction has dimension 6, and the four photon has dimension 8 (using gauge invariance). However the four-fermion box diagram corrects a tree level process in which a four fermion interaction is generated by one-photon exchange. The four-photon interaction, on the other hand, is purely an effect of the quantum corrections, without classical analogue.

The corrections to the basic classical interactions contain the ultraviolet divergences, and the renormalization procedure results in a modification of the input parameters that makes them scale dependent. In QED, quantum effects modify the electron-photon vertex (vertex corrections), the photon propagator (vacuum polarization), the electron propagator (wave function correction) and the electron mass (mass correction). The separation of vertex and vacuum polarization is gauge invariant only in Abelian theories such as QED.

Consider the vacuum polarization diagram; it corrects the photon propagator in a simple way to give the coupling "constant" (electric charge) a momentum dependence.

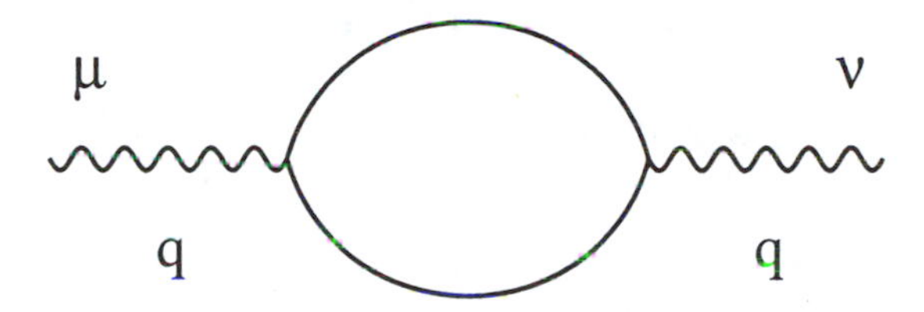

At one loop, using dimensional regularization, it is calculated (details can be found in any primer on field theory) to be

$$\Pi_{\mu\nu}(q) = (\frac{q_\mu q_\nu}{q^2} - \delta_{\mu\nu})\Pi(q^2) \ , \tag{6.1}$$

where

$$\Pi(q^2) = \frac{e^2}{12\pi^2}q^2 \left\{\Delta - \int_0^1 dx x(1-x)\ln\left(\frac{2m_e^2 + 2q^2x(1-x)}{\mu^2}\right)\right\} \ , \tag{6.2}$$

with

$$\Delta = \frac{2}{4-n} - \gamma + \ln 4\pi \ , \tag{6.3}$$

which contains the divergence; n is the dimension of space-time, and γ is the Euler-Mascheroni constant. The diagram diverges in the limit $n \to 4$, with the divergence absorbed in the input parameter.

The appearance of the arbitrary scale μ is a by-product of the regularization, in this case dimensional regularization. Physical quantities cannot depend on μ, which implies that the input parameters develop specific scale dependences of their own.

We note that this correction is purely transverse, and proportional to q^2. This feature continues to be true even at higher loops, enforced by the quantum BRST symmetry, a powerful remnant of the gauge invariance of the classical Lagrangian. This symmetry (through the Ward-Takahashi identities), reduces the degree of divergences of certain diagrams, and makes QED renormalizable.

In QED, it ensures that the photon remains massless even after quantum corrections. To see this, it suffices to examine the corrections to the photon propagator (two-point function). The diagrammatic expansion yields

$$\frac{e^2}{q^2} \quad \to \quad \frac{e^2}{q^2} + \frac{e^2}{q^2}\Pi(q^2)\frac{e^2}{q^2} + \cdots \equiv \frac{e^2(q^2)}{q^2} \ , \tag{6.4}$$

since Π is always proportional to q^2: the photon stays massless, and the only effect of the vacuum polarization diagrams is to make the coupling momentum-dependent.

The running coupling obeys the renormalization group differential equation

$$\frac{de}{dt} = \beta(e) \ , \tag{6.5}$$

where $t = \ln(q^2/\mu^2)$. The β function can be extracted from the coefficient of the divergence in Π. In dimensional regularization, it is natural to use the "minimal substraction" (MS) renormalization scheme where the finite parts of the counterterms are chosen to be zero. However since the divergence always occurs in combination with $\gamma - \ln 4\pi$, it is better to use the modified minimal subtraction ($\overline{MS}$), where this combination is the only momentum independent finite part of the counterterms. These renormalization prescriptions are chosen for their mathematical convenience, and the running parameters they produce must be carefully compared with measurable quantities.

In either scheme, the β function is zero below the electron threshold, when $t < t_e = \ln(4m_e^2/\mu^2)$; above the electron threshold, for $t > t_e$, it can be read off the coefficient of the divergent part of Π. At the one loop level, it is given by

$$\beta(e) = \frac{e^3}{16\pi^2}\frac{4}{3} \ .$$

The arbitrary scale μ is fixed by measurement. In QED, it is traditional to use a different (on-shell) renormalization scheme, based on a direct comparison with the Thomson scattering cross-section; it yields the famous numerical value for the gauge coupling

$$\alpha = \frac{e^2}{4\pi}(q^2 = 2m_e^2) \approx \frac{1}{137} \ .$$

This numerical identification can be seen as setting the scale μ for QED. Fortunately, the QED coupling is smallest at large distances (that is after all the reason we recognize electrons as free particles). At the other end, the renormalization group equation implies that the effective gauge coupling increases in the deep ultraviolet, eventually reaching infinity at the Landau pole at an extraordinary large value of energy. This happens well beyond the domain of validity of this formula which assumes α to be small. It is not known if the singularity is really there; one can only relate that, when last seen, the gauge coupling was increasing with energy-what it does beyond that energy is unknown, beyond the reach of our puny perturbative methods. Inclusion of higher order effects does not alter this trend: there is a natural scale associated with QED, roughly speaking that of the Landau pole. Fortunately it is ridiculously small compared to those at which we operate, which offers some justification for ignoring it.

More QED infinities lurk in the remaining one-loop diagrams. The first

corrects the fermion propagator. It is calculated to be

$$\begin{aligned}\Sigma(p) =& -i\Delta\frac{e^2}{16\pi^2}[\not p + 4m_e] + i\frac{e^2}{16\pi^2}[\not p + 2m_e] \\ &+ i\frac{e^2}{8\pi^2}\int_0^1 dx[\not p\,(1-x) + 2m_e]\ln\left(\frac{p^2x(1-x)+m_e^2x}{\mu^2}\right) \ . \qquad (6.6)\end{aligned}$$

Note that the divergence appears along both the kinetic term and the mass term. It is absorbed by a redefinition of the electron field and of the mass term. It follows that the mass, like any other parameter in the Lagrangian also becomes scale-dependent; its dependence on scale is dictated by the renormalization group equation

$$\frac{dm_e(t)}{dt} = m_e(t)\gamma_m(e) \ . \qquad (6.7)$$

Its one loop expression is

$$\gamma_m = -6\frac{e^2}{(4\pi)^2} \ . \tag{6.8}$$

Finally, we note that the physical mass of the electron, M_e, is to be distinguished from this running mass. It is natural to make the identification

$$m_e(q^2 = M_e^2) = M_e[1 + \mathcal{O}(e^2)] \ .$$

The scale dependence of the mass may also be interpreted in terms of the dimension of the two-fermion operator, and thus as the anomalous dimension of the fermion field.

This renormalization group equation tells us that if the running mass is zero at any scale, it will remain so at all other scales. This is a reflection of the added symmetry gained by setting the mass equal to zero. Indeed if $m_e = 0$, the QED Lagrangian becomes invariant under the chiral transformation

$$\Psi \to e^{i\alpha\gamma_5}\Psi \ , \tag{6.9}$$

which forbids a mass term for the electron to all orders of perturbation theory.

The one-loop correction to the interaction vertex is described by the diagram,

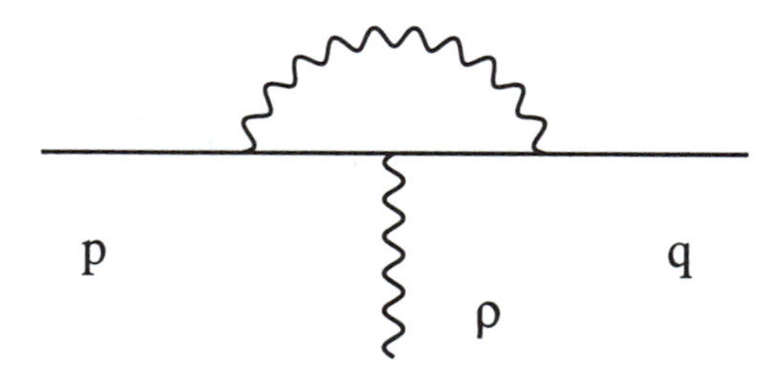

It is written in the form

$$\Gamma_\rho(p,q) = \Gamma_\rho^{(1)}(p,q) + \Gamma_\rho^{(2)}(p,q) \ ,$$

where the first part contains the divergence and the second part is finite. The first term has exactly the same matrix structure as the interaction term, specifically

$$\Gamma_\rho^{(1)}(p,q) = -\, ie\mu^{2-\frac{n}{2}}\gamma_\rho\frac{e^2}{16\pi^2}\Big\{\Delta - 1 - \int_0^1 dx \int_0^{1-x} dy$$

$$\ln \left[\frac{m_e^2(x+y) + p^2 x(1-x) + q^2 y(1-y) - 2p \cdot qxy}{\mu^2} \right] \Bigg\} \tag{6.10}$$

The second term is more complicated. It yields contributions along both γ_ρ and $\sigma_{\rho\mu}k_\mu$, with $k_\mu = q_\mu - p_\mu$. They are both ultraviolet finite, but the γ_ρ term diverges in the infrared when the external particles are on their mass-shells.

$$\Gamma_\rho^{(2)}(p,q) = \frac{e^2}{16\pi^2} \int_0^1 dx \int_0^{1-x} dy \;\times$$
$$\left\{ \frac{2m_e^2 \gamma_\rho [(x+y)^2 - 2(1-x-y) - 2k^2(1-x)(1-y)]}{m_e^2(x+y) + p^2 x(1-x) + q^2 y(1-y) - 2p \cdot qxy} + \right.$$
$$\left. + \frac{8im_e \sigma_{\rho\sigma} k_\sigma [x - y(x+y)]}{m_e^2(x+y) + p^2 x(1-x) + q^2 y(1-y) - 2p \cdot qxy} \right\} . \tag{6.11}$$

The infrared divergence in the first term results from integration over the Feynman parameters. It can be shown, on the grounds of relativistic invariance that the vertex corrections can be cast in the form

$$\Gamma_\rho(p,q) = F_1(k^2)\gamma_\rho + F_2(k^2)\sigma_{\rho\nu}k_\nu . \tag{6.12}$$

The one-loop diagram contributes to both $F_{1,2}$. The first term is the form factor which corrects the basic electron-photon vertex, while the second contributes to the electron's magnetic moment.

We have said that the effect of the quantum corrections is to generate in the effective Lagrangian terms of higher dimensions that respect all the symmetries of the classical Lagrangian which survive quantization. The finite quantum corrections generate in $\mathcal{L}_{eff}$ interaction terms with dimension higher than four. Divided by the appropriate power of the only dimensionful parameter, in this case the electron mass, they decouple in the limit of large electron mass. The reader is cautioned that it does not mean that the theory is trivial at energies below that scale: the electron can still contribute as a virtual particle, say in the scattering of light by light.

We now turn to the uses of dimensional analysis and symmetries, which are powerful tools in drawing a catalog of the finite quantum corrections of QED.

• *Dimension Five Interactions.* We begin the catalog by enumerating all possible interactions of dimension five, made out solely of the covariant derivative and the electron field. There are no Lorentz-invariant term with five D_μ alone. This leaves only Lorentz-invariant combinations of two fermion fields and two D_μ's:

$$\mathcal{O}_5^{(1)} = \overline{\Psi}\sigma^{\mu\nu}\Psi F^{\mu\nu} \; ; \qquad \mathcal{O}_5^{(2)} = \overline{\Psi}D_\mu D_\mu \Psi \; , \tag{6.13}$$

as well as

$$\mathcal{O}_5^{(1)\prime} = \overline{\Psi}\gamma_5\sigma^{\mu\nu}\Psi F^{\mu\nu} \; ; \qquad \mathcal{O}_5^{(2)\prime} = \overline{\Psi}\gamma_5 D_\mu D_\mu \Psi \; . \tag{6.14}$$

The last two terms cannot be generated by QED alone (see problem), leaving only the first two terms. Both **must** appear in $\mathcal{L}_{eff}$ divided by the electron mass, and with a dimensionless prefactor which is **finite** and **computable** in perturbation theory. The first term describes the famous correction to the gyromagnetic ratio of the electron. It is generated by the one-loop vertex diagram. Eventually, comparison with Eq. (6.11) yields the following effective interaction

$$\mathcal{L}_1 = (\frac{\alpha}{\pi})\frac{1}{m_e}\overline{\Psi}\sigma_{\mu\nu}\Psi F_{\mu\nu} \; , \tag{6.15}$$

where α is the fine structure constant. Contributions to the magnetic dipole of the electron are also generated in $\mathcal{L}_n$, with a coefficient of nth order in the fine structure constant.

The second term in (6.13) breaks fermion chirality, and thus cannot contribute to the kinetic term. By expanding the covariant derivative, we see that it contains three different terms. The first just provides a correction proportional to the momentum squared to the electron mass. The second generates a momentum-dependent chirality-breaking correction to the electron photon vertex, and the third yields a two-electron-two-photon vertex.

• *Dimension Six Interactions.* There are three types of terms of dimension six, containing four Ψ's, two Ψ's and three D_μ's, and six D_μ's; they will be divided by the square of the electron mass, with a finite prefactor calculated in perturbation theory. It is easy to see that the only possible terms with two fermion fields are

$$\mathcal{O}_6' = \epsilon_{\rho\sigma\mu\lambda}\overline{\Psi}\gamma_\rho F_{\sigma\mu}D_\lambda\Psi \; , \tag{6.16}$$

which is odd under parity, and does not appear in pure QED, and the parity-invariant interactions

$$\mathcal{O}_6^{\pm} = \overline{\Psi}(\gamma^\rho D^\sigma \pm \gamma^\sigma D^\rho)D^\rho D^\sigma \Psi \; . \tag{6.17}$$

The antisymmetric combination corrects the basic vertex. After integration by parts, it yields the interaction

$$\mathcal{O}_6^{(1)} = \overline{\Psi}\gamma_\rho\Psi\partial_\sigma F_{\rho\sigma} \; . \tag{6.18}$$

In momentum space, it is the coefficient of the term linear in momentum squared of the form factor for the electron-photon vertex; it is called the charge radius of the electron. It appears in the expansion of the form factor F_1.

The symmetric combination yields a finite correction to the fermion kinetic term

$$\mathcal{O}_6^{(2)} = \overline{\Psi}\gamma_\rho D_\rho D_\mu D_\mu \Psi \; , \tag{6.19}$$

that is linear in the momentum squared, as well as more complicated corrections to the interaction of one electron with one, two, and three photons.

The combination with six covariant derivatives can appear in many different ways. A possible term with three field strengths vanishes identically because one cannot make a Lorentz invariant out of three field strengths. Another can be made up of two field strengths and two covariant derivatives, such as

$$\mathcal{O}_6^{(3)} = \partial_\mu F_{\mu\nu}\partial_\rho F_{\rho\nu} \; , \tag{6.20}$$

which gives a finite correction to the photon propagator.

The analysis of the four fermion combinations is more complicated. Using Lorentz invariance, we arrange the 16 fermion bilinears in their Lorentz-covariant form (S,P,V,A,T). The allowed four-fermion interactions are their Lorentz-invariant, charge conjugation-even combinations. They are of the form SS, PP, VV, AA, and TT, but they are not all independent. The Fierz identities yield the following relations

$$SS - PP = -\frac{1}{2}(VV - AA) \; , \tag{6.21}$$

$$SS + PP = -\frac{1}{3}TT \; , \tag{6.22}$$

which leave only the three independent interactions

$$\mathcal{O}_6^{(4)} = \overline{\Psi}\Psi\overline{\Psi}\Psi \; , \qquad \mathcal{O}_6^{(5)} = \overline{\Psi}\gamma_5\Psi\overline{\Psi}\gamma_5\Psi, \qquad \mathcal{O}_6^{(6)} = \overline{\Psi}\gamma_\mu\Psi\overline{\Psi}\gamma_\mu\Psi \; . \tag{6.23}$$

The finite contributions to these interactions are generated by the box diagram

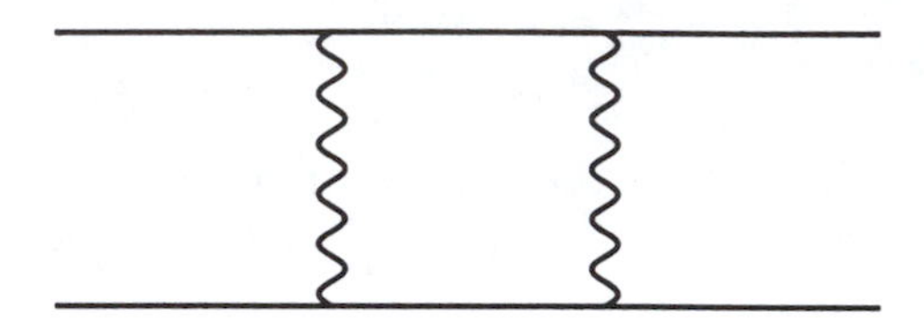

When weak interactions are included, we expect other interactions to be generated, such as SP and VA, because of parity violation.

It is important to emphasize that there already are one-particle-reducible contributions which yield four-fermion interactions, generated by one-photon exchange. In QED, the dimension six terms in the effective Lagrangian yield finite corrections to these processes, but do not generate new types of interaction. As we shall see later, in the standard model, some new four-fermion interactions, forbidden at tree level, do appear at one-loop in the effective Lagrangian. In QED, we have to go to terms of dimension eight to see a similar effect.

• *Dimension Eight.* We leave it to the reader to list all possible terms of dimension eight. Rather we focus on the one term that does not correspond to any interaction of lower dimensions. We can indeed form a term of dimension eight by putting together eight covariant derivatives, which is the same as four field strengths. It produces a direct interaction between photons; it is the famous scattering of light by light, or Delbrück scattering. It is generated at the lowest order by a box diagram

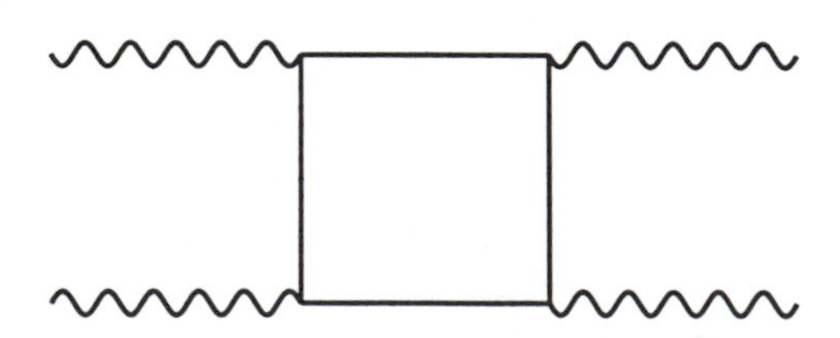

where the internal lines are electrons, and the external lines are photons. Naive power counting implies this diagram to be logarithmically divergent in the ultraviolet, which would spoil the renormalizability of the theory.

However, in gauge theories, there are magic cancellations, and the diagram is finite. In the static approximation, it yields the finite interaction

$$\mathcal{L}_1 = \frac{\alpha^2}{90m_e^4}[(F_{\mu\nu}F_{\mu\nu})^2 + \frac{7}{12}(\epsilon_{\mu\nu\rho\sigma}F_{\mu\nu}F_{\rho\sigma})^2] \, .$$

This important example shows that the effective Lagrangian can contain totally new interactions. In particular, imagine a world where the electron mass is so large that it has not yet been produced in the laboratory. At low energy, its presence would still be felt indirectly through the observation of photon-photon scattering! However, in the limit of very large electron mass, these effects become negligible. This is an example of the decoupling theorem (See T. Appelquist and J. Carazzone, *Phys. Rev.* **D11**, 2856(1975)), which says that all the quantum effects of massive particles become insignificant as their masses become infinite. The important exception, is for particles that get their masses through vacuum expectation values.

PROBLEMS

A. Identify the symmetries that forbid QED from generating the terms in Eq. (6.14).

B. Show that the box diagram that describes Delbrück scattering is ultraviolet finite.

C. Find the Lorentz structure of the QED four-fermion one-loop box diagram, to determine which of the interactions in the text it generates. Then compare these with interactions generated by one photon exchange.

D. Enumerate all the possible operators of dimension seven, and interpret their contributions physically.

E. Identify the lowest order Feynman diagrams which contribute to $\mathcal{O}_5^{(2)}$ and to $\mathcal{O}_6^{(3)}$.

6.2 THE STANDARD MODEL

The organization of the one-loop corrections of the standard model is much more challenging. Their detailed analysis is complicated by its non-Abelian gauge symmetries, and their spontaneous breaking. In spite of these technical difficulties, the corrections have the same structure as those encountered in our study of QED: they cause the parameters of the theory to run with scale as prescribed by the renormalization group, and they also generate interactions; some appear as corrections to those already present in the classical Lagrangian, others generate entirely new interactions.

To distinguish between these two types of corrections, it is useful to analyze in detail the symmetries of the standard model. We start with the electroweak part of the theory, leaving QCD aside for the moment.

Partial Symmetries of the Standard Model

From Chapter 2, we recall the symmetries of the standard model classical Lagrangian: besides the gauged symmetries, $SU(2) \times U(1) \times SU(3)$, it contains four global continuous symmetries, the three lepton numbers, L_e, L_μ, L_τ, and the total quark number, $B/3$, and no discrete symmetries. Non-perturbative quantum effects, associated with the anomaly of the non-Abelian weak $SU(2)$, break these down to two relative lepton numbers, $L_e - L_\mu$, $L_\mu - L_\tau$, and $B - L$, where L is the total lepton number. These effects are negligible and they have no practical effect for our present purposes, although they are important in the study of standard model cosmology.

When we consider a subset of all its interactions, the standard model displays a much richer structure than implied by these symmetries alone. One reason is that particular subset of the interactions may display a larger symmetry than in the whole Lagrangian. We call these partial symmetries, or *accidental* symmetries. For example, QED, which is part of the standard model, preserves P and C, while the weak interactions do not. Of course the rest of the interactions break such symmetries, but by terms which have specific covariance properties with respect to the accidental symmetries. As with the Wigner-Eckhardt theorem, this generally implies selection rules amenable to experimental tests.

For instance, tree-level processes which involve interactions with the larger symmetry, will reflect that symmetry either by producing (tree-level) relations among parameters, or by the absence of some processes. To these must be added quantum corrections (one-loop or beyond), which use interactions from other parts of the Lagrangian that break the partial symmetry. It may even happen that one-loop corrections break the accidental symmetry only to a lesser accidental symmetry, and so on. The radiative corrections may generate new interactions forbidden by the accidental tree level symmetries, or correct the tree level relations implied by the partial symmetries. An example is the electric dipole of the electron, which is forbidden by the symmetries of QED; in the standard model, it is no longer protected, and we expect it to be generated by weak quantum corrections.

Clearly, the study of accidental symmetries will prove very useful in the understanding of the radiative structure of the standard model. Let us now apply this analysis for the whole electroweak model. Its Lagrangian can be split into different parts,

$$\mathcal{L}_{SM} = \mathcal{L}_{YM} + \mathcal{L}_{WD} + \mathcal{L}_{Y} + \mathcal{L}_{H} \ ,$$

each part characterized by different global symmetries that are generally larger than those of $\mathcal{L}_{SM}$. We have already encountered simple examples: the classical Yang-Mills part, $\mathcal{L}_{YM}$ is not only invariant under the gauge groups, but also under the discrete space-time symmetries P and C; the anomaly associated with QCD generates non-perturbatively interactions that break C and CP; the QED part which emerges from spontaneous breaking is invariant under parity.

Lepton Symmetries

The leptonic Weyl-Dirac Lagrangian $\mathcal{L}_{WD}$ displays a much larger global invariance. With three chiral families, it is invariant under $U(3)_L \times U(3)_R$, where $U(3)_L$ acts on the three lepton doublets, and $U(3)_R$ acts on the lepton singlets.

This leptonic global chiral symmetry is explicitly broken by Yukawa interactions, leaving only the three lepton number symmetries. The leptonic bilinears which appear in the Yukawa couplings are of the form $L_i\bar{e}_j$ and transform as $(\mathbf{3}, \overline{\mathbf{3}})$ under this global symmetry. If these bilinears were coupled to a Higgs matrix $\mathbf{H}_e$, itself transforming as a $(\overline{\mathbf{3}}, \mathbf{3})$, invariance could be preserved, but this would require nine Higgs doublets. The standard model contains only one doublet, with the Higgs matrix that couples to the lepton bilinear given by

$$\mathbf{H}_e(x) = \mathbf{Y}_e \tau_2 H^*(x) \ , \tag{6.24}$$

where $\mathbf{Y}_e$ is the lepton Yukawa matrix which breaks the symmetry. If all three leptons had equal mass, the remaining symmetry would be the diagonal $U(3)_{L+R}$, but since the lepton masses are different, this vectorial

$U(3)$ is broken to its three diagonal generators, yielding the three lepton numbers. At high energies, where we can neglect the electron and muon masses, the remaining global symmetry is $U(2)_L \times U(2)_R \times U(1)_{L+R}$.

Quark Symmetries

The same analysis applied to the quark Weyl-Dirac Lagrangian is more complicated. The quark gauged kinetic terms are invariant under the global chiral symmetry generated by $U(3)_L \times U(3)_R \times U(3)_R$, the first acting on the three quark left-handed weak doublets, the other two on the charge 2/3 and -1/3 right-handed singlets, respectively. The Yukawa terms involve two types of weak doublets quark bilinears, $\widehat{\mathbf{Q}}_i\bar{\mathbf{u}}_j$ and $\widehat{\mathbf{Q}}_i\bar{\mathbf{d}}_j$, transforming as $(\mathbf{3}, \bar{\mathbf{3}}, \mathbf{1})$ and $(\mathbf{3}, \mathbf{1}, \bar{\mathbf{3}})$ under the global chiral symmetry, respectively. With only one Higgs doublet, these bilinears couple to the Higgs matrices

$$\mathbf{H}_u(x) = \mathbf{Y}_u H(x) \ ; \qquad \mathbf{H}_d(x) = \mathbf{Y}_d \tau_2 H^*(x) \ . \tag{6.25}$$

Since the Yukawa matrices are different for the up and down quark sectors, there remains only one unbroken symmetry, the (vectorial) quark number. It is nevertheless very instructive to keep track of the individual quark numbers.

In the spontaneously broken vacuum of the standard model, the different quark mass eigenstates, each of which carries its own quark number, mix with one another at tree level *only by emitting a charged W-boson.* These interactions change both quark number and electric charge at the same time. It is traditional to attribute the different quarks with a *flavor* F of their own, usually named after the quark itself; thus the strange quark has *strangeness* (for strange historical reasons, the strange quark has strangeness -1); the charmed quark has *charm*, the bottom quark has *beauty*, and the top quark carries *truth.* The up and down quarks are not assigned special flavor names. W-exchange imply the important **tree-level** selection rules

$$|\Delta F| = 1 \ , \ |\Delta Q| = 1 \ ; \qquad |\Delta F| = 0 \ , \ |\Delta Q| = 0 \ , \tag{6.26}$$

where F is any of the quark flavors. In words, there are no flavor-changing interactions among quarks of the same electric charge at tree level. As these selection rules do not come from the symmetries of the full Lagrangian, we expect radiative corrections to break them. It follows that the standard model flavor-changing neutral interactions among quarks arise purely from quantum effects; they are strictly **predicted** in terms of the input parameters, and thus provide important experimental checks of the standard model. This situation is similar to that in QED where scattering of light by light occurs only through quantum effects, and is predicted in terms of QED's input parameters, the electron charge and mass.

Higgs Symmetry

The Higgs sector of the standard model has a global $SU(2)_R$ symmetry of its own, which we have already discussed. It is convenient for the analysis that follows to write the complex Higgs doublet in terms of four real fields

$$H = \begin{pmatrix} h_1 + ih_2 \\ h_3 + ih_4 \end{pmatrix} . \tag{6.27}$$

The same doublet with opposite hypercharge is just

$$\overline{H} \equiv -i\tau_2 H^* = \begin{pmatrix} -h_3 + ih_4 \\ h_1 - ih_2 \end{pmatrix} . \tag{6.28}$$

We form the matrix

$$\mathcal{H} = (H, \overline{H}) = \begin{pmatrix} h_1 + ih_2 & -h_3 + ih_4 \\ h_3 + ih_4 & h_1 - ih_2 \end{pmatrix} . \tag{6.29}$$

The weak gauged $SU(2)_L$ acts on the two rows, the global $SU(2)_R$ on the two columns; $SU(2)_R$ clearly violates hypercharge, since the first and second rows have opposite hypercharges. The Higgs matrix transforms as $(\mathbf{2}, \mathbf{2})$ under this symmetry

$$\mathcal{H} \to \mathcal{H}' = U_L \mathcal{H} U_R , \tag{6.30}$$

where the unitary matrices $U_{L,R}$ represent $SU(2)_{L,R}$, respectively. These two groups combine to an $SO(4)$ acting on the four real components of the Higgs field. It is easy to check that the Higgs potential involves only the $SO(4)$-invariant combination

$$H^\dagger H = \frac{1}{2} \operatorname{Tr} (\mathcal{H}^\dagger \mathcal{H}) = \det \mathcal{H} = h_1^2 + h_2^2 + h_3^2 + h_4^2 . \tag{6.31}$$

The gauged weak $SU(2)_L$ is clearly preserved, but what happens to $SU(2)_R$ in the rest of the Lagrangian? The Higgs kinetic term clearly preserves the full symmetry, but since the two columns of the matrix have opposite unit values of hypercharge, hypercharge interactions violate $SU(2)_R$ by two units.

The quark Yukawa interactions violate $SU(2)_R$ as well, but in an interesting way, best seen by rewriting the quark Yukawa coupling in the suggestive form

$$\mathcal{L}_Y = \widehat{\mathbf{Q}} \left[\left(\frac{\mathbf{Y}_u + \mathbf{Y}_d}{2} \right) (\overline{\mathbf{u}} H + \overline{\mathbf{d}} \overline{H}) + \left(\frac{\mathbf{Y}_u - \mathbf{Y}_d}{2} \right) (\overline{\mathbf{u}} H - \overline{\mathbf{d}} \overline{H}) \right] . \tag{6.32}$$

The first term is obviously invariant if the combination $(\overline{\mathbf{u}}, \overline{\mathbf{d}})$ transforms as a doublet under $SU(2)_R$ (this is the reason for the subscript R). The second term violates $SU(2)_R$, with the quantum numbers of a triplet. In analogy with the Wigner-Eckhardt theorem, this results in additional sum rules. The contributions from this term are significant because of the sizeable value of the mass *difference* between the top and bottom quarks, which dwarfs the contributions from the lighter families. The gauged kinetic term of the $(\overline{\mathbf{u}}, \overline{\mathbf{d}})$ doublet is not $SU(2)_R$ invariant, since its components have different hypercharges.

A similar reasoning applied to the lepton Yukawa couplings shows that they are not symmetrical in any limit, because the standard model contains no electroweak-singlet leptons to serve as the $SU(2)_R$ partner of $\overline{e}_i$. However these partners do appear in many extensions of the standard model.

The standard model is therefore symmetric under a global $SU(2)_R$ only in the limits

$$\mathbf{Y}_e = \mathbf{Y}_u - \mathbf{Y}_d = 0 \; ; \qquad g_1 = 0 \; . \tag{6.33}$$

Our analysis has not taken into account the spontaneous breakdown of the standard model symmetry. As the Higgs gets its vacuum value, the $SO(4)$ symmetry is spontaneously broken to an $SO(3)$ subgroup. The three broken symmetries yield Nambu-Goldstone bosons, which are eaten by the three gauge vector bosons. The surviving symmetry preserves the trace of the Higgs matrix; it is the vectorial diagonal subgroup $SU(2)_{L+R}$, the so-called *custodial* $SO(3)$.

In the limit $g_1 = 0$, $\cos\theta_W = 1$, and electromagnetism decouples. The three massive gauge bosons W^+_μ, W^-_μ, Z_μ transform as a custodial triplet, and have the same mass. The three currents to which they couple also transform as a custodial triplet. One can see this directly by noting that both left- and right-handed quarks transform as custodial doublets.

Does this symmetry manifest itself at tree level? In the static limit, the exchange of the gauge bosons produces with a current-current four-fermi interaction among these currents. The custodial symmetry simply requires that neutral and charged current interactions appear with the same strength at this level of approximation.

When hypercharge interactions are restored ($g_1 \neq 0$), this situation does not change at tree-level, since the strengths of the charged and neutral current-current interactions are still equal

$$\frac{g_2^2}{m_W^2} = \frac{g_1^2}{m_Z^2} = \frac{g_2^2}{m_Z^2 \cos^2\theta_W} , \tag{6.34}$$

although the form of the neutral current changes by acquiring the electromagnetic current multiplied by $\sin^2\theta_W$. The above relation is often expressed by introducing a parameter ρ which is the ratio of these two interaction strengths

$$\rho \equiv \frac{m_W^2}{m_Z^2 \cos^2\theta_W} . \tag{6.35}$$

It trivially satisfies the tree-level relation

$$\rho - 1 = 0 . \tag{6.36}$$

Since the rest of the Lagrangian violates the custodial symmetry, one expects quantum corrections to this relation. This situation is analogous to $g-2$ in QED, which is zero at tree-level, but is calculably corrected by quantum effects. The custodial symmetry is most badly broken by the large mass difference between the top and bottom quark masses divided by the W-mass (to make it dimensionless). There will also be smaller contributions involving the $W-Z$ mass difference, and electromagnetic corrections, with strengths proportional to $\sin^2\theta_W$.

Running the Standard Model Parameters

As a result of quantization, all the parameters of the standard model become scale dependent. In this section we only state the resulting equations for its parameters, and refer the reader interested in the calculational details to standard texts on quantum field theory.

Gauge Couplings

All three gauge couplings are scale-dependent. In the one-loop approximation, their evolution is governed by the equations

$$\frac{d\alpha_l^{-1}}{dt} = \frac{1}{2\pi} b_l , \tag{6.37}$$

where $\alpha_l = g_l^2/4\pi$, $t = \ln\mu$ and $l = 1, 2, 3$, corresponding to the gauge groups $U(1) \times SU(2) \times SU(3)$. For any gauge group, the coefficients are given by

$$b = \frac{11}{3}C_{\text{adj}} - \frac{2}{3}\sum_f C_f - \frac{1}{6}\sum_h C_h \,, \tag{6.38}$$

where C_{adj} is the Dynkin index of the adjoint representation of the gauge group, C_f is the Dynkin index of the representation of the left-handed Weyl fermions, and C_h is that of the representation of the (real) Higgs field. Applying this formula for one Higgs doublet and n_{fam} chiral families, we find

$$\begin{aligned} b_1 &= -\frac{4}{3}n_{\text{fam}} - \frac{1}{10}\,, \\ b_2 &= \frac{22}{3} - \frac{4}{3}n_{\text{fam}} - \frac{1}{6}\,, \\ b_3 &= 11 - \frac{4}{3}n_{\text{fam}}\,. \end{aligned} \tag{6.39}$$

The coefficient of the hypercharge has been normalized in such a way that

$$b_i = -\frac{3}{20}\left\{\frac{2}{3}\sum_L (Y^2)_L + \frac{1}{3}\sum_H (Y^2)_H\right\}. \tag{6.40}$$

In the standard model, with $n_{\text{fam}} = 3$, the numerical values of these coefficients are just

$$(b_1, b_2, b_3) = (-\frac{41}{10}, \frac{19}{6}, 7)\,.$$

These equations are modified by higher loop effects, but as long as the couplings are reasonably small, they should suffice. All three gauge couplings are perturbative over an enormous range of energies. The QCD coupling becomes strong in the infrared, where we need to include higher order effects, and the hypercharge coupling tends towards a Landau pole in the deep ultraviolet.

Yukawa Couplings

The one-loop renormalization group equations of the Yukawa couplings are of the form

$$\frac{d\mathbf{Y}_{u,d,e}}{dt} = \frac{1}{16\pi^2}\mathbf{Y}_{u,d,e}\boldsymbol{\beta}_{u,d,e}\,. \tag{6.41}$$

The matrix coefficients are given by

$$\begin{aligned}
\beta_u &= \frac{3}{2}(\mathbf{Y}_u{}^\dagger\mathbf{Y}_u - \mathbf{Y}_d{}^\dagger\mathbf{Y}_d) + T - (\frac{17}{20}g_1^2 + \frac{9}{4}g_2^2 + 8g_3^2) \ , \\
\beta_d &= \frac{3}{2}(\mathbf{Y}_d{}^\dagger\mathbf{Y}_d - \mathbf{Y}_u{}^\dagger\mathbf{Y}_u) + T - (\frac{1}{4}g_1^2 + \frac{9}{4}g_2^2 + 8g_3^2) \ , \\
\beta_e &= \frac{3}{2}\mathbf{Y}_e{}^\dagger\mathbf{Y}_e + T - \frac{9}{4}(g_1^2 + g_2^2) \ ,
\end{aligned} \tag{6.42}$$

with

$$T = \mathrm{Tr}\{3\mathbf{Y}_u{}^\dagger\mathbf{Y}_u + 3\mathbf{Y}_d{}^\dagger\mathbf{Y}_d + \mathbf{Y}_e^\dagger\mathbf{Y}_e\} \ . \tag{6.43}$$

The structure of these equations reflects the fact that chiral symmetry that appears in the limit where these couplings are zero, is not broken by (perturbative) radiative corrections. The origin of the different terms can be understood in terms of Feynman diagrams. In the following, we show only diagrams with physical particles (unitary gauge), but in any calculationally-friendly gauge, these diagrams must be supplemented by those involving the longitudinal gauge bosons and various associated ghosts.

The universal factor T comes from the fermion-loop renormalization of the Higgs line

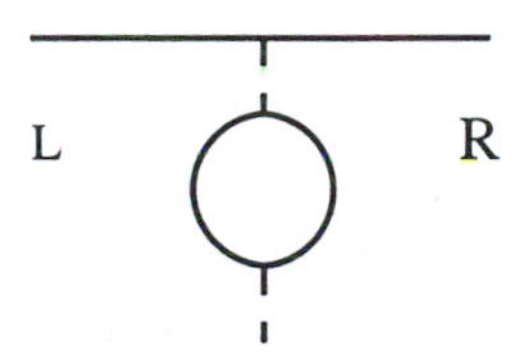

The pure Yukawa coupling contributions come from diagrams of the form

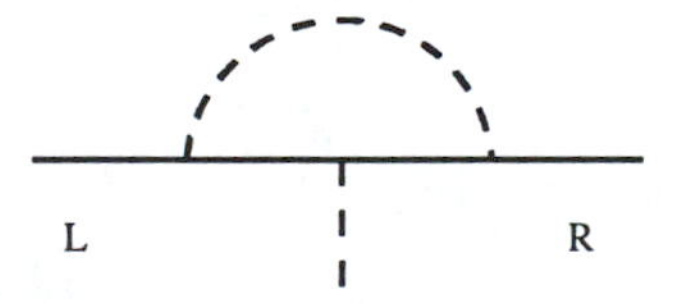

The gauge couplings contributions come from corrections to the fermion and Higgs lines (not shown here) as well as from the one-particle irreducible diagrams

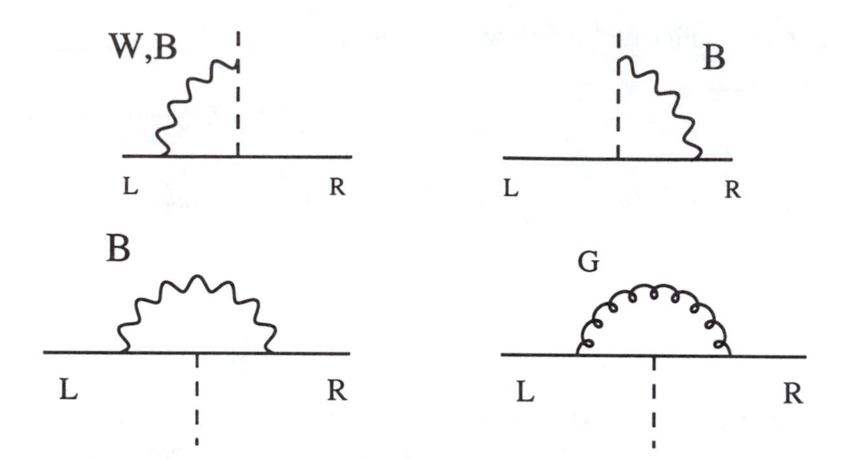

The last diagrams applies only when the external fermions are quarks. We observe that the Yukawa couplings contributions tend to make the couplings blow up in the ultraviolet, while those of the gauge couplings tend towards asymptotic freedom. In order to see which of these two effects prevails, let us look at the coefficients in the (not unrealistic) limit where we keep only the top quark Yukawa, and neglect g_1 and g_2. Then the lepton Yukawa couplings evolve as

$$\frac{dy_\tau}{dt} \approx \frac{3}{16\pi^2} y_\tau y_t^2 \ , \tag{6.44}$$

resulting in a Landau pole in the ultraviolet. The mass of the τ lepton, and thus its Yukawa coupling, is sufficiently small that y_τ blows up beyond the Planck energy. The *bottom* quark Yukawa obeys

$$\frac{dy_b}{dt} \approx \frac{1}{16\pi^2} y_b (\frac{3}{2} y_t^2 - 8g_3^2) \ . \tag{6.45}$$

As the larger QCD coupling dominates, y_b is asymptotically free. Similarly, the *top* Yukawa coupling, which varies according to,

$$\frac{dy_t}{dt} \approx \frac{1}{16\pi^2} y_t (\frac{9}{2} y_t^2 - 8g_3^2) \ , \tag{6.46}$$

is still asymptotically free, although less so than that of the *bottom* quark. For representative values $y_t \approx .7$ and $g_3^2 \approx 1.5$, in the neighborhood of M_Z, we both quark Yukawa couplings decrease at short distances.

Higgs self-coupling

The Higgs quartic coupling has a complicated scale dependence. It evolves according to

$$\frac{d\lambda}{dt} = \frac{1}{16\pi^2}\beta_\lambda, \tag{6.47}$$

where the one loop contribution is given by

$$\beta_\lambda = 12\lambda^2 - (\frac{9}{5}g_1^2 + 9g_2^2)\lambda + \frac{9}{4}(\frac{3}{25}g_1^4 + \frac{2}{5}g_1^2g_2^2 + g_2^4) + 4T\lambda - 4H\ , \tag{6.48}$$

in which

$$H = \mathrm{Tr}\{3(\mathbf{Y}_u{}^\dagger\mathbf{Y}_u)^2 + 3(\mathbf{Y}_d{}^\dagger\mathbf{Y}_d)^2 + (\mathbf{Y}_e{}^\dagger\mathbf{Y}_e)^2\}\ . \tag{6.49}$$

We note that since λ is not protected by symmetry, β_λ is not proportional to λ. Hence setting to zero the Higgs self coupling does not enhance symmetry. This is to be contrasted with the Yukawa couplings whose absence generates chiral symmetries.

The first two terms involving λ come from diagrams of the form

while the pure gauge terms, and H are generated by the gauge and fermion loop corrections, respectively

Finally, the renormalization of the Higgs line are all proportional to λ give contributions to the second and fourth groups of terms.

The value of λ at low energies is related the physical value of the Higgs mass according to the tree level formula

$$m_H = v\sqrt{2\lambda}\ , \tag{6.50}$$

while the vacuum value is determined by the Fermi constant G_F of β decay. Since the Higgs mass is not yet known, we do not have a physical boundary condition for Eq. (6.47). Still we can discuss the evolution of λ as a function of the Higgs mass.

We discuss below the qualitative features of its running, leaving datails to the problems. First, for a fixed vacuum value v, let us assume that the Higgs mass, and therefore λ is large. In that case, β_λ is dominated by the λ^2 term, which drives the coupling towards its Landau pole at higher energies. Hence the higher the Higgs mass, the higher λ is and the closest the Landau pole to experimentally accessible regions. This means that for a given (large) Higgs mass, we expect the standard model to enter a strong coupling regime at relatively low energies, losing in the process our ability to calculate. This does not necessarily mean that the theory is incomplete, only that we can no longer handle it. In analogy with the chiral model description of pion physics, it is natural to think that this effect is caused by new strong interactions, and that the Higgs actually is a composite of some hitherto unknown constituents. An example of such a theory is a generalization of the standard model called *technicolor*. The resulting bound on λ is sometimes called the *triviality bound*. The reason for this unfortunate name (the theory is anything but trivial) stems from lattice studies where the coupling is assumed to be finite everywhere; in that case the coupling is driven to zero, yielding in fact a trivial theory. In the standard model λ is certainly not zero.

In the opposite limit of a small Higgs mass, another strange behavior sets in, leading to another interesting constraint. In this regime, λ is small and its β function is dominated by the term coming from fermion loops. This term, proportional to the fourth power of the Yukawa couplings, can becomes dominant for a heavy top quark. Its effect is to decrease the value of λ in the ultraviolet. Since the change is not proportional to λ, it can in fact drive λ to negative values beyond a certain energy. Naively, this implies a negative contribution to the potential, which destabilizes the theory: large field configurations become energetically favored, and the theory tumbles out of control. The standard model description becomes inconsistent above a certain scale. This yields the *instability bound*. Should the Higgs particle prove to be light, this bound means that something must happen to the standard model at that scale, perhaps in the form of new contributions to the renormalization group evolution appear, from particles not in the standard model. In the supersymmetric generalization of the standard model, for instance, new particles appear and λ is not a fundamental coupling constant, but rather the square of gauge coupling constants.

We can summarize these two bounds in one graph showing the scale at which new physics is expected as a function of the Higgs mass for a given value of the top quark mass, which we take to be 180 GeV.

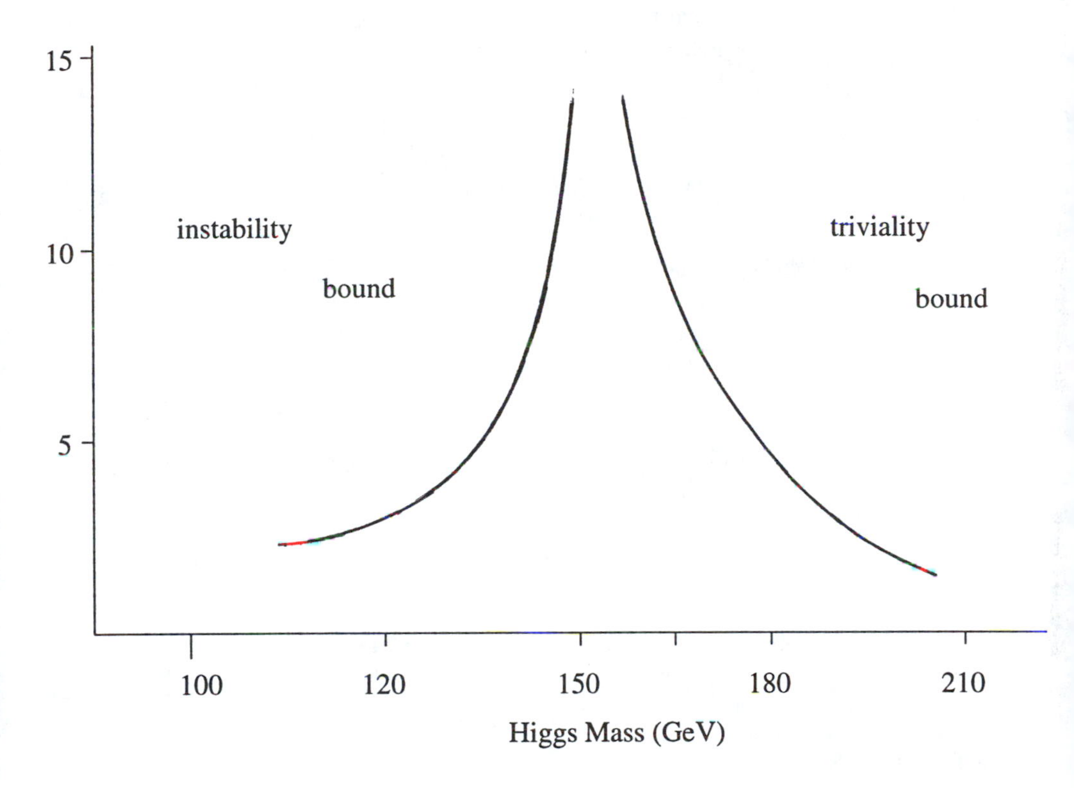

We see that with a (low) Higgs mass of 100 GeV, the instability sets in around 1 TeV; on the other hand, a (large) Higgs mass of 300 GeV implies, through the triviality bound new physics around 10 TeV. This discussion makes it clear that knowledge of the Higgs mass is an important figure of merit for the scale at which new physics will appear. OF course, if the Higgs mass is between 130 and 200 GeV, this analysis does not require new physics below the Planck scale!

At low energy, the effective theory becomes $SU(3) \times U(1)_{EM}$, with only two gauge coupling constants. Their evolution is given by

$$\frac{dg_3}{dt} = \frac{g_3^3}{(4\pi)^2}[\frac{2}{3}(n_u + n_d) - 11] \ , \tag{6.51}$$

and

$$\frac{de}{dt} = \frac{e^3}{(4\pi)^2}[\frac{16}{9}n_u + \frac{4}{9}n_d + \frac{4}{3}n_l] \ , \tag{6.52}$$

where n_u, n_d, and n_l are the number of light fermions, up-like and down-like quarks, and leptons. The remaining parameters of the low energy theory run as well. In the Landau gauge, the vacuum expectation value of the Higgs field runs like its anomalous dimension, that is

$$\frac{d \ln v}{dt} = \frac{1}{16\pi^2}\left(\frac{9}{4}(\frac{1}{5}g_1^2 + g_2^2) - T\right) \ . \tag{6.53}$$

Finally, the fermion masses in the low energy theory evolve as

$$\frac{dm_i}{dt} = m_i \gamma_{(i)} \ , \qquad i = l, q \ , \tag{6.54}$$

where the l and q refer to a particular lepton or quark. At one-loop

$$\gamma_{(i)} = \gamma_{(i)}^{[QED]}\frac{e^2}{(4\pi)^2} + \gamma_{(i)}^{[QCD]}\frac{g_3^2}{(4\pi)^2} \ , \tag{6.55}$$

with one-loop values for fermions of electric charge $Q_{(i)}$

$$\gamma_{(i)}^{[QED]} = -6Q_i^2 \ , \qquad \gamma_{(l)}^{[QCD]} = 0 \ ; \qquad \gamma_{(q)}^{[QCD]} = -8 \ . \tag{6.56}$$

PROBLEMS

A. 1-) Assume that the standard model symmetry is broken by a Higgs field that transforms not as a doublet of the weak $SU(2)$, but according to some arbitrary represntation of weak isospin j. Derive the general formula for the tree-level value of the ρ parameter.
2-) Now assume that you have two Higgs, one is the standard weak iso doublet, and the other is an isotriplet. What constraints does the experimental value of the ρ parameter put on the ratio of their vacuum values? Neglect the effect of quantum corrections.

B. Show that when the standard Higgs doublet gets its vacuum value, the global $SO(4)$ symmetry is broken to an $SO(3)$ subgroup. Identify the surviving symmetry as the diagonal subgroup $SU(2)_{L+R}$. Show that in the limit $g_1 = 0$, the three gauge bosons form a degenerate triplet under that symmetry.

C. 1-) Keeping only the top quark Yukawa and the QCD coupling in the one-loop renormalization equations, find the location of the Landau pole for the τ lepton Yukawa coupling.
2-) Suppose a heavy τ were found. Derive an upper bound on its mass based on Landau pole arguments.

D. Assume one family of quarks and leptons, and neglect g_1 and g_2. Using the RG equations for both the top quark Yukawa coupling and the strong gauge coupling, show that the top Yukawa coupling has an infrared fixed point. Estimate its value, and discuss its significance. For reference, see B. Pendleton and G. G. Ross *Phys.Lett.* **98B** 291(1981), as well as C. T. Hill, *Phys. Rev.* **D24**, 691(1981).

E. Using one-loop expressions, plot the instability and triviality bounds for the measured value of the top quark mass. Suppose the Higgs particle weighs in at 110 GeV. At what energy scale does the standard model cease to be valid?

6.3 HIGHER DIMENSION ELECTROWEAK OPERATORS

We now apply to the standard model what we have learned in organizing the QED quantum corrections. Our procedure was to find all possible invariant field combinations of a given dimension. In perturbation theory, all these combinations are generated in the effective quantum Lagrangian, suppressed by inverse powers of the electron mass, M_e^{d-4}, where d is the engineering dimension of the invariant. The coefficients in front of each combination are determined in perturbation theory, by calculating the appropriate Feynman diagrams.

The same technique also provides an elegant and efficient method to describe and organize the radiative structure of the standard model. It is of course more complicated simply because there are more symmetries to keep track of, and there is an essential complication we have not encountered in QED: the electroweak symmetry is spontaneously broken. In spite of these, the basic ideas we have introduced in our study of QED still apply, with some caveats we proceed to discuss.

Higgs Polynomials

The spontaneous breaking of the electroweak symmetry introduces some subtelty, as we can perform the analysis either in the broken or in the unbroken formulation of the theory. It is more economical to list invariants under the full electroweak symmetry, rather than under its broken remnants. However, invariants under the full symmetry will in general contain polynomials in the Higgs field, which must be evaluated in the electroweak vacuum. As a result, an infinite number of operators with arbitrarily high dimensions in the unbroken formulation can be expected to contribute to one operator of much lower dimension in the broken theory. Nevertheless, as long as we stick to identifying operators by their quantum numbers, this method provides a powerful way to identify interactions in the broken theory. This technique cannot be used for the perturbative calculation of the coefficients in front of the operators, since the false and true vacua are not perturbatively related. Calculations make sense only in the true electroweak vacuum. Since spontaneous breaking brings in another scale, v, masses of the particles will not necessarily appear in the effective Lagrangian as inverse powers (as in QED), but also as positive powers, logarithms, etc... .

To summarize, invariant interactions in the broken theory formulation with a given dimension can be generated either from electroweak invariants of the same dimension, and/or from invariants of higher dimensions that contain polynomial combinations of the Higgs doublet that do not vanish in the electroweak vacuum. The difference in dimension is the order of the Higgs polynomial. Fortunately, these polynomials have a limited set of electroweak quantum numbers, which keeps our method practical.

It is not very hard to list all possible Higgs polynomials which do not vanish in the electroweak vacuum. The first is of course the Higgs doublet itself (or its conjugate) which can be set equal to its vacuum value, v. Any combination of fields of dimension d that transforms with the conjugate quantum numbers of the Higgs doublet stems from a full invariant of dimension $d+1$.

There are several Higgs polynomials of second order. Of the two Higgs binomials with $Y=2$, the isoscalar combination $H^t\tau_2 H$ vanishes identically, since there is only one Higgs doublet. The second is the weak isovector

$$H^t\tau_2\vec{\tau}H \sim (\mathbf{1},\mathbf{1};\mathbf{3},\mathbf{1}^c)_2 \ , \tag{6.57}$$

where the first two entries refer to the Lorentz group, $SU(2)\times SU(2)$, the third to the weak isospin, the fourth to color, and the subscript is the hypercharge. The same combination with H replaced by $\overline{H}$ is the conjugate isovector with hypercharge -2. Full invariants of dimension $d+2$ that contain this polynomial yield interactions of dimension d along the electrically neutral component of this weak isotriplet. There are two $Y = 0$ combinations,

$$H^\dagger H \sim (\mathbf{1}, \mathbf{1}; \mathbf{1}, \mathbf{1}^c)_0 \; ; \qquad H^\dagger \vec{\tau} H \sim (\mathbf{1}, \mathbf{1}; \mathbf{3}, \mathbf{1}^c)_0 \; , \tag{6.58}$$

The first, with no quantum numbers, plays no role in the listing of invariants, as it can appear in any power. The second is another isovector.

There are of course other polynomials quadratic in the Higgs doublet, but they vanish in the electroweak vacuum. For instance, by adding the covariant derivative acting on the Higgs doublet, we obtain a Lorentz vector polynomial

$$H^T \tau_2 \vec{\tau} \mathcal{D}_\mu H \sim (\mathbf{2}, \mathbf{2}; \mathbf{1} \oplus \mathbf{3}, \mathbf{1}^c)_2 \; . \tag{6.59}$$

It vanishes in the Lorentz invariant electroweak vacuum. This vector polynomial does appear in higher dimension polynomials, coupled to another with conjugate quantum numbers. Evaluated in the electroweak theory, it gives rise to interactions that involve Higgs scalars.

The reader is encouraged to show that, with one Higgs doublet, there is only one new cubic Higgs polynomial with isospin $3/2$ and $Y = \pm 3$, and that there are no new quartic polynomials. Hence all higher order Higgs polynomials with electroweak vacuum values are made up of the combinations we have already listed, leaving polynomials with four possible quantum numbers

$$(\mathbf{1}, \mathbf{1}; \mathbf{2}, \mathbf{1}^c)_{\pm 1} \; ; \; (\mathbf{1}, \mathbf{1}; \mathbf{3}, \mathbf{1}^c)_{\pm 2} \; ; \; (\mathbf{1}, \mathbf{1}; \mathbf{3}, \mathbf{1}^c)_0 \; ; \; (\mathbf{1}, \mathbf{1}; \mathbf{4}, \mathbf{1}^c)_{\pm 3} \; , \tag{6.60}$$

together of course with their Kronecker products. This enables us to proceed with our main task: building electroweak-invariant polynomials of a given dimension, using the basic building blocks of the standard model: the left-handed Weyl fermions $f(d = \frac{3}{2})$, the Lorentz vector covariant derivatives $\mathcal{D}(d = 1)$, and the scalar Higgs doublet $H(d = 1)$.

Dimension-Five Interactions

We proceed to list those operators which are invariant under the full symmetry of the standard model, *as well as* those which have the quantum numbers of the Higgs polynomials with electroweak vacuum values. We start by discussing the invariants.

It is not difficult to enumerate all invariants with $d = 5$. Dimension-five invariants must necessarily contain one fermion bilinear: without fermions, the weak doublet Higgs must appear in pairs to conserve weak isospin, so that $d = 5$ combinations must contain either one or three covariant derivatives, which are not possible without losing Lorentz invariance. Hence the possible dimension-five invariants are restricted to the forms $ffHH$, $ff\mathcal{D}H$, and $ff\mathcal{D}\mathcal{D}$, where the covariant derivatives can act on any of the fields including themselves. Here f denote fermions of either chirality (f or $\bar{f}$).

It is useful to recall the Lorentz properties of Weyl fermion bilinears. The products of two left- or right-handed Weyl fermions transform as a linear combination of scalar and tensor, and the products of left and right fermions transforms as a vector and/or axial vector

$$f\ f \sim \bar{f}\ \bar{f} \sim (\mathbf{1},\mathbf{1}) \oplus (\mathbf{3},\mathbf{1})\ ; \qquad f\ \bar{f} \sim (\mathbf{2},\mathbf{2})\ . \tag{6.61}$$

• Consider terms of the form $ffHH$. Lorentz invariance requires the ff combination to be a Lorentz scalar, with zero color triality; thus both f must be leptons of the same chirality. Also, the two Higgs combination must be an isotriplet, restricting each lepton to be isodoublet. The quantum numbers of the antisymmetrized product of two lepton doublets are

$$L_{(i}L_{j)} \sim (\mathbf{1},\mathbf{1};\mathbf{3},\mathbf{1}^c)^{(ij)}_{-2} \oplus (\mathbf{3},\mathbf{1};\mathbf{1},\mathbf{1}^c)^{(ij)}_{-2}\ , \tag{6.62}$$

when symmetrizing over the family indices, and

$$L_{[i}L_{j]} \sim (\mathbf{3},\mathbf{1};\mathbf{3},\mathbf{1}^c)^{[ij]}_{-2} \oplus (\mathbf{1},\mathbf{1};\mathbf{1},\mathbf{1}^c)^{[ij]}_{-2}\ , \tag{6.63}$$

when antisymmetrizing over the family indices. We have a match for the family-symmetric combination, and the dimension-five operator

$$L^T_{(i}\sigma_2\tau_2\vec{\tau}L_{j)} \cdot H^T\tau_2\vec{\tau}H\ , \tag{6.64}$$

is invariant under the gauge groups of the standard model. Unfortunately, it is not invariant under its global symmetries: it violates total and relative lepton numbers by two units. It is not generated in perturbation theory.

Any interaction forbidden only by global symmetries deserves further analysis. Evaluate in the electroweak vacuum, it yields $v^2\widehat{\nu}_{(i}\nu_{j)}$, which we recognize as Majorana mass terms for the neutrinos. This is an example of a dimension-five operator which produces a dimension-three operator in the electroweak vacuum. This analysis shows with hardly any calculation that the standard model neutrinos stay massless to all orders of perturbation theory, not because of its gauge symmetries, but only because of the global lepton number symmetries.

• There are no invariants of the form $ff\overline{H}H$ in the absence of standard model fermion bilinears with zero hypercharge.

• Consider terms of the form $ff\mathcal{D}H$. Simple invariance considerations restrict the fermion bilinear to be a color singlet or octet, a weak doublet, and a Lorentz vector. For quarks, the only possible combinations necessarily have color triality two. For leptons, the only Lorentz vector isodoublet combination has hypercharge ± 3. We conclude that there are no dimension-five invariants of this form.

• The last combination to consider is of the form $ff\mathcal{D}\mathcal{D}$. Each covariant derivative has the following quantum numbers

$$\mathcal{D}_\mu \sim (\mathbf{2}, \mathbf{2}; \mathbf{1} \oplus \mathbf{3}, \mathbf{1}^c \oplus \mathbf{8}^c)_0 \ . \tag{6.65}$$

Since the standard model fermion bilinears with zero hypercharge are Lorentz vectors, no dimension-five invariants of this form can exist.

We conclude that the *unbroken* standard model generates *no invariant dimension-five interactions*. How then did the dimension-five operators of QED come about? They are generated solely from higher dimension operators evaluated in the electroweak vacuum.

This shows that our classification is not complete, and we need to take into account the $d = 5$ combinations with the quantum numbers of Higgs polynomials that take electroweak vacuum values. Coupled with these Higgs polynomials, these produce standard model invariants of higher dimensions that reduce to $d = 5$ interactions in the electroweak vacuum. The dimension-five Higgs polynomial covariants can be several types:

• There are combinations without fermions, of the form $\mathcal{D}^2 H^3$ and $\mathcal{D}^4 H$. We leave it as an exercise to list the Higgs polynomial-covariants of that dimension.

• All combinations of the form $\bar{f}_i f_j \mathcal{D} H$, where $f_i = \mathbf{Q}_i$, $\overline{\mathbf{u}}_i$, $\overline{\mathbf{d}}_i$, L_i, or $\overline{e}_i$, where i, j are the family indices, can have the quantum numbers of the Higgs doublet. This follows because the combinations $\bar{f}_i f_j \mathcal{D}$ always contain electroweak singlets, since they can transform like (singlet) kinetic terms. They yield dimension-six invariants in the unbroken formulation when coupled to the conjugate Higgs. Lepton number conservation requires that $i = j$ for leptons; not so for quarks, leading to flavor-changing decays of the scalar Higgs. These interactions describe chirality-preserving emission and absorption of Higgs scalars from fermions. The flavor-changing processes (unlikely to ever be observed!) such as

$$\overline{\mathbf{s}}^\dagger \sigma_\mu \overline{\mathbf{b}} \mathcal{D}_\mu H \; ; \qquad \overline{\mathbf{d}}^\dagger \sigma_\mu \overline{\mathbf{s}} \mathcal{D}_\mu H \; , \tag{6.66}$$

do not appear at tree level where the Higgs decay is flavor-diagonal. In the electroweak vacuum, they are generated by a diagram of the form,

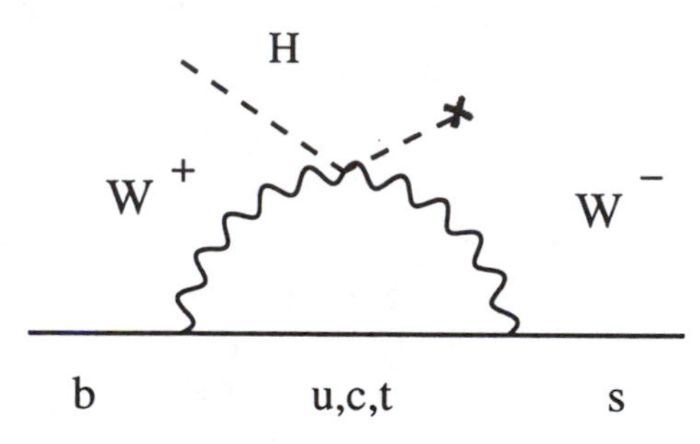

where the cross on the Higgs line means that it is evaluated in the vacuum.

• Terms with one fermion pair and two covariant derivatives, $ff\mathcal{D}\mathcal{D}$, for which Lorentz invariance requires both fermions to be left- or right-handed, yield the operators

$$\mathcal{D}\mathcal{D}\mathbf{Q}_i\overline{\mathbf{u}}_j \; ; \qquad \mathcal{D}\mathcal{D}\mathbf{Q}_i\overline{\mathbf{d}}_j \; ; \qquad \mathcal{D}\mathcal{D}L_i\overline{e}_j \; , \tag{6.67}$$

all with the quantum numbers of the Higgs doublet: they are generated by at least $d = 6$ invariants in the unbroken formulation. Of special interest are the quark magnetic moment interactions

$$\mathbf{Q}_i\sigma_{\mu\nu}\lambda^A\overline{\mathbf{d}}_j\mathbf{G}^A_{\mu\nu} \; ; \qquad \mathbf{Q}_i\sigma_{\mu\nu}\tau^a\overline{\mathbf{d}}_j W^a_{\mu\nu} \; ; \qquad \mathbf{Q}_i\sigma_{\mu\nu}\overline{\mathbf{d}}_j B_{\mu\nu} \; , \tag{6.68}$$

and similar interactions with $\overline{\mathbf{d}}$ replaced by $\overline{\mathbf{u}}$. If $i \neq j$, they generate flavor-changing but charge-preserving interactions. Forbidden at tree-level, these processes provide a direct glimpse into the radiative corrections to the standard model. Evaluated in the electroweak vacuum, they describe rare interactions of the type gluon $\rightarrow \mathbf{s}\overline{\mathbf{d}}$, $Z \rightarrow \mathbf{s}\overline{\mathbf{d}}$, $\mathbf{b} \rightarrow \mathbf{s}\gamma$, or $\gamma \rightarrow \mathbf{s}\overline{\mathbf{d}}$, etc... . These processes occur at the one-loop level through diagrams like

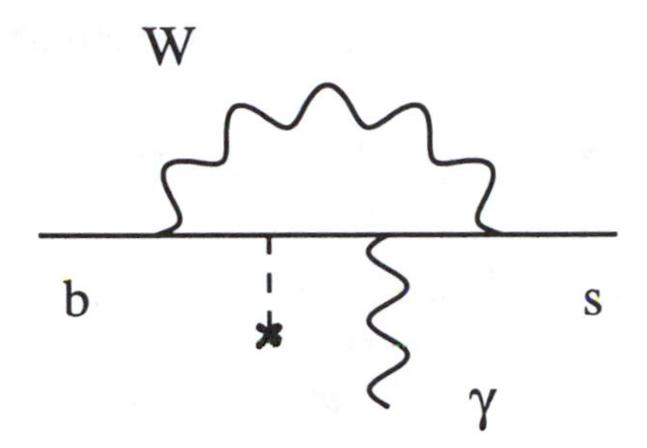

Similar invariants with lepton pairs are allowed only if $i = j$ because of lepton number conservation. There are more subtle covariant combinations involving leptons, but they are forbidden by lepton number conservation. For example,

$$L_i^{\dagger}\sigma_{\mu}\overline{e}_j\mathcal{D}_{\mu}\vec{\tau}\overline{H} \; , \tag{6.69}$$

which transforms as an isovector with $Y = 2$, can be upgraded to a $d = 7$ invariant by adding the isovector Higgs binomial with $Y = -2$. The same remark applies to the weak isovector with $Y = -2$, generically of the form $\mathcal{D}\mathcal{D}L_iL_j$. Neither term is generated in perturbation theory.

Dimension-Six Interactions

They come in many different combinations, $\mathcal{D}^6$, $\mathcal{D}^4H^2$, $\mathcal{D}^2H^4$, H^6, ffH^3, $ff\mathcal{D}H^2$, $ff\mathcal{D}^2H$, $ff\mathcal{D}^3$, and $ffff$, not including operators linear in H, which we have already discussed.

• Invariants of the form $\mathcal{D}^6$ contain either three field strengths or two covariant derivatives of field strengths. In QED, it was not possible to form a symmetric invariant product of three Maxwell field strengths: the symmetric product of two field strengths is a symmetric second rank Lorentz tensor. In the standard model we can escape this restriction by antisymmetrizing on the group indices. This always produces the adjoint representation, leading to invariants of the form

$$\epsilon^{abc}W^a_{\mu\rho}W^b_{\rho\nu}W^c_{\nu\mu} \; ; \qquad f^{ABC}\mathbf{G}^A_{\mu\rho}\mathbf{G}^B_{\rho\nu}\mathbf{G}^C_{\nu\mu} \; , \tag{6.70}$$

involving $SU(2)$ and $SU(3)$ field strengths, respectively. The P and CP violating weak interactions can generate similar interactions with the one field strength replaced by its dual

$$\epsilon^{abc}W^a_{\mu\rho}W^b_{\rho\nu}\widetilde{W}^c_{\nu\mu} \; ; \qquad f^{ABC}\mathbf{G}^A_{\mu\rho}\mathbf{G}^B_{\rho\nu}\widetilde{\mathbf{G}}^C_{\nu\mu} \; , \tag{6.71}$$

There are other operators containing two field strengths

$$\mathcal{D}_\mu W^a_{\nu\rho}\mathcal{D}_\mu W^a_{\nu\rho}\ ; \qquad \partial_\mu B_{\nu\rho}\partial_\mu B_{\nu\rho}\ ; \qquad \mathcal{D}_\mu \mathbf{G}^A_{\nu\rho}\mathcal{D}_\mu \mathbf{G}^A_{\nu\rho}\ , \tag{6.72}$$

which, together with a permutation of the Lorentz indices, provide finite renormalizations to the kinetic terms, and to the higher order vertices. By taking dual field strengths, we generate new interactions

$$\mathcal{D}_\mu W^a_{\nu\rho}\mathcal{D}_\mu \widetilde{W}^a_{\nu\rho}\ ; \qquad \partial_\mu B_{\nu\rho}\partial_\mu \widetilde{B}_{\nu\rho}\ ; \qquad \mathcal{D}_\mu \mathbf{G}^A_{\nu\rho}\mathcal{D}_\mu \widetilde{\mathbf{G}}^A_{\nu\rho}\ . \tag{6.73}$$

• Invariants of the form $\mathcal{D}^4H^2$. Except for operators which are products of invariants, such as $H^\dagger HB_{\mu\nu}B_{\mu\nu}, \ldots$, we have the interesting interaction

$$H^\dagger\tau^a HW^a_{\mu\nu}B_{\mu\nu}\ . \tag{6.74}$$

As before there is a similar term with one dual field strength. Finally there are several types with one field strength and two Higgs derivatives

$$(\mathcal{D}_\mu H)^\dagger\mathcal{D}_\nu HB_{\mu\nu}\ ; \qquad (\mathcal{D}_\mu H)^\dagger\tau^a\mathcal{D}_\nu HW^a_{\mu\nu}\ , \tag{6.75}$$

and

$$(\mathcal{D}_\nu\mathcal{D}_\mu H)^\dagger\mathcal{D}_\nu\mathcal{D}_\mu H\ , \tag{6.76}$$

which describe interactions of gauge and Higgs fields.

• Invariants of the form $\mathcal{D}^2H^4$ are

$$H^\dagger H(\mathcal{D}_\mu H)^\dagger\mathcal{D}_\mu H\ ; \qquad H^\dagger\tau^a H(\mathcal{D}_\mu H)^\dagger\tau^a\mathcal{D}_\mu H\ . \tag{6.77}$$

Other interactions of this type with a different distribution of the weak isospin indices can be obtained from these through $SU(2)$ Fierz transformations.

• Fermion bilinears in invariants of the form $ff\mathcal{D}HH$ are zero triality Lorentz vectors and/or axial vectors. We have already analyzed terms of the form $f_i\overline{f}_j\mathcal{D}H\overline{H}$. An interesting invariant is

$$\overline{\mathbf{d}}^\dagger_i\sigma_\mu\overline{\mathbf{u}}_jH^T\tau_2\mathcal{D}_\mu H\ . \tag{6.78}$$

A similar operator

$$\overline{e}^\dagger_j\sigma_\mu L_iH^T\tau_2\mathcal{D}_\mu H\ , \tag{6.79}$$

with the quantum numbers of the Higgs doublet, will be generated in a term of dimension-seven.

• Invariants of the form $ff\mathcal{D}\mathcal{D}H$ contain only fermions of the same chirality, one being a weak doublet. These are the terms we encountered in constructing dimension-five invariants. They are

$$\mathbf{Q}_i\overline{\mathbf{u}}_j\mathcal{D}\mathcal{D}H \ ; \qquad \mathbf{Q}_i\overline{\mathbf{d}}_j\mathcal{D}\mathcal{D}\overline{H} \ ; \qquad L_i\overline{e}_j\mathcal{D}\mathcal{D}\overline{H} \ . \tag{6.80}$$

They look like Yukawa terms but can also couple to field strengths. They are especially relevant in the quark sector when those with different family indices break the tree-level flavor symmetry of the gauge interactions.

• Invariants of the form $ff\mathcal{D}\mathcal{D}\mathcal{D}$ exist when the fermion bilinear transforms as Lorentz vector and/or axial vector. They must also have zero triality and no hypercharge. The bilinears must therefore be of the form $\overline{f}_i f_j$. The quark family index can be different, leading to flavor-changing charge-preserving interactions. Some interesting examples where two covariant derivatives form into one field strength are

$$\mathbf{Q}_i^\dagger\sigma_\mu\mathbf{Q}_j\mathcal{D}_\rho B_{\rho\mu} \ ; \ \mathbf{Q}_i^\dagger\sigma_\mu\tau^a\mathbf{Q}_j(\mathcal{D}_\rho\cdot W_{\rho\mu})^a \ ; \ \mathbf{Q}_i^\dagger\sigma_\mu\lambda^A\mathbf{Q}_j(\mathcal{D}_\rho\cdot\mathbf{G}_{\rho\mu})^A \ , \tag{6.81}$$

which involve only the quark doublets. We also have

$$\overline{\mathbf{d}}_i^\dagger\sigma_\mu\overline{\mathbf{d}}_j\mathcal{D}_\rho B_{\rho\mu} \ ; \qquad \overline{\mathbf{d}}_i\sigma_\mu\lambda^A\overline{\mathbf{d}}_j(\mathcal{D}_\rho\cdot\mathbf{G}_{\rho\mu})^A \ , \tag{6.82}$$

and others with $\overline{\mathbf{d}}$ replaced by $\overline{\mathbf{u}}$. At the one-loop level, these are generated through chirality-preserving diagrams like

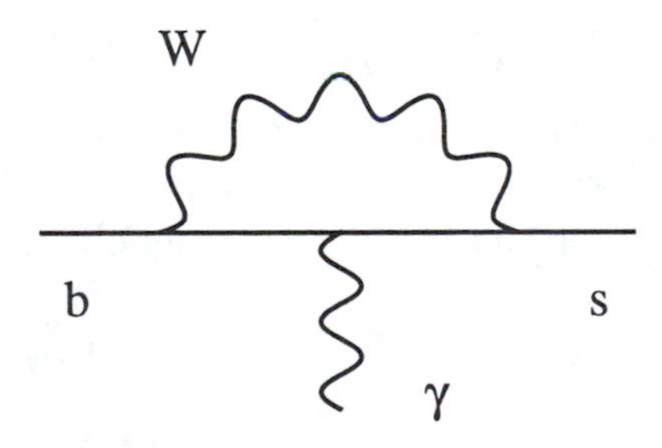

They are called *penguin* diagrams. With different quark flavors, they describe flavor-changing emission of gluons, photons, and decays of the Z boson. Note that we have already encountered flavor-changing interactions

of this type, but they were chirality-changing of the magnetic moment variety, induced by the Higgs vacuum value.

• As expected, gauge invariants of the form $ffff$ have a much richer structure than in QED. Some, which do not violate the tree-level partial symmetries provide finite corrections to tree-level processes. Others, which do not respect the same symmetries, lead to new processes, and allow direct measurements of radiative corrections. Fortunately, many of these operators violate lepton and baryon numbers, and do not appear in the effective Lagrangian.

We can build Lorentz-invariants of this type in two ways. One is $(f\ f)\cdot(f\ f)$, and its conjugate, with bilinears forming scalar and tensor combinations. The other is of the form $(f\ f)\cdot(\overline{f}\ \overline{f})$, with each fermion bilinear in a scalar combination. Other possible invariants, for instance with each bilinear transforming as vector and axial vectors, is Fierz-equivalent to the above. It follows that all invariants can be assembled by first forming the product of any two of the (left-handed) fermion fields of the standard model, $\mathbf{Q}_i$, $\overline{\mathbf{u}}_i$, $\overline{\mathbf{d}}_i$, L_i, or $\overline{e}_i$, and then by contracting them with either themselves or their conjugates. This construction is simplified by assembling the fermion pairs in terms of their triality, hypercharge, baryon and lepton numbers.

- There are 15 different types of fermion bilinears. Multiplied with their conjugates, they yield fifteen types of four-fermion interactions. Some examples are

$$(\mathbf{Q}_i\mathbf{Q}_j)\cdot(\mathbf{Q}_k^\dagger\mathbf{Q}_l^\dagger)\ , \qquad (\overline{\mathbf{u}}_i\overline{\mathbf{u}}_j)\cdot(\overline{\mathbf{u}}_k^\dagger\overline{\mathbf{u}}_l^\dagger)\ \ldots . \tag{6.83}$$

- There are only three types of bilinears with $\Delta B = \Delta L = 0$

$$\mathbf{Q}_i\overline{\mathbf{u}}_j\ , \qquad \mathbf{Q}_i\overline{\mathbf{d}}_j\ , \qquad L_i\overline{e}_i\ . \tag{6.84}$$

Multiplied together, they generate four-fermion interactions that preserve lepton and baryon numbers, but can cause flavor-changing interactions. They enter the standard model effective Lagrangian as

$$\mathbf{Q}_i\mathbf{Q}_j\overline{\mathbf{d}}_k\overline{\mathbf{u}}_l\ ; \qquad \mathbf{Q}_i\overline{\mathbf{u}}_jL_k\overline{e}_l\ ; \qquad \mathbf{Q}_i\overline{\mathbf{d}}_j\overline{L}_ke_l\ .$$

They are generated in one-loop order through box diagrams like

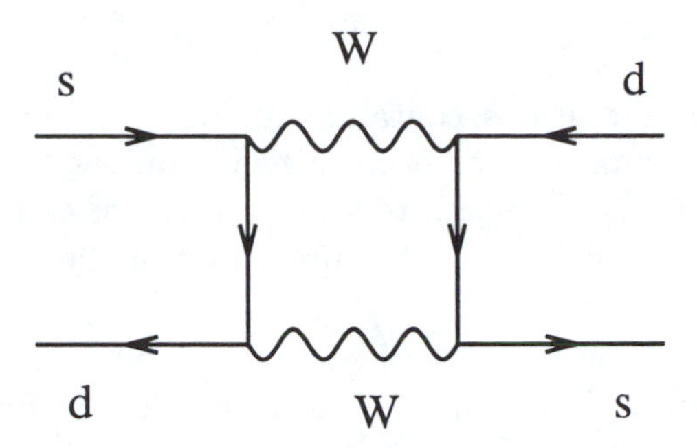

which can violate flavor. This diagram cause charge-preserving interactions that violate strangeness by two units.

Although not generated in the standard model, invariants that violate B and L, but preserve $B - L$ are of interest in many of its extensions. Some of these are, $\mathbf{Q}_i\mathbf{Q}_j\mathbf{Q}_k L_l$, $\mathbf{Q}_i\mathbf{Q}_j\overline{\mathbf{u}}_k^\dagger\overline{e}_l^\dagger$, $\mathbf{Q}_i L_j\overline{\mathbf{u}}_k^\dagger\overline{\mathbf{d}}_l^\dagger$, and $\mathbf{u}_i\mathbf{u}_j\overline{\mathbf{d}}_k^\dagger\overline{e}_l^\dagger$.

Finally, we must also consider four-fermion covariants that contain Higgs polynomials with electroweak vacuum values. To that effect, we list the $Y = 1, 2, 3$ combinations that yield invariants when multiplied by the appropriate Higgs polynomial.

- We do not show the $Y = 1$ combinations since all can be shown to violate either lepton and/or baryon number.
- There are six $Y = \pm 2$ combinations which do not violate any global symmetries:

$$\begin{gathered}\mathbf{Q}_i\overline{\mathbf{u}}_j L_k^\dagger\overline{e}_l^\dagger\ , \qquad L_i L_j\overline{e}_k\overline{e}_l\ , \qquad \mathbf{Q}_i\overline{\mathbf{d}}_j L_k\overline{e}_l\ , \\ \overline{\mathbf{u}}_i\overline{\mathbf{u}}_j\overline{\mathbf{u}}_k^\dagger\overline{\mathbf{d}}_l^\dagger\ , \quad \overline{\mathbf{d}}_i\overline{\mathbf{d}}_j\overline{\mathbf{u}}_k^\dagger\overline{\mathbf{d}}_l^\dagger\ , \quad \mathbf{Q}_i\mathbf{Q}_j\overline{\mathbf{u}}_k\overline{\mathbf{u}}_l\ , \quad \mathbf{Q}_i\mathbf{Q}_j\overline{\mathbf{d}}_k\overline{\mathbf{d}}_l\ .\end{gathered} \tag{6.85}$$

There is in addition one combination with $\Delta B = \Delta L = 1$, of the form $\mathbf{Q}_i\mathbf{Q}_j\overline{\mathbf{d}}_k\overline{\mathbf{d}}_l$. Finally, all the $Y = 3$ four-fermion combinations violate the global symmetries.

The effective Lagrangian contains interactions of arbitrarily high dimensions. Fortunately, we need not list invariants of higher dimensions, as those already listed are sufficient for a thorough discussion of the lowest order radiative structure of the standard model.

PROBLEMS

A. Find the representations contained in the product of two antisymmetric second rank tensors, each transforming as $(\mathbf{1},\mathbf{3})\oplus(\mathbf{3},\mathbf{1})$ of the Lorentz group. Use the result to verify the list of all possible invariants built out of three field strengths presented in the text.

B. Show that with one Higgs doublet, there is only one new cubic Higgs polynomial with $Y = \pm 3$, and no new quartic polynomial. Deduce that all higher order Higgs polynomials that do not vanish in the vacuum are made up of the combinations already listed.

C. Evaluate the operators with two Higgs and one field strength in the electroweak vacuum, and interpret the results.

D. Find the quantum numbers of the symmetric and antisymmetric product of two covariant derivatives in the standard model. Interpret the results physically.

E. Evaluate the mixed interaction (6.74) in the electroweak vacuum. Show that it apparently mixes the photon and the Z boson. What do you conclude?

F. Show that the list of operators with four Higgs and two covariant derivatives is complete, up to integrations by parts and Fierz transformations.

CHAPTER 7

STANDARD MODEL

ONE-LOOP PROCESSES

In the previous chapter we analyzed the rich radiative structure of the standard model, using group-theoretical methods, together with the identification of explicit one-loop Feynman diagrams. Quantum corrections were split in two types; those which renormalize the input parameters of the theory and correct the interactions already present at tree level, and those which produce new interactions. No calculational details were given. This chapter presents explicit computations for some of these effects.

7.1 GENERAL STRUCTURE

The organization of the radiative corrections in the standard model for calculations is complicated by the spontaneous breakdown of the electroweak symmetry. We follow D. Kennedy and B. Lynn (*Nucl. Phys.* **B322**, 1(1989)) who use the Lorentz and gauge symmetries of the Green's functions to separate the radiative corrections in two classes, *oblique* (universal) corrections which alter the gauge boson propagators, and other *straight* (direct) corrections which modify the vertices and the box diagrams. Oblique corrections do not depend on the external fermion legs are not process-specific, while the straight corrections depend on the external states. Although in non-Abelian gauge theories, this separation of the corrections is not quite gauge-invariant, gauge invariance can be restored order by order without affecting the structure of the equations presented below.

Structure of the Oblique Corrections

The radiative corrections to the gauge propagators are universal corrections which appear in all processes that involve the exchange of virtual gauge bosons, irrespective of the external particles. They are to be distinguished from the rest of the quantum corrections, which are in general process-specific.

As a simple first step, consider how such a separation comes about in pure QED where the one-photon exchange amplitude is of the form

$$\mathcal{M}_{QED} = \frac{e^2}{q^2} . \tag{7.1}$$

We want to examine the propagator corrections to this amplitude. The photon two-point function obeys the Schwinger-Dyson equation

$$G_{\gamma\gamma} = \frac{1}{q^2} + \frac{1}{q^2} D_{\gamma\gamma} \Pi_{\gamma\gamma} G_{\gamma\gamma} , \tag{7.2}$$

where $\Pi_{\gamma\gamma}$ is the one-loop vacuum polarization, which is purely transverse

$$\Pi_{\mu\nu}(q^2) = (g_{\mu\nu} - \frac{q_\mu q_\nu}{q^2}) \Pi_{\gamma\gamma}(q^2) . \tag{7.3}$$

The $q^\mu q^\nu$ tensor, when contracted on the external currents, yield terms proportional to the masses of the external fermions, which we neglect. In pictures

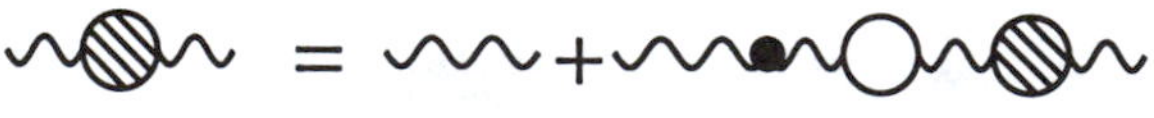

[diagram: dressed propagator = bare propagator + bare propagator, vertex, loop, dressed propagator]

Its solution, which includes the propagator corrections to all orders, has the simple form

$$G_{\gamma\gamma} = \frac{1}{q^2 - \Pi_{\gamma\gamma}(q^2)} . \tag{7.4}$$

In QED, an additional simplification comes from the fact the photon two-point function is proportional to q^2, as a result of the Ward identities which insures that the photon stays massless to all orders of perturbation theory. Thus we can set

$$\Pi_{\gamma\gamma}(q^2) = q^2 e^2 \widetilde{\Pi}_{\gamma\gamma}(q^2) \ . \tag{7.5}$$

This enables us to rewrite the photon exchange amplitude in the suggestive form

$$\mathcal{M}_{QED} = \frac{e_*^2(q^2)}{q^2} \ , \tag{7.6}$$

where the effective coupling constant is

$$\frac{1}{e_*^2(q^2)} = \frac{1}{e^2} - \widetilde{\Pi}_{\gamma\gamma}(q^2) \ . \tag{7.7}$$

We can express the starred coupling as a function of the fine structure constant measured in Thomson scattering $\alpha \equiv e_*^2(0)/4\pi$, and of ultraviolet-finite quantities

$$e_*^2(q^2) = \frac{4\pi\alpha}{1 - 4\pi\alpha[\widetilde{\Pi}_{\gamma\gamma}(q^2) - \widetilde{\Pi}_{\gamma\gamma}(0)]} \ . \tag{7.8}$$

The effect of the propagator corrections has been to replace the gauge coupling constant by its scale-dependent starred counterpart. Of course, a complete accounting of the radiative corrections, requires the other quantum corrections to be added.

We can now apply the same analysis to the standard model, where spontaneous breaking of the electroweak symmetry (softly) invalidate the Slavnov-Taylor identities, allowing for terms proportional to the mass squared of particles circulating in the loops. This is easiest to see in the propagator for the charged W boson. Its vacuum polarization tensor can be written as

$$\Pi^{\pm}_{\mu\nu}(q^2) = (g_{\mu\nu} - \frac{q_\mu q_\nu}{q^2})\Pi^T_{WW}(q^2) + \frac{q_\mu q_\nu}{q^2}\Pi^L_{WW}(q^2) \ . \tag{7.9}$$

The longitudinal term reflects the loss of conservation of the isospin current. As long as one considers processes with all external particles much lighter than the W, it can be reliably neglected in one-loop diagrams. Coupling to the divergence of the isospin current, it generates contributions proportional to the mass of the particles in the currents. In one-loop diagrams, these are external particles, yielding additional suppression by small factors of the order of $(m_f/m_W)^2$. Thus, in discussing one-loop corrections to processes that involve external particles much lighter than the gauge bosons, we can consistently disregard the longitudinal contribution to the W and Z propagators, and keep only the vacuum polarization functions that multiply $g_{\mu\nu}$.

We start again with the (generalized) recursive Schwinger-Dyson equations of the form

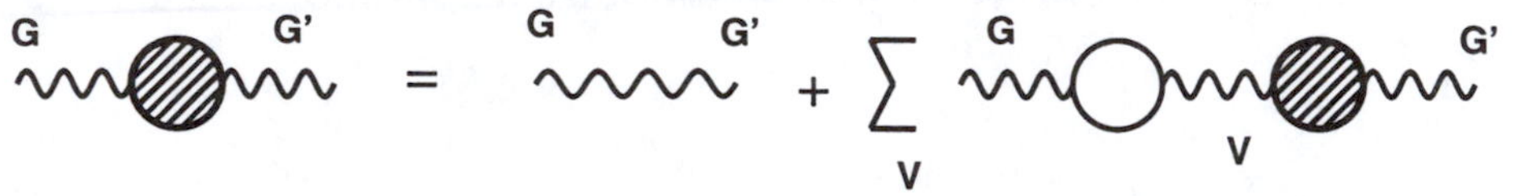

where the bare propagator is present only if G and G' are the same, and the sum in the internal leg is over all the relevant gauge *and* Nambu-Goldstone propagators. Calculations are most easily performed in non-unitary gauges, such as the R_ξ gauge, most often with $\xi = 1$, where the Nambu-Goldstone bosons appear explicitly. As the longitudinal modes of the gauge bosons, the Nambu-Goldstone bosons must be included in this sum.

The structure of the Schwinger-Dyson equation for the charged W boson is rather simple,

$$G_{WW} = D_{WW} + D_{WW}\Pi_{WW}G_{WW} \ , \tag{7.10}$$

where

$$D_{WW} = \frac{1}{q^2 - m_W^2} \ , \tag{7.11}$$

is the bare propagator, and Π_{WW} is the vacuum polarization. Its solution

$$G_{WW} = \frac{1}{q^2 - m_W^2 - \Pi_{WW}} \ , \tag{7.12}$$

includes the vacuum polarization corrections to all orders, where m_W is the W-boson tree-level mass ($s \equiv \sin\theta_{\rm w}$)

$$m_W^2 = \frac{1}{4}g_2^2 v^2 = \frac{e^2}{4s^2}v^2 \ . \tag{7.13}$$

It follows that, neglecting vertex corrections, the matrix elements for processes involving $W^\pm$ exchange can be written in the form

$$\mathcal{M}_{CC} = \frac{e^2}{2s^2}\frac{I_+ I_-}{q^2 - m_W^2 - \Pi_{WW}(q^2)} \ , \tag{7.14}$$

where $I_\pm$ are the isospin raising and lowering charges. As a first step, we can rewrite this matrix element in the form

$$\mathcal{M}_{CC} = \frac{e^2}{2s^2} I_+ I_- \frac{\mathcal{Z}_{[W]}}{q^2 - M^2_{W*}(q^2)} \, . \tag{7.15}$$

The physical mass of the W boson, M_W, is determined from the *running mass* $M_{W*}(q^2)$ at the pole through

$$M^2_{W*}(M^2_W) = M^2_W \, , \qquad \frac{d}{dq^2} M^2_{W*}(M^2_W)|_{q^2=M^2_W} = 0 \, . \tag{7.16}$$

Note that this definition does not exactly correspond to the physical mass (see problem). The physical width of the W boson is identified with the imaginary part of the vacuum polarization at the pole

$$\sqrt{q^2}\Gamma_{W*} \equiv \mathcal{I}m\Pi_{WW}(M^2_W) \, . \tag{7.17}$$

The wavefunction renormalization is the residue at the pole, given by

$$\mathcal{Z}^{-1}_{[W]} = 1 - \frac{d}{dq^2}\Pi_{WW}(q^2)|_{q^2=M^2_W} \, . \tag{7.18}$$

We define q^2-dependent starred parameters for the couplings, by setting

$$\frac{e^2}{s^2}\mathcal{Z}_{[W]} \equiv \frac{e^2_*(q^2)}{s^2_*(q^2)}\mathcal{Z}_{[W]*}(q^2) \, , \tag{7.19}$$

which enables us to write the charged current amplitude in its tree-level form

$$\mathcal{M}_{CC} = \frac{e^2_*}{2s^2_*} I_+ I_- \frac{\mathcal{Z}_{[W]*}}{q^2 - M^2_{W*}(q^2)} \, . \tag{7.20}$$

We can identify the Fermi coupling by considering (7.15) near $q^2 = 0$, yielding

$$4\sqrt{2}G_F = \left(\frac{v^2}{4} + \Pi_{\pm}(0)\right)^{-1} , \tag{7.21}$$

where we have defined

$$\Pi_{WW}(q^2) = \frac{e^2}{s^2}\Pi_{\pm}(q^2) \, . \tag{7.22}$$

It contains an ultraviolet divergence, to be absorbed by the definition of the input parameters. We can define a q^2-dependent Fermi "constant" through the equation

$$\frac{\mathcal{Z}_{[W]*}}{q^2 - M^2_{W*}(q^2)} = \left(q^2 - \frac{e^2_*(q^2)}{s^2_*(q^2)} \frac{1}{4\sqrt{2}G_{F*}(q^2)}\right)^{-1} , \tag{7.23}$$

but he determination of the individual functions G_{F*}, e^2_* and s^2_*, requires us to consider the propagator corrections to the Z boson and the photon.

In the last chapter we have seen that radiative corrections generate mixing between hypercharge B and W^3, the third component of the weak gauge boson. This complicates the Schwinger-Dyson equations. Using the tree-level structure, we define the functions

$$\Pi_{AA} = e^2 \Pi_{QQ} , \tag{7.24}$$

$$\Pi_{ZA} = \frac{e^2}{sc}(\Pi_{3Q} - s^2 \Pi_{QQ}) , \tag{7.25}$$

$$\Pi_{ZZ} = \frac{e^2}{s^2c^2}(\Pi_{33} - 2s^2\Pi_{3Q} + s^4 \Pi_{QQ}) , \tag{7.26}$$

in terms of the current correlation functions Π_{QQ}, Π_{3Q}, Π_{33}, where the subscripts $(Q, 3)$ denote the electromagnetic current and the third component of weak isospin, respectively. At one-loop, these functions do not depend on the couplings. The coupled Schwinger-Dyson equations are now

$$\begin{aligned} G_{AA} &= D_{AA} + D_{AA}\Pi_{AA}G_{AA} , \\ G_{AZ} &= D_{ZZ}\Pi_{ZA}G_{AA} , \\ G_{ZZ} &= D_{ZZ} + D_{ZZ}\Pi_{ZZ}G_{ZZ} , \end{aligned} \tag{7.27}$$

where D_{AA}, D_{ZZ} are the bare propagators. The amplitude due to neutral current exchange is given by

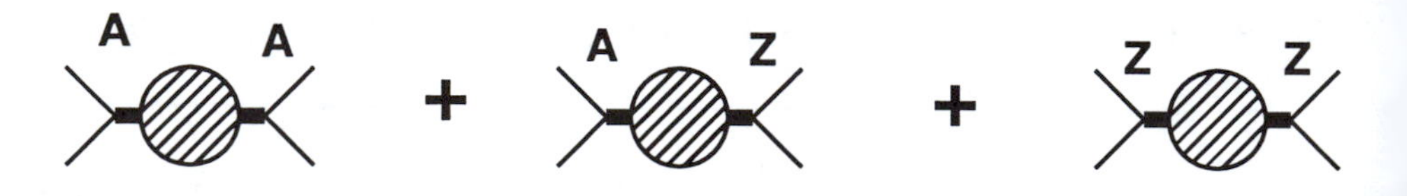

where the solid intermediate line denotes the coupling of the external current to the full propagator

$$\begin{aligned} \mathcal{M}_{NC} =& e^2 QQ' G_{AA} \\ &+ \frac{e^2}{sc}\Big[Q(I'_3 - s^2 Q') + (I_3 - s^2 Q)Q'\Big] G_{ZA} \\ &+ \frac{e^2}{s^2c^2}(I_3 - s^2 Q)(I'_3 - s^2 Q') G_{ZZ} . \end{aligned} \tag{7.28}$$

The solutions of the Schwinger-Dyson equations can be shown to be

$$
\begin{aligned}
G_{AA} &= \frac{1}{q^2 - \Pi_{AA}} + \frac{\Sigma^2}{q^2 - m_Z^2 - \Pi_{ZZ} - \Pi_{ZA}\Sigma} , \\
G_{ZA} &= \frac{\Sigma}{q^2 - m_Z^2 - \Pi_{ZZ} - \Pi_{ZA}\Sigma} , \\
G_{ZZ} &= \frac{1}{q^2 - m_Z^2 - \Pi_{ZZ} - \Pi_{ZA}\Sigma} ,
\end{aligned}
\tag{7.29}
$$

where

$$
m_Z^2 = \frac{v^2}{4}\frac{e^2}{c^2 s^2} , \qquad \Sigma \equiv \frac{\Pi_{ZA}}{q^2 - \Pi_{AA}} . \tag{7.30}
$$

The neutral current amplitude is dominated by two poles, corresponding to the exchange of the photon and the Z boson. Substitution of these solutions in the neutral current matrix element yields

$$
\begin{aligned}
\mathcal{M}_{NC} &= e^2 \frac{QQ'}{q^2 - \Pi_{AA}} \\
&+ \frac{e^2}{s^2 c^2} \frac{\{I_3 - [s^2 - sc\Sigma]Q\}\{I_3' - [s^2 - sc\Sigma]Q'\}}{q^2 - m_Z^2 - \Pi_{ZZ} - \Sigma\Pi_{ZA}} .
\end{aligned}
\tag{7.31}
$$

The QED Ward identities are still operative since $U(1)_\gamma$ is unbroken, allowing us to set

$$
\Pi_{AA}(q^2) = e^2 q^2 \widetilde{\Pi}_{QQ}(q^2) , \tag{7.32}
$$

from which

$$
e_*^2(q^2) \equiv \frac{e^2}{1 - e^2 \widetilde{\Pi}_{QQ}(q^2)} , \tag{7.33}
$$

We obtain an equation of the same form as in QED, although the vacuum polarization is different because of W-boson loops,

$$
e_*^2(q^2) \equiv \frac{e_*^2(0)}{1 - e_*^2(0)[\widetilde{\Pi}_{QQ}(q^2) - \widetilde{\Pi}_{QQ}(0)]} , \tag{7.34}
$$

taking the real part of the two-point functions. All ultraviolet infinities and renormalization prescriptions have cancelled out from this equation, leaving us with concrete testable predictions. It gives the scale dependence of the fine structure constant in terms of its boundary value at the Thomson limit.

A similar construction obtains for the Z boson contribution, but with some subtelties. It follows from Eq. (7.30) that Σ has a pole at $q^2 = 0$, since now

$$\Sigma(q^2) = \frac{1}{q^2}\frac{\Pi_{ZA}}{1 - e^2\widetilde{\Pi}_{QQ}(q^2)} . \tag{7.35}$$

unless Π_{ZA} is itself proportional to q^2. This is the naive expectation from current conservation. However, calculations show a ξ-dependent part to Π_{ZA} that does not vanish at $q^2 = 0$, indicating that the separation of the corrections is not gauge-invariant. Fortunately, the gauge-independence of the construction is restored by a careful consideration of the vertex and box corrections. These contain process-independent but gauge-dependent parts which, when added to the vacuum polarizations, fully justify the separation of the oblique corrections, at the price of a slight redefinitions of the Π functions, to be described later. Thus reassured, we set for a properly defined Π_{ZA},

$$\Pi_{ZA} = q^2\widetilde{\Pi}_{ZA} , \qquad \Pi_{3Q} = q^2\widetilde{\Pi}_{3Q} . \tag{7.36}$$

Comparison with the tree-level current warrants the further definition

$$s_*^2(q^2) \equiv s^2 - sc\Sigma(q^2) , \tag{7.37}$$

so that

$$g_{2*}^{-2}(q^2) \equiv \frac{s_*^2(q^2)}{e_*^2(q^2)} = \frac{s^2}{e^2} - \widetilde{\Pi}_{3Q}(q^2) . \tag{7.38}$$

We can use this equation to express the weak gauge coupling at any scale in terms of its value at the Z-mass

$$g_{2*}^2(q^2) = \frac{g_{2*}^2(m_Z^2)}{1 - g_{2*}^2(m_Z^2)[\widetilde{\Pi}_{3Q}(q^2) - \widetilde{\Pi}_{3Q}(m_Z^2)]} . \tag{7.39}$$

This equation contains no divergences and is prescription-independent. We can now extract the starred Fermi "constant"

$$\frac{1}{4\sqrt{2}G_{F*}(q^2)} = \frac{v^2}{4} + [\Pi_{\pm}(q^2) - \Pi_{3Q}(q^2)] . \tag{7.40}$$

In terms of *the* real Fermi constant we obtain

$$\frac{1}{4\sqrt{2}G_{F*}(q^2)} = \frac{1}{4\sqrt{2}G_F} - \Delta_1(q^2) , \tag{7.41}$$

introducing the finite quantity (assuming the naive Slavnov-Taylor identity $\Pi_{3Q}(0) = 0$)

$$\Delta_1(q^2) \equiv \Pi_{3Q}(q^2) - \Pi_\pm(q^2) + \Pi_\pm(0) . \tag{7.42}$$

Finally the wavefunction renormalization and mass of the Z are identified through the function

$$M^2_{Z*}(q^2) \equiv m^2_Z + \Pi_{ZZ}(q^2) + \Pi_{ZA}(q^2)\Sigma(q^2) , \tag{7.43}$$

so that the mass of the Z is at the pole

$$M^2_{Z*}(M^2_Z) = M^2_Z , \qquad \frac{d}{dq^2}M^2_{Z*}(q^2)|_{q^2=M^2_Z} = 0 . \tag{7.44}$$

In terms of the wavefunction renormalization

$$\mathcal{Z}^{-1}_{[Z]} \equiv 1 - \frac{d}{dq^2}[\Pi_{ZZ}(q^2) + \Pi_{ZA}(q^2)\Sigma(q^2)]|_{q^2=M^2_Z} , \tag{7.45}$$

we define the starred wavefunction renormalization through

$$\frac{e^2}{c^2 s^2}\mathcal{Z}_{[Z]} \equiv \frac{e^2_*(q^2)}{c^2_*(q^2)s^2_*(q^2)}\mathcal{Z}_{[Z]*}(q^2) , \tag{7.46}$$

where we have defined

$$c^2_*(q^2) \equiv 1 - s^2_*(q^2) . \tag{7.47}$$

The result is the expression for the neutral current amplitude

$$\begin{aligned}\mathcal{M}_{NC} =& e^2_*(q^2)\frac{QQ'}{q^2} \\ &+ \frac{e^2_*(q^2)}{s_*(q^2)c_*(q^2)}(I'_3 - s^2_*(q^2)Q')\frac{\mathcal{Z}_{[Z]*}(q^2)}{q^2 - M^2_{Z*}(q^2)}(I_3 - s^2_*(q^2)Q) .\end{aligned} \tag{7.48}$$

These redefinitions have enabled us to rewrite the charged and neutral current amplitudes in their tree-level form, by introducing q^2-dependent couplings, masses, and residues. The starred parameters enter all processes that involve the exchange of the gauge bosons in exactly the same way. For comparison with a real process, other (vertex and box) corrections need to be taken into account, but those are in general process-dependent.

From (7.31), we can express the strength of the neutral current interaction at $q^2 = 0$ in the form $G_F \rho_*(0)/\sqrt{2}$, with G_F identified in (7.21). A simple calculation yields

$$\rho_*(0) = \frac{\frac{v^2}{4} + \Pi_{33}(0)}{\frac{v^2}{4} + \Pi_\pm(0)} = 1 - 4\sqrt{2} G_F [\Pi_\pm(0) - \Pi_{33}(0)] \ , \tag{7.49}$$

since Π_{ZA}, Π_{3Q} and Π_{QQ} all vanish at $q^2 = 0$. This combination is not divergent. It gives the relative strength of the charged to neutral current processes, corrected from its tree-level value. We introduce a q^2-dependent ρ parameter in a similar fashion, that is, through

$$\frac{\mathcal{Z}_{[Z]*}(q^2)}{q^2 - M^2_{Z*}(q^2)} = \left(q^2 - \frac{e^2_*}{s^2_* c^2_*} \frac{1}{4\sqrt{2} G_{F*}(q^2) \rho_*(q^2)} \right)^{-1} , \tag{7.50}$$

which leads to

$$\frac{1}{\rho_*(q^2)} = 1 - 4\sqrt{2} G_{F*}(q^2) \Delta_\rho(q^2) \ , \tag{7.51}$$

where

$$\Delta_\rho(q^2) \equiv \Pi_\pm(q^2) - \Pi_{33}(q^2) \ , \tag{7.52}$$

is the second finite quantity which appears in our discussion of the oblique corrections. Like $g - 2$, the deviation of the ρ parameter from one is a calculable prediction of the standard model. Thus Δ_ρ is indicative of custodial symmetry breaking.

One can also define a derived quantity

$$\Delta_3(q^2) \equiv \Delta_1(q^2) - \Delta_\rho(0) + \Delta_\rho(q^2), \tag{7.53}$$

which appears in the running of the Z mass. The next step is to find the most convenient way to relate these corrections to measurements. In the gauge boson sector, the standard model contains three bare parameters, g_1, g_2, the hypercharge and weak isospin gauge couplings, and v, the Higgs vacuum expectation value, and three measurements are needed to specify their values. Once this is done, all oblique contributions to the starred parameters at arbitrary scales are fully determined, in a way that involves neither ultraviolet infinities nor the prescription for the finite parts of the counterterms: the starred functions are independent of the renormalization prescription (up to truncation errors) since they are directly related to the bare quantities.

The most accurately-known measured parameters are the value of the fine structure constant extracted from Thomson scattering $4\pi/e_*^2(0) \equiv \alpha^{-1} = 137.0359895(61)$, the Fermi coupling constant $G_\mu = 1.16637(2) \times 10^{-5}\ (\text{GeV})^{-2}$, determined from muon decay, and the mass of the Z boson, $M_Z = 91.187 \pm 0.007$ GeV. However these may not always be the most convenient to use, as many quantities are presently measured at the Z mass, and above. This has resulted in the appearance of many choices for the three experimental inputs.

- A popular scheme (B. Lynn, M. Peskin and R. Stuart in *Physics at LEP*, J. Ellis, R. Peccei eds. CERN 86-02 (1986)I, 90) uses as input parameters these three experimental quantities. Thus M_W and the Z and W widths are calculable.
- Another, the *on-shell* scheme, introduced by W. Marciano and A. Sirlin (*Phys. Rev.* **D22**, 2695(1980)), uses as input parameters, the fine structure constant, and the W and Z masses. In this scheme, the Weinberg angle is a derived quantity, fixed by $\sin_W^2 \equiv 1 - m_W^2/m_Z^2$.
- A third is particularly suited to the extrapolation of the standard model to the deep ultraviolet. Introduced by A. Sirlin (*Phys. Lett.* **232B**, 123(1989)), it is called the $\overline{MS}$ scheme, uses as inputs the Fermi constant and the values of the electromagnetic and weak gauge couplings at some scale. These couplings are related to their starred equivalent through a specific renormalization prescription.

Calculation of the Oblique Corrections

The oblique corrections are generated by the two-point functions of the gauge bosons, which are ultraviolet divergent and prescription-dependent although they show up in the starred parameters in ultraviolet-finite combinations.

The calculations of oblique corrections are simplest for the charged W boson. To obtain the W-boson parameters, we need to calculate its vacuum polarization, which is no longer proportional to q^2, because of spontaneous breakdown:

$$\Pi_\pm^T(q^2) = \Pi_\pm^{(0)}(q^2) + q^2\Pi_\pm^{(1)}(q^2)\ . \tag{7.54}$$

The calculations are performed in Euclidean space, using dimensional regularization, where we encounter the ubiquitous divergent one-loop integrals of G. Passarino and M. Veltman, (*Nucl. Phys.* **B160**, 151(1979))

$$\{B_0, B_\mu, B_{\mu\nu}\}(q^2, m_1^2, m_2^2) = \int \frac{d^{2\omega}l}{(2\pi)^{2\omega}}\, \frac{\{1, l_\mu, l_\mu l_\nu\}}{[l^2 + m_1^2][(l+q)^2 + m_2^2]}\ , \tag{7.55}$$

where ω is half the dimension of space-time. Using Lorentz covariance, we set

$$B_\mu = q_\mu B_1 \ , \quad B_{\mu\nu} = q_\mu q_\nu B_2 + \delta_{\mu\nu} B_3 \ . \tag{7.56}$$

After integration over the momenta and introduction of the Feynman parameters, these integrals reduce to

$$B_n(q^2, m_1^2, m_2^2) = c_n \Delta + \int_0^1 dx\ x^n \ln[x(1-x)q^2 + m_1^2(1-x) + x m_2^2] \ , \tag{7.57}$$

where the ultraviolet divergence is contained in

$$\Delta = \frac{1}{2-\omega} - \gamma - \ln\pi \ , \qquad (c_0, c_1, c_2) = (1, -\frac{1}{2}, \frac{1}{3}) \ . \tag{7.58}$$

The quadratically divergent B_3 does not appear in our calculations. In terms of the combinations

$$B \equiv B_0 + 2B_1 \ , \quad C \equiv 2B_0 - 8(B_2 + B_1) \ , \ D \equiv B_2 + B_1 + \frac{1}{4} B_0 \ , \tag{7.59}$$

we find (in the $\xi = 1$ gauge)

$$\Pi_\pm^{(1)} = \frac{1}{16\pi^2} \Big\{ \frac{2}{3} + s^2 C(W,0) + c^2 C(W,Z) - D(W,Z) - D(W,H) \Big\} \ , \tag{7.60}$$

where the arguments refer to the masses ($C(W,Z) \equiv C(M_W^2, M_Z^2)$, etc...), and the contribution that breaks the Slavnov-Taylor identities as a result of the spontaneous breaking of the symmetry

$$\begin{aligned}\Pi_\pm^{(0)} = \frac{-1}{16\pi^2} \Big\{ & (m_Z^2 - 3m_W^2) B_0(W,Z) + (m_W^2 - m_Z^2)(2c^2 + \frac{1}{4}) B(W,Z) \\ & + \frac{(m_W^2 - m_H^2)}{4} B(W,H) + m_W^2 B_0(W,H) + 2s^2 m_W^2 B(W,0) \Big\} \ .\end{aligned} \tag{7.61}$$

We must add to these functions the fermion contributions

$$\Pi_\pm^T = \frac{1}{16\pi^2} \sum_{\text{doublets}} \Big\{ 2q^2 [B_2(f,f') + B_1(f,f')] + m_f^2 B_1(f',f) + m_{f'}^2 B_1(f,f') \Big\} \ . \tag{7.62}$$

These expressions explicitly show the violation to the decoupling theorem: the masses of the particles that originate with the spontaneous breaking of the electroweak symmetry appear in the numerator of the vacuum polarization. This is to be constrated with QED where the electron mass appears in the denominator of the corrections, and thus decouples in the limit of infinite electron mass.

We now consider the vacuum polarizations that enter processes mediated by the photon and the Z boson. We label the four basic vacuum polarizations in terms of currents, in an obvious notation, Π_{AA}, Π_{ZA}, Π_{ZZ}, Π_{WW}. With the tree level propagators, these enter the coupled Dyson equations to produce the full propagators. The others are more complicated by the mixing Π_{ZA}. Before stating their solution, it will prove very instructive to calculate Π_{ZA} in the R_ξ gauge. The basic diagrams, including the ghost contributions, are

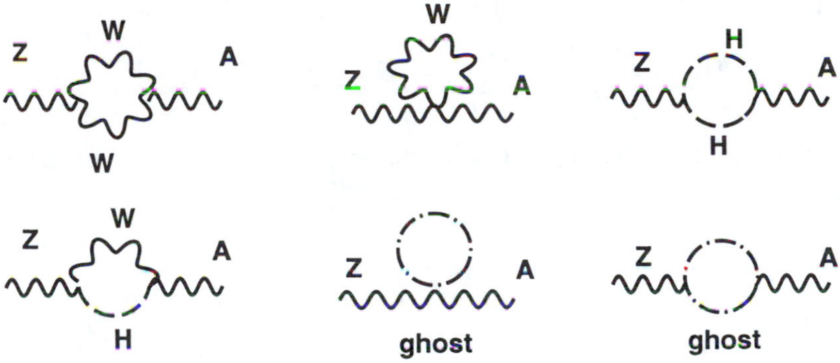

together with the contributions of the leptons and quarks

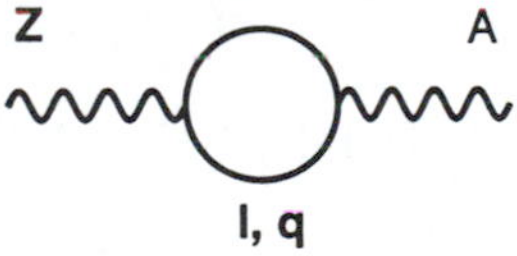

We have indicated only the class of one-loop diagrams we need to consider. As before, we separate the vacuum polarizations in the form

$$\Pi^T_{33}(q^2) = \Pi^{(0)}_{33}(q^2) + q^2\Pi^{(1)}_{33}(q^2) \; , \tag{7.63}$$

to find

$$\Pi_{33}^{(1)} = \frac{1}{16\pi^2}\left[\frac{2}{3} - 9[B_2(W,W) + B_1(W,W)] + \frac{7}{4}B_0(W,W) - D(Z,H))\right] , \tag{7.64}$$

$$\Pi_{33}^{(0)} = \frac{1}{16\pi^2}\left[2m_W^2 B_0(W,W) - \frac{1}{4}(m_Z^2 - m_H^2)B(Z,H) - m_Z^2 B_0(Z,H)\right] , \tag{7.65}$$

with the fermions contributing

$$\Pi_{33f}^{(1)} = \frac{1}{16\pi^2}\frac{1}{2}\sum_{\text{doublets}}\left[2q^2[B_2(f,f) + B_1(f,f)] + m_f^2 B_0(f,f)\right] . \tag{7.66}$$

The photon two-point function is, as expected from the Ward identities, proportional to q^2, and vanishes at $q^2 = 0$, with contributions from the fermions, as in QED,

$$\Pi_{QQf} = \frac{q^2}{16\pi^2}8q^2\sum_f Q_f^2[B_2(f,f) + B_1(f,f)] , \tag{7.67}$$

where Q_f are the fermion electric charge, and contributions from the charged W bosons

$$\Pi_{QQ} = \frac{q^2}{16\pi^2}\left[\frac{2}{3} - 12[B_2(W,W) + B_1(W,W)] + B_0(W,W)\right] . \tag{7.68}$$

Naively one would expect the same from the two-point function that relates the photon to the neutral weak boson, because of current conservation. One finds contributions that vanish at $q^2 = 0$,

$$\Pi_{3Q}^{(1)} = \frac{1}{16\pi^2}\left[\frac{2}{3} - 10[B_2(W,W) + B_1(W,W)] + \frac{2}{3}B_0(W,W)\right] , \tag{7.69}$$

from the gauge bosons, and from the fermions,

$$\Pi_{3Qf}^{(1)} = \frac{1}{16\pi^2}\sum_f Q_f I_f^3[B_2(f,f) + B_1(f,f)] , \tag{7.70}$$

where I_f^3 is the fermion third component of weak hypercharge. However we also find a part that does not vanish at $q^2 = 0$,

$$\Pi_{3Q}^{(0)} = \frac{1}{16\pi^2}2m_Z^2 B_0(W,W) . \tag{7.71}$$

As we discussed earlier, this odd behavior generates a pole at $q^2 = 0$ in Σ, which induces new massless states in the solutions of the Schwinger-Dyson equations.

To see what is amiss, let us retrace our steps. Our procedure has singled out the oblique corrections, neglecting the vertex and box corrections, on the grounds that, unlike the vacuum polarizations, they depend on the external legs. We have also tacitly assumed the gauge-invariance of this procedure, and we lied: in non-Abelian theories, the vertex and box corrections contains parts which are independent of their legs! Furthermore our separation is not gauge-invariant. Gauge-invariance is restored only after taking into account of these contributions from the vertex and box diagrams.

The existence of *process-independent* parts in the vertex and box diagrams creates an ambiguity in the extraction of the oblique corrections from the amplitudes. To see this, write any amplitude in the form

$$\mathcal{M} = (1+\Gamma)D(1+\Gamma) + \Theta \ , \tag{7.72}$$

where Γ and Θ are the process-independent contributions from the vertices and boxes, respectively. We can rewrite the amplitude in a propagator form, by introducing a modified vacuum polarization, $\widehat{\Pi}$, so that in lowest order

$$\mathcal{M} \approx D_0 + D\widehat{\Pi}D_0 \ , \tag{7.73}$$

where

$$\widehat{\Pi} \approx \Pi + 2D_0^{-1}\Gamma + D_0^{-2}\Theta + \cdots \ , \tag{7.74}$$

contains the vertex and box corrections. Applied to Π_{3Q}, the gauge invariant propagator is found to be

$$\widehat{\Pi}_{3Q} = \Pi_{3Q} + q^2\Gamma_3 + (q^2 - m_W^2)\Gamma_3 + \text{box corrections} \ , \tag{7.75}$$

where Γ_3 is the neutral current non-Abelian vertex. Explicit one-loop calculations yield

$$\Gamma_3 = \frac{1}{16\pi^2}\left[2B_0(q^2=0, W, W) + \mathcal{O}(q^2)\right] . \tag{7.76}$$

Magically, this restores the Ward identity, as now we have

$$\widehat{\Pi}_{3Q}(0) = 0 \ . \tag{7.77}$$

For more details we refer the reader to D. C. Kennedy's TASI lectures, *Perspectives in the Standard Model*, R. K. Ellis, C. T. Hill and J. Lykken eds. (World Scientific, Singapore, 1991).

The effect of heavy particles on the oblique corrections is summarized by calculating their effect on the finite combinations $\Delta_{1,\rho}$. The Higgs and top quark contributions are calculated to be

$$\begin{aligned}\rho_\star(0)-1 \;=\; & \frac{3G_\mu}{8\sqrt{2}\pi^2}\Big[m_t^2+m_b^2-\frac{2m_t^2m_b^2}{m_t^2-m_b^2}\ln\left(\frac{m_t^2}{m_b^2}\right) \\ & +m_W^2\ln\left(\frac{m_H^2}{m_W^2}\right)-m_Z^2\ln\left(\frac{m_H^2}{m_Z^2}\right)\Big]\,.\end{aligned} \tag{7.78}$$

As we expect from our analysis of custodial symmetry, this correction vanishes in the limit $m_t = m_b$, and $m_Z = m_W$. Note the dependence on the Higgs mass which enters only through the spontaneous breaking, albeit only logarithmically. In this sense the Higgs mass also escapes the decoupling theorem, but only weakly.

Similarly, the top and Higgs masses contribute to the other finite quantity as

$$\Delta_3(q^2)=\frac{1}{16\pi^2}\frac{q^2}{12}[\ln\frac{m_h^2}{m_Z^2}-2\ln\frac{m_t^2}{m_Z^2}]\,. \tag{7.79}$$

Finally let us remark that the effect of heavy fermions on these corrections can be described, using the decoupling theorem. One can envisage two types of fermions; those that get their masses through spontaneous breaking of electroweak symmetry, and those that have vector-like masses. Since the masses of the vector-like fermions enter at tree level, like the electron mass in QED, their effects on radiative corrections will come in as inverse power of their masses, and effectively decouple for large vector-like masses. As we have seen, the situation is different for chiral masses, which enter radiative corrections in the numerator of the corrections. Such is the case for the top quark. The effect of heavy ($> m_Z$) chiral fermions on the radiative corrections yields rapidly convergent series in q^2. The dependence of the corrections on the scalar Higgs mass is logarithmic. Hence it suffices to consider only their lowest order contributions in q^2. It is convenient to define the three dimensionless quantities

$$\alpha S \equiv -4e^2\frac{d}{dq^2}\Delta_3(q^2)|_{q^2=0}\ , \tag{7.80}$$

$$\alpha T \equiv \frac{e^2}{s^2c^2m_Z^2}\Delta_\rho(0)\ , \tag{7.81}$$

$$\alpha U \equiv -4e^2 \frac{d}{dq^2} \Delta_\rho(q^2)|_{q^2=0} \ , \tag{7.82}$$

in terms of which the effect of heavy chiral particles on the oblique corrections can be conveniently discussed.

PROBLEMS

A. Verify the solutions (7.29) of the Schwinger-Dyson equations for the neutral boson propagators.

B. Express the residues $\mathcal{Z}_{[W]*}$ and $\mathcal{Z}_{[Z]*}$ in terms of $\Delta_1(q^2)$, and $\Delta_\rho(q^2)$.

C. The extraction of the physical parameters such as mass and widths of unstable particles, such as the Z or W bosons, from their propagators can be tricky. Compare the definitions in the text (*on-shell* mass) with the physical definition: mass as the real part of the pole of the propagator in the energy plane. Derive an approximate relation between these masses, Then, use unitarity to show that

$$M_{\text{phys}} = M_{on-shell}[1 - \frac{3}{8}\Big(\frac{\Gamma_{\text{phys}}}{M_{on-shell}}\Big)^2 + \cdots]$$

For details, see S. Willenbrock and G. Valencia *Phys. Let.* **259B**, 373(1991).

D. Suppose the standard model contains a fourth chiral family of quarks and leptons, with masses $> m_Z$ (disregarding the problem caused by its massless neutrino). Calculate their contributions to the S, T, and U parameters. See M. Peskin and T. Takeuchi, *Phys. Rev.* **D46**, 381(1992), and G. Altarelli and R. Barbieri, *Phys. Lett.* **253B**, 161(1991).

E. Calculate the one-loop oblique electroweak corrections to $\Gamma(Z \to \mathbf{b}\overline{\mathbf{b}})$. Identify the one-loop straight corrections, and discuss their strength relative to the oblique ones. In reality, where do you expect the main corrections to come from?

F. Derive the expression for the gauge invariant propagator $\widehat{\Pi}_{3Q}$. you might want to consult M. Kuroda, G. Moultaka, D. Schildknecht, *Nucl. Phys.* **B350**, 25(1991).

7.2 $\Delta F = 1$ PROCESSES

These interactions take place solely through quantum corrections, so that their occurence and absolute strength is a prediction of the standard model.

They are conveniently described in the language of the low energy effective field theory, in which the effect of heavy particles such as W-bosons on the lower energy theory is to generate non-renormalizable interactions of calculable strength. The celebrated and trivial example is the current-current four-fermion interaction originally proposed by Fermi. It applies here not only to tree-level processes but also to interactions generated in higher loops. The flavor-changing neutral current interactions have no tree-level counterpart, and their four-fermion interactions directly measure the radiative structure of the standard model. In the operator language, the strengths of these higher-order operators will not only evolve with scale according to the renormalization group, but also mix with the strengths of other interactions.

The strength of any operator calculated in perturbation theory serves as a boundary condition at the scale at which the effect of heavy particles have been "integrated-out", in our case, the W-mass. This will have to be compared with its strength at the experimental scale. Since in that region the color coupling becomes larger, perturbative and (at lower scales) non-perturbative QCD effects become important. They must be accounted for, through both perturbative and non-perturbative techniques. After heroic efforts, the perturbative part can be calculated reliably down to a surprisingly low energy. Estimating non-perturbative contribution is still at the artisanal level, with the possible exception of lattice calculations that are becoming more reliable with time.

In short, the calculation of these processes proceeds in three steps: first the calculation of the quantum interactions in electroweak theory produces, at the W-boson scale, effective "short-distance" interactions. Second, using the renormalization group, the strengths of these interactions are calculated at experimental scales, and thirdly, the "long-distance" non-perturbative evaluation of the matrix elements is performed. Only then can the result be compared with experiment.

In running from M_W down to experimental scale m, the "short-distance" strengths will be modified in two ways. The first, due to operator-mixing, will add contributions from other operators; the second, due to the anomalous dimension of the operators, will result in multiplication by

$$\eta \equiv \frac{\alpha_{QCD}(m_W)}{\alpha_{QCD}(m)} , \tag{7.83}$$

raised to the appropriate power. In the following, we do not compute these effects, but refer the reader to the many excellent references on the subject.

One may construct many different flavor-changing interactions between quarks of the same electric charge. They appear as four-fermion operators, and operators that involve two fermions and one neutral gauge boson (gluon, Z, γ), called "penguin" vertices. Combined with a fermion line coupled to the neutral gauge boson, they also generate four-fermion interactions. all are generated by loop diagrams involving virtual W-bosons.

Of the many flavor-changing neutral electroweak processes, we begin with the electroweak prediction for the very rare (branching ratio of the order of 10^{-10}!) semileptonic decay $K \to \pi + \overline{\nu}\nu$, which in the quark picture occurs through the process $\overline{\mathbf{s}} + \mathbf{d} \to \nu + \overline{\nu}$. It occurs via the standard model effective Lagrangian

$$\mathcal{L}_{\text{eff}}^{\Delta S=1} = 4\frac{G_F}{\sqrt{2}} g_2^2 \; \mathbf{s}_L^\dagger \sigma_\mu \mathbf{d}_L \sum_{i=1}^{3} \nu_{Li}^\dagger \sigma^\mu \nu_{Li} \; \mathcal{E}_i \; . \tag{7.84}$$

This effective interaction comes from two types of one-loop diagrams. One comes from Z-boson exchange, and the other is from box diagrams

where the $Z\overline{\mathbf{d}}\mathbf{s}$ vertex is due to

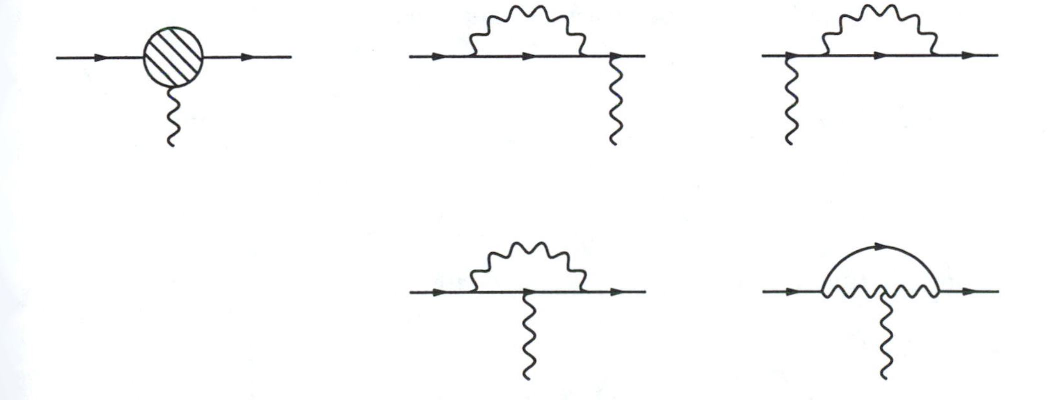

There are more diagrams with each internal gauge line replaced by a charged Higgs (R_ξ gauge) and also their retinue of ghosts. The resulting one-loop calculations can be cast in the form

$$\mathcal{E}_i = \sum_{j=\mathbf{c},\mathbf{t}} \mathcal{U}^*_{j\mathbf{s}}\mathcal{U}_{j\mathbf{d}}\Big[\mathcal{E}_Z(x_j) + \mathcal{E}_{\text{box}}(x_j, y_i)\Big] , \tag{7.85}$$

where $x_{\mathbf{c},\mathbf{t}} = m^2_{\mathbf{c},\mathbf{t}}/m^2_W$, and $y_j = m^2_{e,\mu,\tau}/m^2_W$, after using

$$\mathcal{U}^*_{\mathbf{us}}\mathcal{U}_{\mathbf{ud}} = -\sum_{j=\mathbf{c},\mathbf{t}} \mathcal{U}^*_{j\mathbf{s}}\mathcal{U}_{j\mathbf{d}} = 0 , \tag{7.86}$$

which follows from the unitarity of the CKM matrix. We have also separated the two gauge-dependent contributions coming from Z exchange and box diagrams, respectively. Their sum is of course gauge-invariant. In the R_ξ gauge this manifests itself by opposite sign ξ-dependent terms for each. In a general R_ξ gauge, T. Inami and C.S. Lim (*Prog. Theo. Phys.* **65**, 297(1981)) find,

$$\mathcal{E}_Z(x_j) = \frac{x_j}{4} - \frac{3}{8}\frac{1}{x_j - 1} + \frac{3}{8}\frac{2x_j^2 - x_j}{(x_j - 1)^2}\ln x_j + f(x_j, \xi) , \tag{7.87}$$

where the ξ dependence is contained in f. Also

$$\begin{aligned}\mathcal{E}_{\text{box}}(x_j, y_i) = &-\frac{1}{8}\frac{y_i^2 \ln y_i}{y_i - x_j}\Big(\frac{y_i - 4}{y_i - 1}\Big)^2 - \frac{9}{8}\frac{1}{(y_i - 1)(x_j - 1)} - f(x_j, \xi)\\ &+ \frac{x_j \ln x_j}{8}\Big[\frac{x_j}{y_i - x_j}\Big(\frac{x_j - 4}{x_j - 1}\Big)^2 + \frac{x_j - 7}{x_j - 1}\Big] .\end{aligned} \tag{7.88}$$

Their sum yields the gauge-invariant contributions

$$\begin{aligned}\mathcal{E}_Z + \mathcal{E}_{\text{box}} = &\frac{x_j}{4} - \frac{3}{8}\frac{x_j}{x_j - 1}\Big(1 + \frac{3}{y_i - 1}\Big) - \frac{1}{8}\frac{x_j y_i \ln y_i}{y_i - x_j}\Big(\frac{y_i - 4}{y_i - 1}\Big)^2\\ &+ \frac{1}{8}\Big[1 + \frac{3}{(x_j - 1)^2} + \frac{x_j}{x_j - 1}\Big(\frac{x_j - 4}{x_j - 1}\Big)^2\Big] x_j \ln x_j .\end{aligned} \tag{7.89}$$

Applying this formula to our process, we find

$$\mathcal{L}^{\Delta S=1}_{\text{eff}} \approx 2\frac{G_F}{\sqrt{2}} g_2^2\ \mathbf{s}^\dagger_L \sigma_\mu \mathbf{d}_L \sum_{i=1}^{3} \nu^\dagger_{Li}\sigma^\mu \nu_{Li}\ \lambda\Big(\frac{m^2_c}{m^2_W}\Big) \ln\Big(\frac{m^2_c}{m^2_W}\Big) . \tag{7.90}$$

The amplitude for the decay is given by its matrix element between the initial and final states, that is

$$\mathcal{A}(K^+ \to \pi^+ + \nu_i\overline{\nu}_i) = 2\frac{G_F}{\sqrt{2}}g_2^2\ \lambda\left(\frac{m_c^2}{m_W^2}\right)\ln\left(\frac{m_c^2}{m_W^2}\right)$$
$$\langle K^+|\mathbf{s}_L^\dagger \sigma_\mu \mathbf{d}_L|\pi^+\rangle\langle 0|\nu_{Li}^\dagger \sigma^\mu \nu_{Li}|\nu_i\overline{\nu}_i\rangle\ . \quad (7.91)$$

The evaluation of the neutrino matrix element can be easily performed, but our present theoretical tools are not adequate for a complete calculation of the hadronic matrix element: the overlap between a quark-antiquark state and a meson, $\langle meson|\mathbf{q}\overline{\mathbf{q}}\rangle$, is determined by the strong QCD interaction at large distances. This is a general problem when calculating processes that involve hadrons: we can compute the short distance behavior where all interactions, including QCD, are perturbative. However their contribution to physical states requires a knowledge of the overlap between quark states and mesons and hadrons. At present we can only measure these overlaps or else calculate them in some (usually more inspired than justified) approximation, such as the bag model, lattice, etc... .

In this case, however, we are lucky, because $\langle K^+|\mathbf{s}_L^\dagger \sigma_\mu \mathbf{d}_L|\pi^+\rangle$ is closely related to an amplitude that occurs in the charge-changing K_{l3} decay $K^+ \to \pi^0 + e^+ + \nu_e$. Since we do not observe the neutrino species, the amplitudes add incoherently. At the end of the day, one obtains

$$\frac{\Gamma(K^+ \to \pi^+ + \nu\overline{\nu})}{\Gamma(K^+ \to \pi^0 + e^+ + \nu)} = \frac{\alpha_{QED}^2}{2\pi^2 \sin^4\theta_{\rm w}}\left(\frac{m_c^2}{m_W^2}\right)\ln\left(\frac{m_c^2}{m_W^2}\right)\ . \quad (7.92)$$

This result does not include the *perturbative* QCD corrections. Their effect is to dress the diagrams with gluon lines. They can be discussed in a very beautiful way, using the language of operators in an effective field theory where high mass particles have been integrated out.

Another rare process that has been recently observed is the flavor-changing decay $B^+ \to K^{*+} + \gamma$. The short-distance behavior to its amplitude is due to the transition $\mathbf{b} \to \mathbf{s} + \gamma$. This process is generated by the diagrams

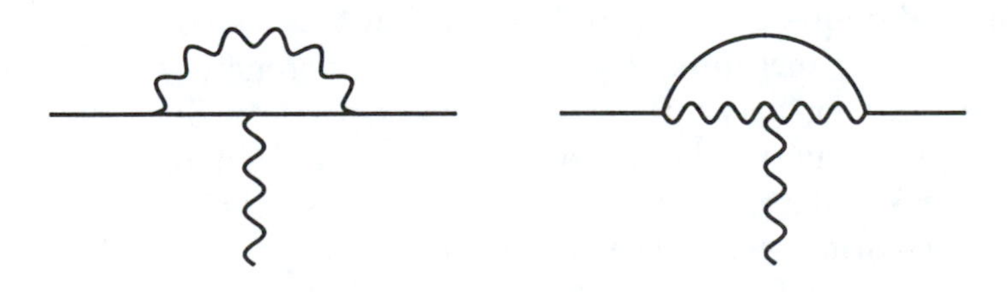

The $\mathbf{b\bar{s}}\gamma$ vertex can only be of the magnetic moment form, since to preserve gauge invariance, the photon must enter in the field strength combination (the photon is on its mass-shell). As these diagrams are clearly homotopic to penguin shapes, they are called (magnetic) penguins. Indeed we find

$$\mathcal{L}_{\text{eff}}^{\Delta F=1} = \frac{G_F}{\sqrt{2}}\frac{e}{4\pi^2}F^{\mu\nu}\left[m_b\mathbf{b}_R^\dagger\sigma_{\mu\nu}\mathbf{s}_L + m_s\mathbf{s}_R^\dagger\sigma_{\mu\nu}\mathbf{b}_L\right]\sum_{j=u,c,t}\mathcal{U}_{jb}^*\mathcal{U}_{js}\mathcal{G}_j \ , \tag{7.93}$$

where

$$\mathcal{G}_j = \frac{x_j}{6}[z_j - 3z_j^2 - 6z_j^3] + z_j^4x_j^2\ln x_j + \frac{1}{2}z_j + \frac{9}{4}z_j^2 + \frac{3}{2}z_j^3 - \frac{3}{2}z_j^4x_j^3\ln x_j \ , \tag{7.94}$$

in which we have set $z_j = (x_j - 1)^{-1}$. It is easy to see that the main effect comes from the virtual top quark contribution. This is the easy part, but our result is still incomplete. For one we have not included the effects of QCD. As we discussed, our result will be modified by perturbative QCD in the form of a multiplicative function of η, evaluated at the *bottom* quark mass.

These corrections are best accounted for in the language of the operator product expansion, but this is beyond the scope of this book.

PROBLEMS

A. Write the most general gauge-invariant expression for the $\mathbf{b\bar{s}}\gamma$ vertex, and evaluate its strength in the electroweak theory.

B. Write the most general gauge-invariant interaction for the QCD penguin. Draw the relevant Feynman diagrams, and evaluate them.

7.3 $\Delta F = 2$ PROCESSES

This is another class of effects which are purely radiative and constitute absolute predictions of the standard model. The celebrated example is the very precisely measured $K_L - K_S$ mass difference, which arises from electroweak mixing between K^0 and its antiparticle. Such flavor-mixing provide a powerful test laboratory for the standard model. After presenting their kinematical description, we discuss how these effects find a natural explanation within the standard model.

Weak Mixing Phenomenology

Some quark-antiquark bound states differ from one another only by quantum numbers associated with global flavor symmetries. Although preserved by QCD and QED interactions these symmetries are violated by the weak interactions. As a result, these QCD-distinct states oscillate into one another through the weak interactions. The study of their time evolution and decay can then be treated using standard quantum mechanics techniques. With three chiral families, QCD allows for this kind of mixing in six different systems:

• **the K-system**: There are two states, $K^0 \sim (\mathbf{d}\overline{\mathbf{s}})$, and its antiparticle $\overline{K}^0 \sim (\overline{\mathbf{d}}\mathbf{s})$, which are similar in all respects, except for their *strangeness* (for historical reasons, the s-quark is given strangeness -1). Both K^0 and $\overline{K}^0$ can decay weakly to the same channel

$$K^0 \to \pi^+\pi^- , \qquad \overline{K}^0 \to \pi^+\pi^- , \tag{7.95}$$

changing strangeness by one unit ($|\Delta S| = 1$). Their decay into the same final state indicates that they can mix through second order perturbation theory. It follows that K^0 and $\overline{K}^0$ are not mass eigenstates. According to quantum mechanics, the mass eigenstates must be linear combinations of these two states. This is more than a standard mixing problem because both particles decay: the study of the $K^0 - \overline{K}^0$ system requires a Hamiltonian which is neither diagonal (to allow for mixing) nor real (to allow for the decays). Thus there are two fundamental time scales in the problem, the decay time and the mixing time; for mixing to be observed, the particles must live long enough. This is clearly true for the $K^0 - \overline{K}^0$ system where this effect was first observed. With three chiral families, the standard model predicts a similar behavior in other systems:

• **the D-system**: $D^0 \sim (\mathbf{u}\overline{\mathbf{c}})$ and $\overline{D}^0 \sim (\overline{\mathbf{u}}\mathbf{c})$ are distinguished only by their *charm*. Since the weak interactions cause $\Delta c \neq 0$ transitions, mixing is expected as well.

• **the B-system**: $B^0 \sim (\mathbf{d}\overline{\mathbf{b}})$ and $\overline{B}^0 \sim (\overline{\mathbf{d}}\mathbf{b})$, distinguished only by their *beauty*.

• **the B$_s$-system**: $B_s^0 \sim (\mathbf{s}\overline{\mathbf{b}})$ and $\overline{B}_s^0 \sim (\overline{\mathbf{s}}\mathbf{b})$, distinguished by their *strange beauty*.

• **the T-system**: $T^0 \sim (\mathbf{u}\overline{\mathbf{t}})$ and $\overline{T}^0 \sim (\overline{\mathbf{u}}\mathbf{t})$, distinguished by their *truth*, and finally

• **the $\mathbf{T_c}$-system:** $T_c^0 \sim (\mathbf{c\bar{t}})$ and $\overline{T}_c^0 \sim (\mathbf{\bar{c}t})$, distinguished in *charm* and in *truth.*

These two-state systems are all described by the same physics, but the value of their parameters dictate their relevance to experiments. The top quark does not live long enough to form bound states, so that the effect in the $\mathbf{T}$ and $\mathbf{T_c}$ is irrelevant. The B-meson and K-meson systems are the ones directly measured by today's experiments, which also set limits on the parameters of the $\mathbf{B_s}$ and $\mathbf{D}$ systems. For illustrative purposes, we focus the discussion on the $K^0 - \overline{K}^0$ system.

Suppose that at time $t = 0$, QCD scattering produces a particle, such as $|K^0\rangle$. It is not an eigenstate of the weak interactions, and will mix with $|\overline{K}^0\rangle$. We want to compute the probability that at some time t later one observes a $|\overline{K}^0\rangle$, as tagged by its decay products.

Since these particles are heavy, we use the formalism of non-relativistic quantum mechanics, with time replaced by proper time. Two-state systems can be described by a two-state vector which obeys a Schrödinger equation

$$i\frac{d\psi}{d\tau} = \mathbf{H}\psi \ , \tag{7.96}$$

where $\mathbf{H}$ is the Hamiltonian. To account for decays, it is not hermitian, but of the form

$$\mathbf{H} = \mathbf{M} - i\mathbf{\Gamma} \ , \tag{7.97}$$

where $\mathbf{M}$ and $\mathbf{\Gamma}$ are hermitian matrices,

$$\mathbf{H} = \begin{pmatrix} M_{11} - i\Gamma_{11} & M_{12} - i\Gamma_{12} \\ M_{12}^* - i\Gamma_{12}^* & M_{22} - i\Gamma_{22} \end{pmatrix} \ , \tag{7.98}$$

in which the diagonal entries M_{ii}, Γ_{ii}, are real. In this basis the QCD eigenstates are identified as

$$|K^0\rangle \leftrightarrow \begin{pmatrix} 1 \\ 0 \end{pmatrix} \ , \qquad |\overline{K}^0\rangle \leftrightarrow \begin{pmatrix} 0 \\ 1 \end{pmatrix} \ . \tag{7.99}$$

Under the combined operation of CPT, particles are transformed into antiparticles with space and time inverted, and the diagonal entries are mapped into one another. The standard model is a local field theory, and therefore CPT-invariant, allowing us to set

$$m - \frac{i}{2}\gamma \ \equiv \ M_{11} - i\Gamma_{11} = M_{22} - i\Gamma_{22} \ , \tag{7.100}$$

and rewrite the Hamiltonian in the form

$$\mathbf{H} = \begin{pmatrix} m - \frac{i}{2}\gamma & p^2 \\ q^2 & m - \frac{i}{2}\gamma \end{pmatrix} , \tag{7.101}$$

where p and q are arbitrary *complex* numbers. The eigenstates of $\mathbf{H}$, given by

$$\begin{aligned} |\chi_1\rangle &= \frac{1}{\sqrt{|p|^2 + |q|^2}} \begin{pmatrix} p \\ q \end{pmatrix} , \\ |\chi_2\rangle &= \frac{1}{\sqrt{|p|^2 + |q|^2}} \begin{pmatrix} p \\ -q \end{pmatrix} . \end{aligned} \tag{7.102}$$

satisfy

$$\mathbf{H}|\chi_{1,2}\rangle = (m - \frac{i}{2}\gamma \pm pq)|\chi_{1,2}\rangle . \tag{7.103}$$

The masses and width of the eigenstates are defined by

$$m_{S,L} - \frac{i}{2}\gamma_{S,L} \equiv m - \frac{i}{2}\gamma \pm pq , \tag{7.104}$$

so that

$$pq = \frac{\Delta m}{2} - \frac{i}{4}\Delta\Gamma , \tag{7.105}$$

where Δm is the mass difference and $\Delta\Gamma$ is the difference of the widths. In another language, we can write the mass eigenstates as

$$|K_S\rangle \equiv \frac{1}{\sqrt{|p|^2 + |q|^2}} \left\{ p|K^0\rangle + q|\overline{K}^0\rangle \right\} , \tag{7.106}$$

$$|K_L\rangle \equiv \frac{1}{\sqrt{|p|^2 + |q|^2}} \left\{ p|K^0\rangle - q|\overline{K}^0\rangle \right\} , \tag{7.107}$$

where we have tagged the states with S(hort) and L(ong) labels according to which has the shortest and longest lifetimes. Thus a state, initially taken to be the pure QCD eigenstate $|K^0\rangle$ will become (in its proper time frame)

$$\begin{aligned} |K^0(\tau)\rangle =& \frac{1}{2} \left\{ e^{-im_S\tau - \frac{\gamma_S}{2}\tau} + e^{-im_L\tau - \frac{\gamma_L}{2}\tau} \right\} |K^0\rangle \\ &+ \frac{1}{2}\frac{q}{p} \left\{ e^{-im_S\tau - \frac{\gamma_S}{2}\tau} - e^{-im_L\tau - \frac{\gamma_L}{2}\tau} \right\} |\overline{K}^0\rangle . \end{aligned} \tag{7.108}$$

Similarly a state, initially $|\overline{K}^0\rangle$, becomes later

$$|\overline{K}^0(\tau)\rangle = \frac{1}{2}\frac{p}{q}\left\{e^{-im_S\tau-\frac{\gamma_S}{2}\tau} - e^{-im_L\tau-\frac{\gamma_L}{2}\tau}\right\}|K^0\rangle + \frac{1}{2}\left\{e^{-im_S\tau-\frac{\gamma_S}{2}\tau} + e^{-im_L\tau-\frac{\gamma_L}{2}\tau}\right\}|\overline{K}^0\rangle \,. \tag{7.109}$$

As a result, the probability of observing a $|\overline{K}^0\rangle$ at a (proper) time τ is given by

$$|\langle\overline{K}^0|K^0(\tau)\rangle|^2 = \frac{1}{4}|\frac{q}{p}|^2[e^{-\gamma_S\tau} + e^{-\gamma_L\tau} - 2e^{-\frac{\gamma_L+\gamma_S}{2}\tau}\cos\Delta m\tau] \,. \tag{7.110}$$

Although derived here in the context of the $K^0 - \overline{K}^0$ system, this analysis applies to any of the other two-state oscillatory systems.

We note the appearance of several time scales in the problem, the two decay rates, the inverse mass difference, in addition to the duration of the experimental observation. Clearly, a necessary condition for observing any effect is that the decays not be too fast compared to ΔT, the duration of the observation: if both $\gamma_{S,L}\Delta T \gg 1$, there will be nothing to observe. Even if ΔT is small compared to the lifetime of either particle, one may still not be able to observe any effect: the time scale of oscillations, given by $(\Delta m)^{-1}$, must also be short compared to the decay time. The figure of merit for detecting oscillatory behavior is the ratio

$$x = \frac{\Delta m}{\Gamma}\,, \qquad \Gamma = \gamma_S + \gamma_L \,. \tag{7.111}$$

Assuming that the observation times are short enough, we can envisage several situations. If $x \sim 1$, oscillations between two states can be observed, with at least one particle quasi-stable on the time scale appropriate to the measurement. If $x \ll 1$, the oscillatory amplitude will decay before oscillations have a chance to develop, and no effect will be observed. The x parameters can be calculated for each system in the standard model, although not with complete reliability, because of the strong matrix element $\langle q\bar{q} \mid \text{Meson}\rangle$. To see how p and q relate to measured quantities, set

$$p = |p|e^{i\alpha_p}\,, \qquad q = |q|e^{i\alpha_q}\,, \tag{7.112}$$

yielding from the eigenvalues

$$|p||q|\cos(\alpha_p + \alpha_q) = \frac{\Delta m}{2}$$
$$|p||q|\sin(\alpha_p + \alpha_q) = -\frac{\Delta\Gamma}{4} . \tag{7.113}$$

Only one combination of angles appears, which reflects the phase arbitrariness of the form

$$|K^0\rangle \to e^{i\chi}|K^0\rangle , \qquad |\overline{K}^0\rangle \to e^{-i\chi}|\overline{K}^0\rangle , \tag{7.114}$$

$$p \to e^{i\chi}p , \qquad q \to e^{-i\chi}q . \tag{7.115}$$

It originates in the phase arbitrariness inherent in the definition of the CPT transformation. The other phase, $\alpha_p - \alpha_q$, is therefore convention-dependent and not physical.

The width and mass differences fix all the parameters except the absolute value of the ratio q/p. Measurements in the neutral kaon system yield

$$\frac{1}{\gamma_S} = 0.8934 \pm 0.0008 \times 10^{-10}\ s , \qquad \frac{1}{\gamma_L} = 5.17 \pm 0.04 \times 10^{-8}\ s ,$$

$$\Delta m = 3.489 \pm 0.009 \times 10^{-12}\ \text{MeV} ,$$

so that the oscillation figure of merit is $x = .4773 \pm 0.0023$. The $K^0 - \overline{K}^0$ is therefore an ideal laboratory for observing this effect: the mass and width differences are comparable, and one state is sufficiently long-lived. The sum of the phases and the absolute value of the product of p and q are then

$$\alpha_p + \alpha_q = 46.33^\circ , \qquad |p||q| = 2.5496 \times 10^{-12}\text{MeV} .$$

Since $\alpha_p - \alpha_q$ is convention-dependent, there remains only to determine the ratio $|q|/|p|$.

CP-violation

We note that CP conservation implies

$$\frac{|q|}{|p|} = 1 , \qquad (CP \text{ conservation}) . \tag{7.116}$$

Indeed, under CP,

$$CP|K^0\rangle = e^{i\eta}|\overline{K}^0\rangle \ , \qquad CP|\overline{K}^0\rangle = e^{-i\eta}|K^0\rangle \ , \tag{7.117}$$

where η is an arbitrary phase, so that if CP were exact, the CP eigenstates would also be the eigenstates of the Hamiltonian, $|K_L\rangle$ and $|K_S\rangle$, with $|p| = |q|$. Thus the determination of this ratio lies with the observation of CP-violation

In reality, CP is not a good symmetry, and the ratio $|q|/|p|$ deviates slightly from one. Indeed $|K_S\rangle$ is observed to decay 68.61% of the time into $\pi^+\pi^-$ and 31.39% of the time into $\pi^0\pi^0$, while $|K_L\rangle$ decays mostly into $\pi^0\pi^0\pi^0$ (21.5%), $\pi^+\pi^-\pi^0$ (12.4%), $\pi^\pm\mu^\mp\overline{\nu}$ (27.1%), and $\pi^\pm e^\mp\nu$ (38.7%). The two- and three-pion S-wave neutral combinations are CP eigenstates

$$CP|\pi\pi\rangle = |\pi\pi\rangle \ , \qquad CP|\pi\pi\pi\rangle = -|\pi\pi\pi\rangle \ , \tag{7.118}$$

leading to the conclusion (up to 1964) that $|K_S\rangle$ is CP-even and $|K_L\rangle$ is CP-odd. Unexpectedly, $|K_L\rangle$ was found to decay into the CP-even $\pi^+\pi^-$ 0.2% of the time, showing that CP is not a good (conserved) quantum number.

The deviation of $|p/q|$ from one is a measure of CP violation. The extraction of this ratio from experiments with hadronic final states is rather complicated. First define ratios of CP-violating to CP-conserving amplitudes

$$\eta_{+-} \equiv \frac{A(K_L \to \pi^+\pi^-)}{A(K_S \to \pi^+\pi^-)} \ , \qquad \eta_{00} \equiv \frac{A(K_L \to \pi^0\pi^0)}{A(K_S \to \pi^0\pi^0)} \ . \tag{7.119}$$

Their absolute values and phases are directly measured to be

$$|\eta_{+-}| = (2.285 \pm 0.019) \times 10^{-3} \ , \qquad \arg \eta_{+-} = 43.5 \pm 0.6^\circ \ ,$$

$$|\eta_{00}| = (2.275 \pm 0.019) \times 10^{-3} \ , \qquad \arg \eta_{00} = 43.4 \pm 1.0^\circ \ .$$

Since π is an isovector ($I = 1$), the two-pion state is either an isoscalar ($I = 0$) or an isotensor ($I = 2$), while K^0 is a component of an isospinor ($I = 1/2$). It follows that the decay $K^0 \to 2\pi$ can be caused by $\Delta I = \frac{1}{2}, \frac{3}{2}, \frac{5}{2}$ interactions. Of these three posibilities, only the $\Delta I = \frac{1}{2}$ part is experimentally dominant. For example $K_S \to 2\pi$ decays twice as often into $\pi^+\pi^-$ than to $\pi^0\pi^0$, which means that there is hardly any I=2 component in the final state. We define the amplitudes for transition to a given isospin channel as

$$\langle (2\pi)_I | H_{\rm weak} | K^0 \rangle \equiv A_I e^{i\delta_I} \;, \quad I = 0, 2 \;, \tag{7.120}$$

where we have explicitly shown δ_I, the $\pi - \pi$ phase shift in the I channel. The antiunitary CPT transformation yields

$$\begin{aligned} \langle (2\pi)_I^{\rm out} | H_{\rm weak} | K^0 \rangle &= \langle (2\pi)_I^{\rm in} | H_{\rm weak} | \overline{K}^0 \rangle^* \;, \\ &= e^{2i\delta_I} \langle (2\pi)_I^{\rm out} | H_{\rm weak} | \overline{K}^0 \rangle^* \;, \end{aligned} \tag{7.121}$$

using the phase convention $CPT|K^0\rangle = |\overline{K}^0\rangle$. The phase shifts relate the "in" to the "out" states since CPT changes "in" states into "out" states, and vice versa. Thus

$$\langle (2\pi)_I | H_w | \overline{K}^0 \rangle = A_I^* e^{i\delta_I} \;. \tag{7.122}$$

Note that the A_I have convention-dependent phases, since under the previously considered phase transformation,

$$|K^0\rangle \to e^{i\chi} |K^0\rangle \;, \qquad A_I \to e^{i\chi} A_I \;. \tag{7.123}$$

In this case the origin of the phase dependence in the definition of CPT is apparent. It follows that the ratios

$$z \equiv \frac{q A_0^*}{p A_0} \;, \qquad w \equiv \frac{A_2}{A_0} \;, \tag{7.124}$$

are independent of the phase convention. CP-conservation implies $|z| = 1$, which suggests we introduce a new parameter, ϵ_p, to directly measure the deviation of the ratio $|q/p|$ from one,

$$\epsilon_p \;\equiv\; \frac{1 - z}{1 + z} \;. \tag{7.125}$$

Two other parameters, ϵ_p', ζ_p, measure CP-violation in the two-pion decay modes

$$\zeta_p - i\epsilon_p' \;\equiv\; \frac{w}{\sqrt{2}} e^{i(\delta_2 - \delta_0)} \;; \tag{7.126}$$

with an exact $\Delta I = 1/2$ rule, they are identically zero. All these quantities ϵ_p, ϵ_p', and ζ_p are independent of phase conventions. This practice, alas, is not always followed in the literature!

It is but simple (and messy) algebra to express η_{+-} and η_{00} in terms of these quantities,

$$\eta_{00} = \left(\epsilon_p - \frac{2\epsilon'_p}{1-2\zeta_p}\right)\frac{1}{(1+\frac{2\epsilon_p\epsilon'_p}{1-2\zeta_p})}\ , \tag{7.127}$$

$$\eta_{+-} = \left(\epsilon_p + \frac{\epsilon'_p}{1+\zeta_p}\right)\frac{1}{(1-\frac{\epsilon_p\epsilon'_p}{1+\zeta_p})}\ . \tag{7.128}$$

Using both the $\Delta I = \frac{1}{2}$ rule, which suggests that we can neglect $\zeta_p \ll 1$, and the small value of $\epsilon_p \ll 1$, we arrive at the approximate expressions

$$\eta_{00} \simeq \epsilon_p - 2\epsilon'_p\ , \qquad \eta_{+-} \simeq \epsilon_p + \epsilon'_p\ , \tag{7.129}$$

$$\left|\frac{\eta_{00}}{\eta_{+-}}\right|^2 \simeq 1 - 6\mathcal{R}e\left(\frac{\epsilon'_p}{\epsilon_p}\right)\ , \tag{7.130}$$

so that any deviation η_{+-}/η_{00} from one measures the ratio ϵ'_p/ϵ_p. The most recent measurement at FermiLab (Alavi-Harati *et al*, hep-ex/9905060) yields

$$\mathcal{R}e\left(\frac{\epsilon'_p}{\epsilon_p}\right) = 28.0 \pm 3.0(\text{stat}) \pm 2.8(\text{syst}) \times 10^{-4}\ , \tag{7.131}$$

to be compared with CERN's (G. D. Barr *et al*, *Phys. Lett.* **B317**, 233(1997)) earlier result of $23 \pm 6.5 \times 10^{-4}$. These values are of the same sign, but somewhat larger than the standard model predictions, perhaps providing a glimpse of physics beyond the standard model. However, the standard model calculation of this ratio is very tricky because of the strong matrix elements, and the predictions range from 4 to 17×10^{-4}. Its evaluation is beyond the scope of this book but we refer the reader to A. J. Buras and J. Fleischer in *Heavy Flavors II*, A. J. Buras and M. Lindner eds, (World Scientific, Singapore, 1998).

CP-violation is also measured in the semileptonic decays of K_L. In the standard model, flavor violation comes about only through the emission or absorption of a charged W boson, so that at the quark-W vertex, we always have $\Delta S = \Delta Q$. In semileptonic decays, ΔQ can be measured directly as the charge of the lepton in the final state, since the lepton pair is produced by the decay of a virtual W-boson

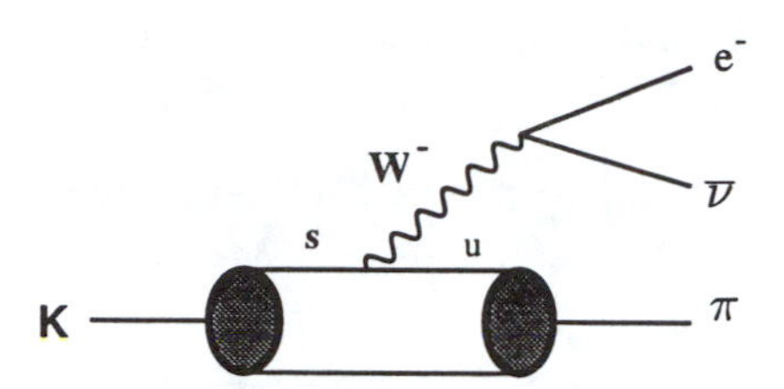

It follows that the comparison of $K_L \to \pi^- \ell^+ \nu_\ell$ and $K_L \to \pi^+ \ell^- \overline{\nu}_\ell$ can tell us the relative amount of K^0 and $\overline{K}^0$ in $|K_L>$. Specifically

$$
\begin{aligned}
A(K_L \to \pi^- \ell^+ \nu_\ell) &= \frac{p^*}{\sqrt{|p|^2+|q|^2}} \langle K^0 | H_{\text{weak}} | \pi^- \ell^+ \nu_\ell \rangle \, , \\
A(K_L \to \pi^+ \ell^- \overline{\nu}_\ell) &= \frac{-q^*}{\sqrt{|p|^2+|q|^2}} \langle \overline{K}^0 | H_{\text{weak}} | \pi^+ \ell^- \overline{\nu}_\ell \rangle \, .
\end{aligned}
\tag{7.132}
$$

Using CPT, we conclude that the parameters

$$
\begin{aligned}
\delta_\ell &= \frac{\Gamma(K_L \to \pi^- \ell^+ \nu_\ell) - \Gamma(K_L \to \pi^+ \ell^- \overline{\nu}_\ell)}{\Gamma(K_L \to \pi^- \ell^+ \nu_\ell) + \Gamma(K_L \to \pi^+ \ell^- \overline{\nu}_\ell)} \, , \\
&= \frac{1-|q/p|^2}{1+|q/p|^2} \, ,
\end{aligned}
\tag{7.133}
$$

give a direct measurement of CP-violation. Experimentally

$$
\delta_\mu = (3.04 \pm 0.25) \times 10^{-3} \, , \qquad \delta_e = (3.33 \pm 0.14) \times 10^{-3} \, ,
$$

which are consistent with the value obtained from $|\eta_{+-}|$.

Standard Model Calculation

To calculate the eigenvalues of the (complex) Hamiltonian, as well as p and q, we observe that the mixing comes about from loop effects. We also know that CP-violation can only occur through processes that involve all three quark families. Thus in the K system, $|p| \neq |q|$ must be due to virtual top quark contributions.

The decays of K^0 and $\overline{K}^0$ are $\Delta S = 1$ transitions while the mixing matrices are $\Delta S = 2$ processes. They can only be caused by the weak interactions, and treated as perturbations on the QCD and QED Hamiltonians.

Using again the $K^0 - \overline{K}^0$ system as an example, we write

$$\langle f|(\mathbf{M} - i\mathbf{\Gamma})|i\rangle = \langle f|H|i\rangle + \sum_n \frac{\langle f|H_{\text{weak}}|n\rangle\langle n|H_{\text{weak}}|i\rangle}{M_K - E_n + i\epsilon} + \cdots \,, \tag{7.134}$$

where $H = H_{QCD} + H_\gamma + H_{\text{weak}}$ is the total Hamiltonian, and M_K is the common eigenvalue of the QCD and QED Hamiltonians acting on the K^0 and $\overline{K}^0$ states. We set $i = 1$ for K^0, $i = 2$ for $\overline{K}^0$. The sum, over all possible intermediate states, can be split up into a real and imaginary part, using Cauchy's theorem,

$$\frac{1}{M_K - E_n + i\epsilon} = \mathcal{P}\frac{1}{M_K - E_n} - i\pi\delta(M_K - E_n) \,, \tag{7.135}$$

where $\mathcal{P}$ is Cauchy's principal value. Using CPT invariance, a little algebra gives for both diagonal elements,

$$\begin{aligned} m &= \langle K^0|H|K^0\rangle + \mathcal{P}\sum_n \frac{\langle K^0|H_{\text{weak}}|n\rangle\langle n|H_{\text{weak}}|K^0\rangle}{M_K - E_n} \,, \\ \gamma &= \pi\sum_n \langle K^0|H_{\text{weak}}|n\rangle\langle n|H_{\text{weak}}|K^0\rangle\delta(M_K - E_n) \,. \end{aligned} \tag{7.136}$$

Since we have no lowest order flavor-changing and charge-conserving interactions,

$$\langle K^0|H_{QCD} + H_\gamma|\overline{K}^0\rangle = 0 \,, \tag{7.137}$$

the off-diagonal elements are simply

$$M_{12} = \langle K^0|H_{\text{weak}}|\overline{K}^0\rangle + \mathcal{P}\sum_n \frac{\langle K^0|H_{\text{weak}}|n\rangle\langle n|H_{\text{weak}}|\overline{K}^0\rangle}{M_K - E_n} \,, \tag{7.138}$$

and for the widths

$$\Gamma_{12} = \pi\sum_n \langle K^0|H_{\text{weak}}|n\rangle\langle n|H_{\text{weak}}|\overline{K}^0\rangle\delta(M_K - E_n) \,. \tag{7.139}$$

We can replace the sum over intermediate states by the sum over the different channels, such as $\pi^+\pi^-$, ..., weighted by ρ_{ch}, the density of states in that channel, to obtain

$$\Gamma_{12} = \pi \sum_{\text{ch}} \langle K^0 | H_{\text{weak}} | \text{ch} \rangle \langle \text{ch} | H_{\text{weak}} | \overline{K}^0 \rangle \rho_{\text{ch}} \ . \tag{7.140}$$

These $\Delta F = 2$ interactions are generated by s-channel box diagrams of the form

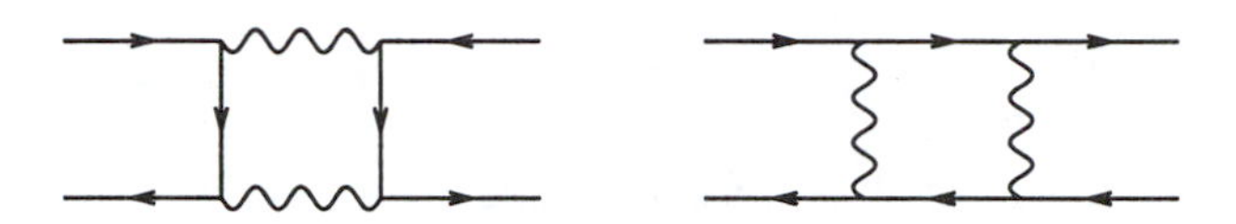

In the R_ξ gauge, every diagram must be replaced with the same diagram with each W-boson line replaced by the charged Higgs line, in addition to their associated ghosts (not shown). These lead to the effective Lagrangian that violate flavor by two units

$$\mathcal{L}_{\text{eff}}^{\Delta F=2} = \frac{G_F}{\sqrt{2}} g_2^2 \mathbf{d}^\dagger_{La} \sigma_\mu \mathbf{d}_{Lb} \mathbf{d}^\dagger_{La} \sigma^\mu \mathbf{d}_{Lb} \ \mathcal{F}_K \ , \tag{7.141}$$

where $a \neq b$, and $\mathbf{d}_{La} = \mathbf{d}_L,\ \mathbf{s}_L,\ \mathbf{b}_L$. Similar interactions with the charge 2/3 quarks on the external lines are also generated in the standard model. The function $\mathcal{F}$ is calculated from the box diagrams, is the sum of four terms, neglecting the mass of the *up*-quark,

$$\mathcal{F}_K = \sum_{j=\mathbf{c},\mathbf{t}} \sum_{k=\mathbf{c},\mathbf{t}} \mathcal{U}^*_{j\mathbf{d}_1} \mathcal{U}_{j\mathbf{d}_2} \mathcal{U}^*_{k\mathbf{d}_1} \mathcal{U}_{k\mathbf{d}_2} F(x_j, x_k) \ , \tag{7.142}$$

where $x_{\mathbf{c},\mathbf{t}} = m^2_{\mathbf{c},\mathbf{t}}/m^2_W$. The unitarity of the CKM matrix has enabled us to cancel the leading contributions; this is the GIM mechanism of S. L. Glashow, J. Iliopoulos, L. Maiani (*Phys. Rev.* **D2**, 1285(1970)). For $a \neq b$,

$$\sum_{j=\mathbf{u},\mathbf{c},\mathbf{t}} \mathcal{U}^*_{j\mathbf{d}_a} \mathcal{U}_{j\mathbf{d}_b} = 0 \ . \tag{7.143}$$

Explicit calculations yield for $j \neq k$ the symmetric function

$$F(x_j, x_k) = -\frac{x_j x_k}{x_j - x_k} \left[\frac{1}{4} - \frac{3}{2} \frac{\ln x_j}{x_j - 1} - \frac{3}{4} \frac{1}{(x_j - 1)^2} \right] - \frac{3}{8} \frac{x_j x_k}{(x_j - 1)(x_k - 1)}) \ , \tag{7.144}$$

plus $(x_j \leftrightarrow x_k)$, and

$$F(x_j, x_j) = -\frac{3}{2} \left(\frac{x_j}{x_j - 1} \right)^3 \ln x_j - x_j \left[\frac{1}{4} - \frac{9}{4} \frac{1}{x_j - 1} - \frac{3}{2} \frac{1}{(x_j - 1)^2} \right] \ . \tag{7.145}$$

Applying the Wolfenstein parametrization to the $K^0-\overline{K}^0$ system, we find the three contributions

$$\mathcal{F}_K = \lambda^2(1-\frac{\lambda^2}{2})^2 F(x_{\rm c},x_{\rm c}) + 2A^2\lambda^6(1-\rho+i\eta)F(x_{\rm c},x_{\rm t}) + A^2\lambda^{10}(1-\rho+i\eta)^2 F(x_{\rm t},x_{\rm t}) \; . \qquad (7.146)$$

As anticipated, it shows that CP-violation occurs only through a virtual top quark. The last term is very small and can be neglected.

The mass difference is related to the real part of the matrix element of the effective Lagrangian between the K^0 and $\overline{K}^0$ states

$$\Delta m = -2\,\mathrm{Re}\,\langle K^0|\mathcal{L}_{\rm eff}|\overline{K}^0\rangle \; , \qquad (7.147)$$

that is,

$$\Delta m = \frac{G_F}{\sqrt{2}} g_2^2 \langle K^0|\mathbf{d}_L^\dagger\sigma_\mu\mathbf{s}_L\mathbf{d}_L^\dagger\sigma^\mu\mathbf{s}_L|\overline{K}^0\rangle\mathcal{F} \; . \qquad (7.148)$$

Originally, M. K. Gaillard and B. W. Lee (*Phys. Rev.* **D10**, 897(1974)), approximated the strong matrix element by using vacuum saturation,

$$\langle K^0|\mathbf{d}_L^\dagger\sigma_\mu\mathbf{s}_L\mathbf{d}_L^\dagger\sigma^\mu\mathbf{s}_L|\overline{K}^0\rangle \approx \langle K^0|\mathbf{d}_L^\dagger\sigma_\mu\mathbf{s}_L|0\rangle\langle 0|\mathbf{d}_L^\dagger\sigma^\mu\mathbf{s}_L|\overline{K}^0\rangle \; , \qquad (7.149)$$

which enabled them to use

$$\langle K^0(p)|\mathbf{d}_L^\dagger\sigma_\mu\mathbf{s}_L|0\rangle = p_\mu\frac{F_K}{\sqrt{2m_K}} \; . \qquad (7.150)$$

where F_K is the kaon decay constant. Nowadays, one simply admits to ignorance by adding a fudge factor to the vacuum saturation amplitude, called B_K. Putting it all together, we find

$$\Delta m = \frac{G_F}{\sqrt{2}} B_K F_K^2 m_K \mathcal{F}_K \; , \qquad (7.151)$$

The same analysis can be applied to the $B^0-\overline{B}^0$ system, with the result

$$\mathcal{F}_B = A^2\lambda^6 F(x_{\rm c},x_{\rm c}) + 2A^2\lambda^6(1-\rho+i\eta)F(x_{\rm c},x_{\rm t}) + A^2\lambda^6(1-\rho+i\eta)^2 F(x_{\rm t},x_{\rm t}) \; . \qquad (7.152)$$

In this case, it is the top quark contribution dominates the mass difference.

We must caution the reader that we have presented only the easy part of these calculations. In the real world, strong QCD rules, and one must develop sophisticated techniques to evaluate its contributions. The best way is to organize the calculation so as to efficiently parametrize the strong QCD effects in terms of unknown constants in the operator product expansion formalism. The ambitious reader is referred to the previously-cited review of Buras and Fleischer, and references therein.

PROBLEMS

A. Use isospin Clebsch-Gordan coefficients to show that $\Delta I = 1/2$ dominance implies that $|A(K_S \to \pi^+ \, \pi^-)|^2 = 2|A(K_S \to \pi^0 \, \pi^0)|^2$.

B. 1-) Assume for a moment that the standard model contains only the up, down and strange quarks. Show that the box contribution of the up quark to the box diagram gives a $K_L - K_S$ mass difference that is unacceptably large.
2-) Show that with the addition of the charm quark, the mass difference gets suppressed. Proceed to evaluate the value of the charm quark mass that yields agreement with experiment. See M. K. Gaillard and B. W. Lee, *Phys. Rev.* **D10**, 897(1974).
3-) Discuss the significance of these results, in view of the long range contribution to these effects.

C. Use diagrams to show that the $\Delta I = 1/2$ is plausible in the standard model.

D. Show that in the standard model $\epsilon' \neq 0$. Estimate the ratio ϵ'/ϵ.

CHAPTER 8

FIRST JOURNEY: *MASSIVE NEUTRINOS*

8.1 NEUTRINOS IN THE STANDARD MODEL

In this chapter we discuss the possibility that neutrinos have masses. For quite some time, many different experiments have reported deficits in the number of solar neutrinos detected on earth. A recent experiment has reported an anomalous angular behavior in the flux of neutrinos produced by cosmic rays. There are also unconfirmed reports of neutrino oscillations in laboratory experiments. While keeping in mind that weak interaction experiments tend to be wrong in the first go, some of these anomalies might survive the test of time, to be explained in terms of small neutrino masses. If corroborated, these effects would represent the first tangible deviation of Nature from the original standard model. Thankfully, additional experiments either in progress or in the building stage should bring this interesting topic to a decisive conclusion in the near future.

We begin by showing how neutrinos come out to be naturally massless in the standard model; we then present generalizations of the standard model designed to accomodate neutrino masses. Finally we discuss neutrino oscillations, the major experimental tool for determining their masses and mixings.

Lorentz Analysis

For completeness, we review some of the salient facts about neutrinos, some of which have been covered in earlier chapters. Standard model neutrinos are purely left-handed and antineutrinos are right-handed. No right-handed neutrinos have ever been observed. Accordingly, neutrinos are represented by left-handed Weyl spinors, $\nu^i_L(x)$ which transform non-unitarily as the $(\mathbf{2},\mathbf{1})$ representation of the Lorentz group, where i denotes their flavor. Antineutrinos are described by the right-handed conjugate fields, $\bar{\nu}^i_L(x) = \sigma_2\nu^{i*}_L(x)$, that transform as the $(\mathbf{1},\mathbf{2})$ representation.

These fields allow three types of bilinears. The first, corresponding to the product $(\mathbf{2},\mathbf{1}) \otimes (\mathbf{1},\mathbf{2}) = (\mathbf{2},\mathbf{2})$, transforms as a four-vector

$$J^{\mu ij}(x) = \nu^{i\dagger}_L(x)\sigma^\mu\nu^j_L(x) \ , \tag{8.1}$$

where as usual $\sigma^\mu = (\sigma^0 \equiv 1,\ \vec{\sigma})$ denote the 2×2 unit and Pauli matrices, respectively. When summed over the flavors, it is the neutrino part of the neutral current. In the following, when confusion is impossible, we omit the L and R subscripts: a spinor field without a bar over it is understood to transform as $(\mathbf{2},\mathbf{1})$, one with a bar over it as $(\mathbf{1},\mathbf{2})$.

The $SU(2)$ decomposition of the product $(\mathbf{2},\mathbf{1}) \otimes (\mathbf{2},\mathbf{1}) = (\mathbf{1},\mathbf{1})_A \oplus (\mathbf{3},\mathbf{1})_S$, indicates that we can form two other bilinears in neutrino fields. The first, given by

$$\widehat{\nu}^{(i}(x)\nu^{j)}(x) \ , \tag{8.2}$$

corresponds to the Lorentz scalar $(\mathbf{1},\mathbf{1})$. Since the ν^i obey Fermi statistics, antisymmetrization on the Lorentz indices requires this bilinear to be symmetric in the flavor indices i, j, as indicated by the round brackets. It is Lorentz invariant, and can appear in the Lagrangian as the Majorana mass term

$$\mathcal{L}_{\text{Maj}} = m_{(ij)}\widehat{\nu}^{(i}\nu^{j)} + \text{ c.c.} \ . \tag{8.3}$$

We can characterize its Majorana mass entries by assigning to each neutrino flavor ν^i a global lepton number L_i. The ii diagonal elements break the lepton number L_i (but preserves a discrete lepton number parity $(-1)^{L_i}$, while the ij entries preserves only the difference $L_i - L_j$. On the other hand, the diagonal part of the current bilinear conserves lepton number, while its off-diagonal element preserves only the sum $L_i + L_j$.

In all generality, masses of both neutral and charged particles can be described by the Majorana formalism. We need (at least) two left-handed fields $N_a(x)$, $a = 1, 2$, to form the symmetric Majorana mass matrix

$$\mathcal{L} = (\widehat{N}_1 \ \widehat{N}_2) \begin{pmatrix} m_1 & m \\ m & m_2 \end{pmatrix} \begin{pmatrix} N_1 \\ N_2 \end{pmatrix} + \text{ c.c.}. \tag{8.4}$$

The entries in the 2×2 symmetric Majorana matrix are distinguished by their "lepton numbers": the diagonal $m_{1,2}$ violate $L_{1,2}$-number by two units, while preserving the $L_{2,1}$-number; the off-diagonal term m violates both L_1- and L_2-number while preserving their difference.

To describe the electron, we identify $N_2 \equiv \sigma_2 e_R^*$ and $N_1 \equiv e_L$, so that $L_1 - L_2$ becomes the electron number (or electric charge). The off-diagonal term of the (Majorana) mass matrix simply becomes the Dirac mass,

$$m(\widehat{N}_2 N_1 + \widehat{N}_1 N_2) = 2m\ e_R^\dagger e_L \ , \tag{8.5}$$

and the diagonal terms are set to zero to conserve electric charge and lepton number. For this reason, the off-diagonal Majorana mass is sometimes called the Dirac mass to distinguish it from the "true" diagonal Majorana mass. For charged particles, this formalism is unnecessary, and in fact quite cumbersome. For neutral fermions however, there is no local quantum number (in this case, electric charge), which forbids the diagonal elements, and this formalism must be used.

To summarize, the Majorana mass matrix is a complex, symmetric matrix, of the form

$$\begin{pmatrix} \text{Maj} & \text{Dirac} \\ \text{Dirac} & \text{Maj} \end{pmatrix} . \tag{8.6}$$

The diagonal elements are of Majorana type, and the off-diagonal elements of Dirac type. It follows that massive neutrinos can be of two types. Either the mass is Majorana and violates lepton number, and there need not be extra fermion degrees of freedom, or it is Dirac and preserves lepton number at the price of introducing for each massive neutrino a right-handed partner with the same lepton number.

The third type of neutrino bilinear is not Lorentz invariant. Antisymmetric in flavor (indicated by square brackets), it is of the form

$$\widehat{\nu}^{[i}(x)\vec{\sigma}\nu^{j]}(x) \ , \tag{8.7}$$

with the Lorentz quantum numbers $(\mathbf{3},\mathbf{1})$, corresponding to an antisymmetric second rank tensor. Together with its conjugate, it represents the electric and magnetic moments of the neutrinos. Recall that an antisymmetric tensor $F_{\mu\nu}$ can be decomposed into two space vectors $E_i = F_{0i}$ and $B_i = \frac{1}{2}\epsilon_{ijk}F_{jk}$, and these two vectors can be arranged as a complex

(Hertz) vector $\vec{E} \pm i\vec{B}$, corresponding to $(\mathbf{3}, \mathbf{1})$ and $(\mathbf{1}, \mathbf{3})$ respectively. One can also think of the $(\mathbf{3}, \mathbf{1})$ as a complex antisymmetric tensor which obeys the complex self-duality condition

$$F_{\mu\nu} = \frac{i}{2}\epsilon_{\mu\nu\rho\sigma}F^{\rho\sigma} \ . \tag{8.8}$$

Fermi statistics requires antisymmetrization in the flavor indices. This is as it should be: there is no electromagnetic moment for only one left-handed field – just as for dancing the tango, it takes two. For charged particles, this is well understood since the electromagnetic magnetic moment interaction causes a transition between left- and right-chiralities. It follows that a single Majorana particle has no electromagnetic moments.

These bilinears can appear in the Lagrangian multiplied by the field strengths because of Lorentz invariance, leading to dimension-five interactions that violate total lepton number by two units. If allowed by symmetries, they would be generated by loop diagrams.

Standard Model Analysis

The analysis of possible neutrino bilinears can now be applied to take into account the quantum numbers of the standard model. First of all, the left-handed neutrinos appear as the upper components of the three flavor isodoublets

$$L_i = \begin{pmatrix} \nu^i \\ e^i \end{pmatrix} \sim (\mathbf{2}, \mathbf{1}; \mathbf{2}, \mathbf{1}^c)_{-1}, \quad i = 1, 2, 3 \ . \tag{8.9}$$

The notation denotes the representations under $(SU_2 \times SU_2;\ SU_2^W \times SU_3^c)_Y$, the first two $SU(2)$ refer to the Lorentz group, Y is the hypercharge, SU_2^W is the weak isospin, and SU_3^c the color group; for example, the left-handed antielectron transforms as $\bar{e}_i \sim (\mathbf{2}, \mathbf{1}; \mathbf{1}, \mathbf{1}^c)_2$.

Majorana masses would therefore be generated by Lorentz invariant bilinears in the fields L_i, with different electroweak quantum numbers, depending on their family values.

The first Lorentz-invariant bilinear is antisymmetric in family indices, with quantum numbers

$$(\mathbf{1}, \mathbf{1}; \mathbf{1}, \mathbf{1}^c)_{-2}^{[ij]} \ . \tag{8.10}$$

Since it is a weak isosinglet with hypercharge -2, it carries electric charge, and cannot be a mass term.

The second Lorentz-invariant is the family-symmetric bilinear that transforms as

$$(\mathbf{1}, \mathbf{1}; \mathbf{3}, \mathbf{1}^c)^{(ij)}_{-2} \ . \tag{8.11}$$

It also carries hypercharge, but it is a weak isotriplet, and one easily checks that one of its components is electrically neutral: it can describe the Majorana mass of the standard model neutrinos. Since the standard model contains no Higgs triplets, we conclude that there are no tree-level masses for the neutrinos.

However, this fermion bilinear has the same electroweak quantum numbers as that of two standard model Higgs doublets, enabling us to form a dimension-five Lorentz invariant which satisfies all the standard model gauge symmetries

$$\frac{m^{ij}}{v^2} \widehat{L}_i \tau_2 \vec{\tau} L_j \cdot H^t \tau_2 \vec{\tau} H \ , \tag{8.12}$$

where the τ-matrices are the weak isospin generators. If it can be generated in the effective Lagrangian, it would lead to neutrino Majorana masses when the Higgs assume their vacuum value $v = G_F^{-1/2}$.

However, this interaction cannot be generated in perturbation theory because of lepton number: in the standard model, the Higgs field carries no lepton number, and this operator violates lepton number by two units. We conclude that Majorana masses are forbidden in the standard model only by a *global chiral symmetry*: lepton number.

Is there any way to avoid this conclusion? After all, lepton number is not a good quantum symmetry, because of the anomaly associated with the electroweak local symmetries. This leads to a violation of the left handed part of *total* lepton number L, while preserving $B - L$. However, this effective interaction also violates $B - L$ by two units, and our conclusion remains the same. To produce an invariant term, we would need to form something like

$$\widehat{L}_i \tau_2 \vec{\tau} L_j \cdot H^t \tau_2 \vec{\tau} H \ \mathcal{O} \ , \tag{8.13}$$

where $\mathcal{O}$ is a weak isosinglet combination of fields with two units of B. If such a construct existed and acquired a non-zero value in the electroweak vacuum, this would surely generate the tiniest of tiny neutrino masses. This invariant would satisfy the $B - L$ symmetry, but it is still not invariant under the *relative* lepton numbers $L_e - L_\mu$, and $L_\mu - L_\tau$, which are exactly conserved (non-anomalous). Hence neutrinos are strictly massless in the standard model. In grand unified models, such as $SU(5)$, the relative lepton numbers are violated and this interaction could generate tiny neutrino masses.

One may still think that it is possible to generate electroweak invariant operators with the quantum numbers of $\mathcal{O}$ out of six quarks, but it is unlikely that they will be non-zero in the electroweak vacuum. Otherwise, it would mean that QCD forces could cause the breaking of baryon number, contrary to the Vafa-Witten theorem which states that QCD does not break vectorial symmetries. This theorem is of very general import, although it may be weakened when the electroweak Yukawa couplings of the quarks are included.

The discussion of neutrino electromagnetic moments proceeds in the same way. One can form two combinations, depending on the symmetry of the family numbers, with the quantum numbers

$$(\mathbf{3}, \mathbf{1}; \mathbf{1}, \mathbf{1}^c)^{(ij)}_{-2} \ , \tag{8.14}$$

for the family symmetric bilinear, and

$$(\mathbf{3}, \mathbf{1}; \mathbf{3}, \mathbf{1}^c)^{[ij]}_{-2} \ , \tag{8.15}$$

for the family antisymmetric case. Again we note that only the combination that is antisymmetric in families can have a neutral component and thus be identified as a magnetic moment. In order to form invariants in which these might appear, it is convenient to introduce the generalized Hertz vectors

$$\vec{W}_m \equiv \vec{W}_{0m} + \frac{i}{2}\epsilon_{mnp}\vec{W}_{np} \ , \tag{8.16}$$

built out of the weak isospin field strengths, and

$$B_m = B_{0m} + \frac{i}{2}\epsilon_{mnp}B_{np} \ , \tag{8.17}$$

made out of the weak hypercharge field strengths. Hypercharge invariance requires two Higgs fields as well as the above to make invariants of dimension seven. With only one Higgs field, the family antisymmetric invariants of lowest dimensions are given by

$$\widehat{L}_{[i}\sigma_m\vec{\tau}L_{j]} \cdot \begin{cases} H^t\tau_2\vec{\tau}HB_m \\ (H^t\tau_2\vec{\tau}H \times \vec{W}_m) \end{cases} . \tag{8.18}$$

If allowed, these would give rise to transition moments among neutrinos, but using arguments detailed earlier for the mass terms, they are forbidden by lepton number conservation.

Finally, we can form invariants with the family-symmetric combinations, such as

$$\widehat{L}_{(i}\sigma_m\tau_2 L_{j)} H^t \tau_2 \vec{\tau} H \cdot \overrightarrow{W}_m \ , \tag{8.19}$$

but they do not lead to static properties for the neutrinos. This lengthy description of electroweak neutrinos has taught us two things:

- Majorana neutrino masses break electroweak symmetry by $\Delta I_w = 1$,
- Majorana neutrino masses are protected only by *global* lepton number symmetries.

By now even the most questioning reader should be convinced that standard model neutrinos are strictly massless. Thus any experimental evidence for neutrino masses would be an indication of the necessity of going *beyond* the standard model.

The present experimental limits on neutrino masses come from laboratory experiments; they are indeed very tiny

$$m_{\nu_e} \leq 10 - 15 \text{ eV} \ ; \quad m_{\nu_\mu} \leq 0.17 \text{ MeV} \ ; \quad m_{\nu_\tau} \leq 18.2 \text{ MeV} \quad .$$

Many more restrictions on neutrino masses, from the early history of the universe, and stellar astrophysics, are beyond the scope of this book. The interested reader can consult *The Early Universe* by R. Kolb and M. Turner, (Addison-Wesley, 1990).

In addition, the absence of lepton-number violating processes, such as neutrinoless double β decay, puts stringent direct limits on lepton-number violation.

Finally, we note that measurements of the ρ parameter, *i.e.* the relative strengths of neutral and charged currents, provide strong evidence that electroweak breaking in the isotriplet channel is very small compared to the main breaking in the $\Delta I_w = 1/2$ channel. Thus it is reasonable to believe that mere breaking of lepton number would not necessarily create large neutrino masses. In the next section, we discuss ways to relax these restrictions, and present various extensions of the standard model with neutrino masses.

PROBLEMS

A. Use CP to build the neutrino bilinears that correspond to their electric and magnetic moments, out of $\widehat{\nu}^{[i}\vec{\sigma}\nu^{j]}$ and their conjugates. Show that invariance under CP forbids the electric dipole moment term.

B. With two Higgs doublets, construct possible operators of dimension seven that can yield neutrino electric and magnetic moments.

C. Can you build a six quark condensate with the quantum numbers of the lepton doublet bilinear?

8.2 ELECTROWEAK MODELS WITH MASSIVE NEUTRINOS

It should be clear from the previous section that the standard model must be extended to accomodate massive neutrinos. As we shall see, the extensions do not put in question the nature of the standard model, but rather add more parameters to it. However, in the context of renormalizable theories, all necessarily involve new degrees of freedom, of either spin zero and/or one-half.

In order to classify these models, we use only the quantum numbers of the standard model, and organize their presentation in terms of the new degrees of freedom and their global lepton number symmetries.

Generically there are two kinds of neutrino mass models. The first preserves at least total lepton number, and contains only neutrino masses of the Dirac type; these require new fermion degrees of freedom to serve as the Dirac partners of the neutrinos. In the second type of models, the neutrino masses can be of both Majorana and Dirac type, and lepton number is necessarily broken; these models differ by the mechanism that break the lepton numbers, which can occur by adding only extra fermions, only extra Higgs fields or both.

Fermion Extensions

These generate neutrino masses at the cost of introducing only fermion degrees of freedom, and they are of two types, depending on whether or not lepton number is broken.

Lepton-number conserving models

To preserve lepton number and massive neutrinos we need to introduce new fermions to serve as the Dirac partners of the left handed neutrinos. These new fermion degrees of freedom can have any electroweak quantum numbers (see problem), but it is simplest to assign them no electroweak quantum numbers. Because they have no weak interactions, they are called "sterile" neutrinos.

In general these new fields $\overline{N}_a(x)$ are labelled by their own flavor index a, which need not be the same (see problem) as the number of chiral families. For simplicity, it is taken to be the same, *i.e.* $a = i = 1, \ldots n_{\text{fam}}$, the number of families. These isosinglet fermions couple to the lepton doublets via the new Yukawa interactions

$$Y_{ij}^{(0)}\widehat{L}_i\overline{N}_j\tau_2 H + \quad \text{c.c.} \;, \tag{8.20}$$

where $Y_{ij}^{(0)}$ is an unknown Yukawa matrix, and H is the Higgs doublet of the standard model. At least one global lepton number is preserved, with $L = -1$ for all the $\overline{N}_i$'s. Electroweak breaking then generates a neutrino Dirac mass of the order of $245 \times Y^{(0)}$ GeV. The Yukawa matrix can be diagonalized by the transformation

$$\mathbf{Y}^{(0)} = \mathbf{U}_\nu^\dagger \mathbf{D}_\nu \mathbf{V}_\nu \;, \tag{8.21}$$

where $\mathbf{U}_\nu$ and $\mathbf{V}_\nu$ are unitary matrices, and $\mathbf{D}_\nu$ is diagonal. This diagonalization produces a mixing matrix in the charged leptonic weak current, in direct analogy to the CKM matrix in the quark sector. It is easy to form the charged current in terms of the mass eigenstates

$$J_\mu^{(+)} = e^\dagger \sigma_\mu \mathcal{U}_{MNS} \nu_m \;, \tag{8.22}$$

where

$$\mathcal{U}_{MNS} \equiv \mathbf{U}_e \mathbf{U}_\nu^\dagger \;, \tag{8.23}$$

is the lepton weak mixing matrix, named here for its inventors, Z. Maki, M. Nakagawa, and S. Sakata (*Prog. Theo. Phys.* **28**, 870(1962)). Here $\mathbf{U}_e$ is the matrix that diagonalizes the charged leptons, obtained from diagonalizing $\mathbf{Y}^{(-1)}$. The extraction of the independent phases from the mixing matrix proceeds as for the CKM matrix, by performing the *Iwasawa* decomposition; for three families, it leaves us with only one CP-violating phase and three mixing angles. As in the quark case, there is no mixing in the neutral current. At the end of the procedure, we are left with invariance under the total lepton number $L = L_e + L_\mu + L_\tau$.

The experimental limits on neutrino masses imply that the $\mathbf{Y}^{(0)}$ coupling constants must themselves be very small, in the range $\mathbf{Y}^{(0)} \leq (10^{-10} - 10^{-4})$. If one accepts such tiny couplings (after all, we already have $m_e = 10^{-6} M_W$!), this represents a viable extension of the standard model, but its drawback is that it falls short of explaining the smallness of the neutrino masses compared to those of the charged leptons. For this reason, it does not bear the mark of a model that can satisfy theorists, many of which feel that this simple extension cannot be the whole story. In the next section we present a lepton number violating generalization which contains a plausible scenario for these small numbers.

PROBLEMS

A. Devise an electroweak model with three right handed neutrino fields that transform as weak isotriplets. Determine their hypercharge and Yukawa couplings that generate neutrino masses. Discuss the phenomenology of this type of model.

B. Construct a standard model model with five extra right-handed sterile neutrinos, with arbitrary mixing to the weak lepton doublets. Assume that two of them form a very massive Dirac pair, discuss the ensuing neutrino mass spectrum and mixing with the standard model neutrinos. What would be the low energy signatures of this model?

Lepton-number violating models

Models that break lepton number face the experimental fact that there are excellent limits against it. For instance, the neutrinoless double β decay

$$^{76}Ge \to {}^{76}Se + e^- + e^- \ , \tag{8.24}$$

which violates electron lepton number by two units has never been observed. This results on an impressive bound on the lifetime $\tau \geq 1.1 \times 10^{25}\mathrm{yr}$. It can be expressed in terms of the bound

$$\sum_i (\pm)_i (\mathcal{U}_{ei})^2 m_{\nu_i} \leq 1 \sim 2 \text{ eV} \ . \tag{8.25}$$

Here the sign depends on the CP property of the neutrino state, $\mathcal{U}_{ei}$ is the element of the MNS mixing matrix between the electron and ith species neutrino of mass m_i. One way to remember the $\pm$ sign is to note that in the event of lepton number conservation there is no effect. Thus a Dirac neutrino enters this formula as two degenerate states, one helicity state with a $+$ sign, the other with a $-$ sign.

It is straightforward to incorporate lepton number violation in the models of the previous section with pure Dirac masses: simply add Majorana mass terms for the right-handed singlets, which break explicitly lepton number. This leads to the extra couplings

$$\mathcal{L}_{\text{extra}} = Y^{(0)}_{ij} \widehat{L}_i \overline{N}_j \tau_2 H + M_{ij} \widehat{\overline{N}}_i \overline{N}_j \ + \ \text{c.c.} \ . \tag{8.26}$$

Here, $\mathbf{Y}^{(0)}$ is the previously discussed Yukawa matrix. The new feature is M_{ij}, the symmetric Majorana mass matrix of the Dirac partners, that does not break any local electroweak symmetry, only global lepton numbers.

In the absence of electroweak breaking, this theory describes three massless neutrinos, interacting with three massive Majorana neutrinos. The values of their Majorana masses are completely arbitrary; they may be very large, limited perhaps only by the Planck mass. The symmetric Majorana matrix $\mathbf{M}$ can be diagonalized by a Schur transformation,

$$\mathbf{M} = \mathbf{U}_0^t \mathbf{D}_0 \mathbf{U}_0 \ , \tag{8.27}$$

where $\mathbf{U}_0$, $\mathbf{D}_0$ are unitary and diagonal matrices, respectively. The entries in $\mathbf{D}_0$ are the physical masses of the extra neutral particles.

With electroweak breaking, the mass matrix becomes more complicated. The three neutrinos, ν_i and their Dirac partners, $\overline{N}_i$ now mix through the Majorana mass matrix,

$$(\widehat{\nu} \ \widehat{\overline{N}}) \begin{pmatrix} \Delta I_w = 1 & \Delta I_w = \frac{1}{2} \\ \Delta I_w = \frac{1}{2} & \Delta I_w = 0 \end{pmatrix} \begin{pmatrix} \nu \\ \overline{N} \end{pmatrix} \ , \tag{8.28}$$

where we have indicated the electroweak breaking properties of the 3×3 block submatrices. Without a Higgs triplet, the $\Delta I_w = 1$ entry is zero at tree level, although it will be corrected by loop-generated radiative effects, since lepton number is violated. The $\Delta I_w = \frac{1}{2}$ off-diagonal entries come from the Dirac-type Yukawa couplings of the previous section. The $\Delta I_w = 0$ entry represents the tree-level Majorana mass term. The new tree-level Majorana mass matrix is given by

$$\mathcal{M} = \begin{pmatrix} 0 & v\mathbf{Y}^{(0)} \\ v\mathbf{Y}^{(0)t} & \mathbf{M} \end{pmatrix} \ . \tag{8.29}$$

Perturbation theory limits the off-diagonal entries of $\mathcal{M}$ to be not much larger than the scale of electroweak breaking, but no such restriction applies to the diagonal $\Delta I_w = 0$ entries. When they are much larger than the electroweak-breaking scale, the diagonalization of this matrix yields a satisfactory explanation for the suppression of neutrino masses over those of their charged partners.

In that limit, it is convenient to introduce a parameter which denotes the ratio of the electroweak breaking scale to the $\Delta I_w = 0$ masses,

$$\epsilon = \frac{\Delta I_w = \frac{1}{2}}{\Delta I_w = 0} \ . \tag{8.30}$$

In many models, the $\Delta I_w = 0$ scale occurs naturally at a much larger value than the electroweak scale, and this parameter is very small. Thus we are led to diagonalize the whole matrix in the limit $\epsilon \ll 1$. This way to generate small neutrino masses has been dubbed the *see-saw mechanism.*

Let us also assume in this discussion that $\mathbf{M}$ has no zero eigenvalues (see problem). We scale out ϵ, and write the full Majorana matrix in the Schur form

$$\mathcal{M} = \mathcal{V}^t \mathcal{D} \mathcal{V} \ , \tag{8.31}$$

where

$$\mathcal{V} = \begin{pmatrix} \mathbf{U}_{11} & \epsilon \mathbf{U}_{12} \\ \epsilon \mathbf{U}_{21} & \mathbf{U}_{22} \end{pmatrix} \ ; \qquad \mathcal{D} = \begin{pmatrix} \epsilon^2 \mathbf{D}_\nu & 0 \\ 0 & \mathbf{D}_0 \end{pmatrix} \ . \tag{8.32}$$

We find three light eigenvalues, suppressed $\mathcal{O}(\epsilon^2)$ over the $\Delta I_w = 0$ scale, *i.e.* further suppressed from the electroweak scale by a factor of ϵ. This mechanism naturally yields tiny neutrino masses, at the cost of introducing a new scale of physics, much larger than the electroweak scale. Since the matrix $\mathcal{V}$ is itself unitary, we have

$$\begin{aligned} \mathbf{U}_{11}\mathbf{U}_{11}^\dagger + \epsilon^2 \mathbf{U}_{12}\mathbf{U}_{12}^\dagger &= 1 \ , \\ \mathbf{U}_{22}\mathbf{U}_{22}^\dagger + \epsilon^2 \mathbf{U}_{21}\mathbf{U}_{21}^\dagger &= 1 \ , \\ \mathbf{U}_{11}\mathbf{U}_{21}^\dagger + \mathbf{U}_{21}\mathbf{U}_{22}^\dagger &= 0 \ , \end{aligned} \tag{8.33}$$

and because ϵ is small, the matrices $\mathbf{U}_{11}$ and $\mathbf{U}_{22}$ are almost unitary.

The charged weak current mixes mass eigenstates, but with subtle differences over the previous case. The weak charged current density can now be expressed in terms of the mass eigenstates

$$J_\mu^{(+)} = e^\dagger \sigma_\mu \mathbf{U}_e (\mathbf{U}_{11}^\dagger \nu_m + \epsilon \mathbf{U}_{21}^\dagger \overline{N}_m) \ . \tag{8.34}$$

The light lepton mixing matrix is simply

$$\mathcal{U}_{MNS} = \mathbf{U}_e \mathbf{U}_{11}^\dagger \ , \tag{8.35}$$

where now $\mathbf{U}_{11}$ is almost unitary, so that $\mathcal{U}_{MNS}$ can be decomposed *à la* Iwasawa

$$\mathbf{U}_e \mathbf{U}_{11}^\dagger = \mathcal{P} \mathcal{U}_{MNS}^{'} \mathcal{P}' \ , \tag{8.36}$$

where $\mathcal{P}$ and $\mathcal{P}'$ are diagonal phase matrices, and $\mathcal{U}_{MNS}^{'}$ contains three mixing angles and one phase. The phase matrix $\mathcal{P}$ can be absorbed in $e_L^\dagger$ and then transferred into e_R, but a new feature is that the $\mathcal{P}'$ phase matrix cannot be absorbed into ν, since it would reappear in the mass term $\epsilon^2 \widehat{\nu}_m \mathbf{D}_\nu \nu_m$. We see that there are two additional CP-violating phases in the lepton sector, which depend on the presence of the $\mathcal{O}(\epsilon^2)$ Majorana mass matrix. We expect these extra phases to be suppressed by ratios of the order of neutrino Majorana masses over the relevant energies, and thus almost inobservable.

The mixing matrix $\mathbf{U}_{11}$ and the Majorana masses of the light neutrinos $\mathcal{M}_\nu = \epsilon^2\mathbf{D}_\nu$ are determined from diagonalizing the matrix

$$\mathcal{Y} = \mathbf{Y}^{(0)}\frac{v^2}{\mathbf{M}}\mathbf{Y}^{(0)t} , \tag{8.37}$$

$$= -\mathbf{U}_{11}^t\epsilon^2\mathbf{D}_\nu^{-1}\mathbf{U}_{11} . \tag{8.38}$$

This formula applies only if $\mathbf{M}$, the $\Delta I_w = 0$ entry in the Majorana mass, has no zero eigenvalue. The neutrino part of the weak neutral current is given by

$$J_\mu = \nu_m^\dagger\sigma_\mu\nu_m + \epsilon\nu_m^\dagger\sigma_\mu\mathbf{U}_{11}\mathbf{U}_{21}^\dagger\overline{N}_m + \epsilon\overline{N}_m^\dagger\sigma_\mu\mathbf{U}_{21}\mathbf{U}_{11}^\dagger\nu_m + \mathcal{O}(\epsilon^2) . \tag{8.39}$$

The neutral mass eigenstates are labelled by the subscript m. The weak neutral current has a small neutrino flavor-changing component between the light and heavy neutrinos, as well as a more suppressed flavor changing interaction among the light degrees of freedom.

This mechanism introduces a new scale, between the electroweak and Planck energies. It naturally accomodates the small value of the neutrino masses by linking them to the ratio of the electroweak scale to this new scale. In models where the renormalization group allows perturbative extrapolation to the deep ultraviolet, it is natural to consider this type of scale.

PROBLEMS

A. Diagonalize and discuss the ensuing physics when the $\Delta I_w = 0$ entry in the Majorana mass matrix has one zero eigenvalue.

B. Suppose there are two right-handed singlet neutrino fields per family. First diagonalize the one-family case. Then generalize your results to three families.

C. Find a diagram that would cause neutrinoless double beta decay, and then express the lifetime in terms of the bound in the text.

8.3 HIGGS EXTENSIONS

These models contain leptonic bosons, spin zero degrees of freedom that carry lepton number, but no extra fermions. Thus neutrino masses can only be of Majorana type, which are possible only if lepton number is violated. These models are further characterized by the mechanism of lepton number violation. In the first type of models, lepton number is violated explicitly by the interactions of the leptonic bosons. The second type of models are invariant under lepton number, but the potential is such as to cause its spontaneous breaking. These contain a Nambu-Goldstone boson, called the *Majoron*.

In a renormalizable theory, the leptonic bosons acquire their lepton number by their Yukawa couplings to standard model lepton bilinears. In the color singlet channel, the only possible lepton bilinears have the following Lorentz and electroweak quantum numbers

$$(\mathbf{1},\mathbf{1};\mathbf{3},\mathbf{1}^c)_{-2}\ , \qquad (\mathbf{1},\mathbf{1};\mathbf{1},\mathbf{1}^c)_{-2}\ , \qquad (\mathbf{1},\mathbf{1};\mathbf{1},\mathbf{1}^c)_{4}\ . \tag{8.40}$$

These allow three types of possible Yukawa interactions, given by

$$\widehat{L}_{(i}\tau_2\vec{\tau}L_{j)}\ \vec{T}_{Y=2}\ , \qquad \widehat{L}_{[i}\tau_2L_{j]}\ S^{+}_{Y=2}\ , \qquad \widehat{\bar{e}}_{(i}\bar{e}_{j)}\ S^{--}_{Y=-4}\ . \tag{8.41}$$

We have denoted the symmetrization properties of the flavor indices in the usual way by curved and square brackets. We see that there can be three different types of leptonic bosons, two have total lepton number $L = -2$, an isotriplet $\vec{T}$ with hypercharge 2, and components $(\varphi^{++},\varphi^{+},\varphi^{0})$, a singly charged isosinglet S^{+}, and the third, a doubly charged isosinglet S^{--} , with $L = 2$. These couplings are used to determine the quantum numbers of the new fields. To see how lepton number can be broken, we must rely on the interactions among these fields and the standard model Higgs doublet.

Theories of this type with massive neutrinos can then be enumerated in terms of the possible renormalizable interactions in the potential. With one Higgs doublet H, and these fields, there are only three types of cubic terms,

$$S^{--}S^{+}S^{+}\ , \qquad \vec{T}\cdot\vec{T}S^{--}\ , \qquad H^{t}\tau_2\vec{\tau}H\cdot\vec{T}\ . \tag{8.42}$$

All three break total lepton number with $\Delta L = 2$, leaving only a discrete subgroup. On the other hand, the quartic Higgs potential contains, in addition to the absolute squares of all the quadratic terms, the interactions

$$H^\dagger \tau^a H \; T^{*b} T^c \epsilon^{abc}, \qquad H^\dagger \tau^a H \; T^{*a} S^+ \; , \qquad H^t \tau_2 \tau^a H \; T^{*a} S^{--} \; , \quad (8.43)$$

None of these terms explicitly violate L; still, they must be included in any renormalizable theory that contains these fields, as they will be generated by radiative corrections with infinite coefficients. We present examples of such models, where explicit breaking comes about through cubic terms in the potential.

Isotriplet Higgs models with explicit breaking

The simplest extension with a leptonic boson adds only one field to the standard model: an isotriplet Higgs with $Y = -2$. The model explicitly breaks lepton number by adding a dimension-three coupling. The extra terms are

$$Y_{ij} \widehat{L}_{(i} \tau_2 \vec{\tau} L_{j)} \cdot \vec{T} + \mu H^t \tau_2 \vec{\tau} H \cdot \vec{T} + \text{c.c.} \; , \quad (8.44)$$

where Y_{ij} are new Yukawa couplings, and the cubic term breaks lepton number, but leaves a lepton number parity symmetry $(-1)^L$. The model contains extra Higgs interactions that break no symmetries

$$m^2 T^{a*} T^a + \lambda_0 H^\dagger \tau^a H \; T^{*b} T^c \epsilon^{abc} + \lambda_1 (T^{a*} T^a)^2 + \lambda_2 H^\dagger H T^{a*} T^a \; , \quad (8.45)$$

With a positive m^2 in the potential, the isotriplet does not get a vacuum value. Since total lepton number is broken by μ, the dimension-five operator that produces neutrino masses is generated by loop diagrams. An attractive feature of this model is that neutrino masses are naturally suppressed by loop effects (see problem).

One could change the sign of the isotriplet mass term in the potential; in that case, the $\vec{T}$ leptonic boson would get a vacuum value, and produce tree-level neutrino masses, without any natural suppression mechanism for the value of their masses.

The family structure of the symmetric Yukawa matrix $\mathbf{Y}$ is rather interesting. While this model does violate some total lepton number, through the Higgs cubic coupling, it may preserve various relative lepton numbers, depending on the form of the matrix $\mathbf{Y}$. When it is of the form

$$\mathbf{Y} = \begin{pmatrix} 0 & 0 & 0 \\ 0 & 0 & y \\ 0 & y & 0 \end{pmatrix} \; , \quad (8.46)$$

two lepton numbers, L_e and $L_\mu - L_\tau$ are left invariant, while the cubic term breaks their sum $L_\mu + L_\tau$. In this case, the electron neutrino stays massless, and the μ and τ neutrinos become Dirac partners of one another (see problem).

When only one combination of lepton numbers is left invariant, it is of the form

$$\mathbf{Y} = \begin{pmatrix} 0 & 0 & x \\ 0 & 0 & y \\ x & y & 0 \end{pmatrix} . \tag{8.47}$$

This Majorana model was first proposed to explain an anomaly (which went away) in the beta decay of Tritium (see problem).

Hybrid Higgs models with explicit breaking

Our analysis of possible Higgs cubic couplings show that there are no models of neutrino masses and explicit breaking of lepton number that contains only one isosinglet. There is however one model with both types of singlets S^{--} and S^{+}. The relevant terms in its Lagrangian are two Yukawa couplings and one cubic Higgs coupling,

$$A_{ij}\widehat{L}_{[i}\tau_2 L_{j]}S^{+} + S_{ij}\widehat{\overline{e}}_{(i}\overline{e}_{j)}S^{--} + \mu S^{--}S^{+}S^{+} + \text{c.c.} \,, \tag{8.48}$$

where $\mathbf{A}$ and $\mathbf{S}$ are the Yukawa matrices. In this model, neutrino masses are also generated at loop level, and thus naturally small (see problem).

With both isosinglet and isotriplet, $\vec{T}$ and S^{--}, there is yet another model with two extra Yukawa couplings, and one cubic coupling. The Lagrangian contains the extra terms

$$S_{ij}\widehat{\overline{e}}_{(i}\overline{e}_{j)}S^{--} + Y_{ij}\widehat{L}_{(i}\tau_2\vec{\tau}L_{j)}\vec{T} + \mu\vec{T}\cdot\vec{T}S^{--} + \text{c.c.} \,, \tag{8.49}$$

together with quadratic and quartic invariant terms in the Higgs potential. The reader is encouraged to work out the resulting neutrino masses (see problem).

It is relatively easy to build new models of neutrino masses by adding to the Higgs sectors. While most such extensions are not motivated except perhaps by the pressure of thesis writing, two deserve a special mention. One is through the addition of another Higgs doublet; the other is the addition of an electroweak singlet Higgs with no hypercharge, which is found in some grand unified models and in the invisible axion model (see chapter 9).

With two Higgs doublets, the most interesting new interaction that breaks lepton number in the cubic term is of the form

$$H^t \tau_2 H' S^+ \ , \qquad (\Delta L = 2) \ . \tag{8.50}$$

The extra quartic term, $H^t \tau_2 H' S^- S^{++}$, does not violate lepton number. The two Higgs have the same hypercharge, but in order to avoid flavor changing neutral current effects they must couple to different channels.

This cubic term allows for another model with one Yukawa coupling and one cubic term,

$$\mathcal{L}_{\text{extra}} = A_{ij} \widehat{L}_{[i} \tau_2 L_{j]} S^+ + \mu H^t \tau_2 H' S^+ + \text{c.c.} \ . \tag{8.51}$$

Noteworthy is the fact that the Yukawa matrix is purely antisymmetric in family space, and the neutrino masses are generated at loop order. Again, we leave it as a problem to explore the consequences of this model, first proposed by A. Zee, *Phys. Lett.* **B93**, 389(1980).

With an electroweak singlet, Φ, cubic couplings can all be turned into quartic couplings, yielding the possible terms $S^{--} S^+ S^+ \Phi$, $\vec{T} \cdot \vec{T} S^{--} \Phi$, $H^t \tau_2 \vec{\tau} H \cdot \vec{T} \Phi$, and $H^t \tau_2 H' S^+ \Phi$. Three of these couplings give Φ six units of lepton number, and the last one two units. With this singlet field one can further consider its own self-interactions. The ones that break lepton number explicitly can appear in the potential in the form $m^2 \Phi^2$, $\mu \Phi^3$, Φ^4, and $\Phi^* \Phi^3$. With two Higgs doublets and one singlet, we contemplate several more quartic interactions, namely $H^t \tau_2 H' S^+ \Phi$, and $H^t \tau_2 H' \Phi^2$.

It is obvious that with the addition of new fields, one can build many new models. Restraint and good taste prevents the author from going any further, but should not hinder the student from doing so.

PROBLEMS

A. In the model with one Higgs isotriplet, identify the lowest order Feynman diagram that generates neutrino masses. Estimate its magnitude.

B. In the model with the Yukawa matrix of the form Eq. (8.47), diagonalize the neutrino mass matrix and describe its consequences.

C. Analyze the possible outcomes of the model of Eq. (8.48).

D. Perform the same analysis for the model of Eq. (8.49).

E. Analyze the neutrino eigenvalues of the model of Eq. (8.51), and estimate their magnitudes.

F. Invent a model of neutrino masses with explicit lepton number violation that has not been covered in this section.

Spontaneous Breaking of Lepton Number

As we mentioned, it could be that lepton number is violated spontaneously. In that case, for each lepton number that is broken, there will be an associated massless Nambu-Goldstone boson, the Majoron, denoted generically by J. Since physics must be invariant under a constant shift in $J(x)$, it couples to the divergence of the current associated with the broken lepton number symmetry with an interaction term

$$\mathcal{L}_{\text{Maj}} = \frac{1}{V} J(x) \partial^\mu \mathcal{J}^l_\mu \ , \tag{8.52}$$

where V is the vev of the field whose vacuum value does the breaking, and $\mathcal{J}^l_\mu$ is the lepton current. This is the only place J appears in the Lagrangian of a spontaneously broken lepton number.

Models with Majorons differ, depending on the electroweak properties of the field that does the breaking. If breaking is generated by a weak isomultiplet, the models run into contradictions with experiments, as we detail below. No such restriction exists when lepton number is broken by a standard model singlet field.

Isotriplet Breaking

The simplest model of this type contains one isotriplet leptonic boson, no cubic interaction terms, and with the potential arranged so that it gets a vacuum expectation value, and therefore breaks lepton number spontaneously. Its extra Yukawa coupling is

$$\mathcal{L}_{\text{extra}} = Y_{ij} \widehat{L}_{(i} \tau_2 \vec{\tau} L_{j)} \vec{T} + \text{c.c.} \ , \tag{8.53}$$

with the Higgs potential chosen to preserve lepton number

$$-m^2 T^{a*} T^a + \lambda_0 H^\dagger \tau^a H \, T^{*b} T^c \epsilon^{abc} + \lambda_1 (T^{a*} T^a)^2 + \lambda_2 H^\dagger H T^{a*} T^a \ , \tag{8.54}$$

with the sign of the mass term judiciously chosen. The neutral component of $\vec{T}$ can, by acquiring a vev, generate tree-level $\Delta I_w = 1$ neutrino masses, although it leaves us with no understanding of the small ratio of triplet to doublet breakings. In addition, experiments tell us that triplet breaking of the electroweak symmetry is suppressed relative to the doublet breaking. These constraints can be met by carefully arranging the Higgs couplings in the potential, while making sure that the vacuum value of $\vec{T}$ does not break electric charge (see problem).

The most interesting feature of this model, of course, is the presence of the Nambu-Goldstone boson, called the triplet Majoron indicating the nature of the breaking mechanism. The lepton current is now

$$\mathcal{J}^{l}_{\mu} = -2\vec{T}^{*}\overleftrightarrow{\partial}_{\mu}\vec{T} + \nu_i^{\dagger}\sigma_{\mu}\nu_i + e_i^{\dagger}\sigma_{\mu}e_i - \bar{e}_i^{\dagger}\sigma_{\mu}\bar{e}_i \ , \tag{8.55}$$

and the triplet Majoron couples with strength $1/V$, where V is the vev of the triplet. It must be much smaller than v, the vev of the doublet, to account for the experimental value of the ρ parameter.

This upper bound on V has a nefarious consequence in a renormalizable theory. A detailed study of the potential shows that if $V \ll v$, there exists a neutral Higgs boson, ρ_J, lighter than the Z boson. Since the triplet Higgs has electroweak quantum numbers, the triplet Majoron and this field will couple to the Z, leading to the decay

$$Z \longrightarrow J + \rho_J \ . \tag{8.56}$$

This decay adds to the decay width of the Z in a way that contradicts present data. We used this example to illustrate the problems inherent to non-singlet Majoron models. We add that because of the small value of V, lepton number cannot be gauged in this type of model. Gauging can occur only in singlet-breaking models.

Isosinglet Breaking

The simplest singlet Majoron model requires, besides the electroweak order parameter for lepton number, a set of right-handed antineutrinos. The singlet Φ couples to the Dirac partners of the neutrinos in the form

$$\mathcal{L}_{\text{extra}} = Y^{(\nu)}_{ij}\widehat{L}_i N_j \tau_2 H + Y^{(0)}_{ij}\widehat{N}_i N_j \Phi \ + \ \text{c.c.} - V(\Phi, H) \ , \tag{8.57}$$

where $\mathbf{Y}^{(0)}$ is another Yukawa matrix. The potential part that contains the new field is given by

$$V(\Phi, H) = -M^2\Phi^*\Phi + \Lambda_0(\Phi^*\Phi)^2 + \Lambda_1\Phi^*\Phi H^{\dagger}H \ , \tag{8.58}$$

which preserves total lepton number. The sign of the mass term is chosen so as to break lepton number when Φ, which carries two units of lepton number, gets a vacuum value. The mass M is an arbitrary mass parameter, of the scale at which lepton number is broken. The singlet Majoron couples to the divergence of the total lepton current, it is given by

$$\mathcal{J}^l_\mu = -2\Phi^* \overleftrightarrow{\partial}_\mu \Phi + \nu_i^\dagger \sigma_\mu \nu_i + e_i^\dagger \sigma_\mu e_i - \bar{e}_i^\dagger \sigma_\mu \bar{e}_i \ . \qquad (8.59)$$

This coupling has many interesting consequences; in particular, using the equations of motion, the Majoron-electron vertex is given by

$$\mathcal{L}_{e-\text{Maj}} = \frac{m_e}{V} J \ \bar{e}\gamma_5 e \ , \qquad (8.60)$$

using Dirac notation. This allows to turn a photon into a Majoron through the process

$$\gamma + e^- \longrightarrow J + e^- \ . \qquad (8.61)$$

In stars, it provides a way to dissipate energy which, for a sufficiently low value of V, can lead to contradiction with the existence of red giants (see problem).

PROBLEMS

A. Analyze the potential of the triplet Majoron model, and determine bounds on its parameters so that electric charge is conserved, and the triplet vev is much smaller than the doublet's. Show that there always exists a neutral Higgs lighter than the Z.

B. Derive the contribution to the Z width in the triplet Majoron model. Use the present value of the width to infer an allowable range for the scale of the triplet breaking. See G. B. Gelmini and M. Roncadelli, *Phys. Lett.* **B99**, 411(1981).

C. Derive a bound on the vacuum value of the singlet Majoron, based on the existence of Arcturus (a red giant), among others.

8.4 NEUTRINO OSCILLATIONS

In 1957, motivated by rumors that neutrinos were sometimes produced in natural beta decay, Pontecorvo proposed that one should consider, in analogy with the $K_0 - \overline{K}_0$ system, that antineutrinos decay might oscillate into neutrinos. The rumors turned out to be just that, but the idea of neutrino oscillations remained. In 1962, when it was experimentally established that there were two types of neutrinos, the Nagoya group (Maki *et al*, *op. cit.*) proposed that oscillations could also take place between different neutrinos flavors, a phenomenon now called flavor oscillations.

Oscillations between neutrinos occur only if they can be differentiated in some way. This in turn requires that the lepton numbers which distinguish them from one another be broken, for instance by having different masses.

The two types of oscillations, flavor and particle-antiparticle, differ in terms of their lepton number properties. Neutrino-antineutrino oscillations, which break lepton number by two units, are very difficult to detect since they involve a spin flip. Their physical consequences are therefore weighted by ratios of (small) Majorana masses to large neutrino momenta.

Flavor oscillations on the other hand do not violate total lepton number, only relative lepton numbers. Accordingly, they may be easier to detect since they will not be suppressed by such small ratios.

For simplicity, we consider only flavor oscillations between two neutrinos. In vacuum, the (tilded) neutrino mass eigenstates satisfy the evolution equation

$$i\frac{d}{dt}\begin{pmatrix}\tilde{\nu}_e \\ \tilde{\nu}_\mu\end{pmatrix} = \begin{pmatrix}E_e & 0 \\ 0 & E_\mu\end{pmatrix}\begin{pmatrix}\tilde{\nu}_e \\ \tilde{\nu}_\mu\end{pmatrix} . \tag{8.62}$$

By definition, these mass eigenstates evolve in time without mixing. However, if there is mixing among the species, the states produced by the Weak Interactions will not be mass eigenstates. Generically, the (tilded) mass eigenstates will be related by a canonical transformation to the **weak** eigenstates, which we label by bareheaded wavefunctions, in this case, ν_e, and ν_μ. We have

$$\begin{pmatrix}\tilde{\nu}_e \\ \tilde{\nu}_\mu\end{pmatrix} = \begin{pmatrix}\cos\theta_0 & \sin\theta_0 \\ -\sin\theta_0 & \cos\theta_0\end{pmatrix}\begin{pmatrix}\nu_e \\ \nu_\mu\end{pmatrix} ,$$

where θ_0 is the vacuum mixing angle.

Now let us assume that at $t = 0$ a muon neutrino state is created in a weak interaction, say from the decay $\pi^+ \to \mu^+ + \nu_\mu$. By inverting the transformation above, we express this initial state as a linear combination of the mass eigenstates

$$|\nu_\mu\rangle = \cos\theta_0|\tilde{\nu}_\mu\rangle + \sin\theta_0|\tilde{\nu}_e\rangle . \tag{8.63}$$

At a later time t, that neutrino state will have evolved into

$$|\nu_\mu(t)\rangle = \cos\theta_0 e^{-iE_\mu t}|\tilde{\nu}_\mu\rangle + \sin\theta_0 e^{-iE_e t}|\tilde{\nu}_e\rangle .$$

We then sample the composition of this beam by using a detector which triggers only on charged current interactions. It is therefore only sensitive to the weak eigenstates. To find the triggering probabilities of these detectors, we rewrite the beam as a linear combination of weak eigenstates,

$$|\nu_\mu(t)\rangle = A(t)|\nu_\mu\rangle + B(t)|\nu_e\rangle \ , \tag{8.64}$$

with amplitudes

$$\begin{aligned} A(t) &= \cos^2\theta_0 e^{-iE_\mu t} + \sin^2\theta_0 e^{-iE_e t} \ , \\ B(t) &= \cos\theta_0 \sin\theta_0 (e^{-iE_e t} - e^{-iE_\mu t}) \ . \end{aligned} \tag{8.65}$$

The absolute value squared of these coefficients denotes the triggering probabilities. Suppose we have a detector that triggers on positrons, indicating the absorption of the weak eigenstate ν_e. Its triggering probability is then

$$\begin{aligned} P_{(\mu\to e)}(t) &= |\langle\nu_\mu(t)|\nu_e\rangle|^2 \ , \\ &= \sin^2 2\theta_0 \sin^2(\Delta E t) \ , \end{aligned} \tag{8.66}$$

where ΔE is the difference between the two energies. Assuming that the two highly relativistic neutrinos have the same momentum, their energies are

$$\begin{aligned} E_e &= \sqrt{p^2 + m_1^2} \approx p + \frac{m_1^2}{2p} \ , \\ E_\mu &= \sqrt{p^2 + m_2^2} \approx p + \frac{m_2^2}{2p} \ , \end{aligned} \tag{8.67}$$

which enables us to recast ΔE in terms of the difference in squared masses. It is natural to introduce the oscillation length L_0

$$L_0 \equiv \frac{2\pi}{\Delta E} = \frac{4\pi p}{\Delta m^2} \ , \tag{8.68}$$

in terms of which the appearance probability is

$$P_{(\mu\to e)}(t) = \sin^2 2\theta_0 \sin^2 \frac{2\pi t}{L_0} \ . \tag{8.69}$$

It peaks whenever the neutrino has travelled a distance equal to $L_0/4$. The distance between peaks is L_0. A convenient way to remember the size of the oscillation length is through the formula

$$L_0(\text{meters}) = 2.47\frac{E(\text{MeV})}{\Delta \text{m}^2(\text{eV}^2)} \quad . \tag{8.70}$$

This treatment of neutrino oscillations which might appear too simple-minded, nevertheless captures the esentials of the phenomenon: the oscillation length is both an experimental and conceptual benchwork. If we place the detector too many oscillation lengths away from the source, the different components of the beam, corresponding to states of different masses, will have separated so much that they can no longer interfere with one another. At the other end of the spectrum, oscillations cannot be detected unless the position of the neutrino is localized well within one oscillation length. This implies through the uncertainty principle that the momentum of the neutrino has a large spread, and that a wavepacket treatment of the neutrino beam is warranted. Some of these subtelties are addressed in the problems.

Laboratory experiments designed to detect neutrino oscillations fall into several categories (for a more comprehensive survey, see the review articles in the bibliography). The first class of experiments uses man-made sources of neutrinos and/or antineutrinos. With antineutrinos beams generated in reactors, experiments are difficult since the antineutrinos are not directional, as the flux falls sharply with distance. Other experiments use neutrino and antineutrino beams produced in the decay of particles generated at particle accelerators. When the detector is close to the birthplace of the neutrinos (short baseline), the beam is highly directional. However when the detector is far away (long baseline) the spreading of the beam becomes a factor. The best arrangement is to monitor the same beam by both near and far detectors. Such an experiment (MINOS) is under construction at FermiLab. So far, the search for flavor oscillations in these types of experiments has not yielded any corroborated positive evidence for oscillations. These experiments have served to rule out certain regions of parameter space in the $\Delta m^2 - \sin^2 2\theta_0$ plane. A reported positive result by LSND, challenged by KARMEN, will soon be tested at FermiLab as well (BooNE). In this field, knowledge of acronyms such as CHORUS, ICARUS, K2K, NOMAD, is half the physics!

The second class of experiments are very sensitive underground detectors that monitor neutrinos emitted by natural sources, notably neutrinos emitted in the core of the Sun, and the neutrinos and antineutrinos emitted as decay products from cosmic ray collisions in the atmosphere. Among these, most remarkable is the Homestake Mine experiment (for a review, see R. Davis, *Prog. Part. Nucl. Phys.* **32**, 13(1994)), the first to see a deficit in solar neutrinos. The new generation of such detectors were constructed to search for proton decay (IMB, Kamiokande), just in time to detect the neutrino burst from supernova SN1987A! The lowest energy solar neutrinos have been detected by other detectors (SAGE, GALLEX), to be further discussed below.

PROBLEMS

A. Suppose that the mass of the neutrino created in the decay of a π^+ is measured to such accuracy that it tells us which neutrino mass eigenstate is being emitted. Use the uncertainty principle to show that in this case, neutrino oscillations will not be observable.

B. Estimate the maximum distance from the source a detector may be placed so as to observe oscillations.

C. Discuss the detectability of neutrino oscillations in terms of the oscillation length L_0 with an initial wavepacket of momentum spread Δp (see B. Kayser, *Phys. Rev.* **D24**, 110(1981)).

8.5 SOLAR NEUTRINOS: PRODUCTION AND DETECTION

According to the theory of the solar engine, electron neutrinos are produced during the so-called *pp* chain which starts with proton fusion and ends up with α particles. Neutrinos are also produced during the so-called CNO cycle as well, but at a much reduced rate, as the *pp* chain is responsible for 98.5% of the energy generated by the sun. The *pp* chain reactions are as follows:

$$p + p \to {}^2H + e^+ + \nu_e, \qquad 0 < E_\nu < 0.420 \text{ MeV} ,$$

$$p + e^- + p \to {}^2H + \nu_e \qquad E_\nu = 1.44 \text{ MeV} ,$$

$${}^2H + p \to {}^3He + \gamma ,$$

$${}^3He + {}^3He \to {}^4He + p + p ,$$

$${}^3He + {}^4He \to {}^7Be + p ,$$

$$^7Be + e^- \to {}^7Li + (\gamma) + \nu_e, \qquad E_\nu = \begin{cases} 0.861 \text{ MeV}(90\%) \\ 0.383 \text{ MeV}(10\%) \end{cases},$$

$$^7Be + p \to {}^8B + \gamma \ ,$$

$$^8B \to {}^8Be + e^+ + \nu_e \ , \qquad 0.81 < E_\nu < 14.06 \text{ MeV} \ ,$$

$$^7Li + p \to {}^4He + {}^4He \ ,$$

$$^8Be \to {}^4He + {}^4He \ .$$

The reason for the two lines in ^7Be capture is that 10% of the time the ^7Li is produced in a metastable state which then radiatively decays. So there are three neutrino lines, one from the *pep* reaction, and two from Be capture, as well as two continuous bands, one at low energy coming from the original fusion reaction, the other at higher energies coming from the decay of Boron. The flux of the lower energy neutrinos is directly related to the proton density at the core. On the other hand, the flux of the higher energy neutrinos depends on the abundance of ^8B. Clearly the latter is much more environmentally sensitive than that of the primal protons. This translates into the dependence of the neutrino flux on the Sun's temperature. The flux of neutrinos from Boron decay depends on the T_C^{18}, where T_C is the Sun's core temperature. This is to be compared with the T_C^8 dependence of ^7Be, and that for *pp* neutrinos produced in the fusion reaction which depend only on $T_C^{-1.2}$. In the last few years, the theoretical understanding of the standard solar model, (J. N. Bahcall, S. Basu, and M. H. Pinsonneault, *Phys. Lett.* **B433**, 1(1998); A. Brun, S. Turck-Chièze, and P. Morel, *Ap. J.* **506**, 913(1998)) has received spectacular corroboration by the measurement of seismic waves on the Sun's surface at the predicted level.

• The first experiment, performed deep underground in the Homestake mine by R. Davis and collaborators, monitors the reaction

$$\nu_e + {}^{37}Cl \to e^- + {}^{37}Ar \ . \tag{8.71}$$

Since it takes neutrinos with at least .814 MeV of energy to produce Argon, only neutrinos coming from the *pep* capture, Beryllium capture and the Boron decay can trigger this transition. Radiochemical methods are used to count the produced ^{37}Ar atoms in a large vat of cleaning fluid. This experiment, originally suggested by B. Pontecorvo in 1945, has been performed for over twenty five years albeit with funding interruptions. It has consistently detected fewer neutrinos than expected from theory by roughly a factor of two.

• The second experiment is of the same type, but uses a cleverly chosen transition that can be triggered by much lower energy neutrinos. These Gallium detectors use the raction

$$\nu_e + {}^{71}Ga \to e^- + {}^{71}Ge \ . \tag{8.72}$$

It can be triggered by neutrinos with as little energy as .23 MeV. Thus it is sensitive to neutrinos that come from the fusion reaction, as well of course as to those that come from all the other sources. However, the flux of the *pp* neutrinos is much larger than the flux emanating from Boron decay. This type of experiment has been performed at two underground detectors, SAGE (V. Gavrin *et al*, in *Neutrino 96*, K. Huitu, K. Enqvist, and J. Maalampi, eds. (World Scientific, Singapore, 1997)), near Baksan in the Caucasus mountain range, and GALLEX (P. Anselmann *et al Phys. Lett.* **B342**, 440(1996); W. Hampel *et al ibid* **B388**, 364(1996)) in the Gran Sasso tunnel, not too far from Hadrian's villa. Both detectors have been calibrated with a radioactive ^{51}Cr source, and both detect less than the expected flux of neutrinos.

• The third type of experiment, conducted in the US by the IMB collaboration, and in Japan at Kamioka (Y. Fukuda *et al*, *Phys. Rev. Lett.* **77**, 1683(1994)), uses a water Čerenkov detector, originally built for proton decay, and used to detect neutrinos coming from the SN1987A supernova. A neutrino entering the detector can scatter with an electron

$$\nu_e + e^- \to \nu_e + e^- \ . \tag{8.73}$$

and the electron recoils mostly in the direction it was hit. Solar theory predicts an excess number of recoil electrons opposite the direction of the sun at the time of impact. Since Kamiokande can measure recoil electrons with an energy greater than 7.3 MeV, it is sensitive only to neutrinos generated in Boron decay. The newer and larger SuperKamiokande detector has also observed the same deficit in solar neutrinos from ^{8}B, although not with sufficient energy to determine the shape of the spectrum.

In the convenient Solar Neutrino Unit (SNU), which equals 10^{-36} captures per target atom per second, the approximate theoretical expectations from the standard solar model are listed in the following table

Reaction	^{37}Ar	^{71}Ge
pp	0	69.6
pep	0.2	2.8
^{7}Be	1.15	34.4
^{8}B	5.9	12.4
CNO	0.5	9.7
Total	7.6	128.9

These are to be compared with the experimental results

$$\begin{aligned} \text{Homestake :} &\quad 2.1 \pm 0.3 \\ \text{Gallex :} &\quad 79 \pm 10 \pm 6 \\ \text{Sage :} &\quad 74 \pm 12 \pm 6 \end{aligned} \tag{8.74}$$

The Kamiokande experiment which detects only ^{8}B neutrinos also reports solar neutrinos in less than expected numbers. In units of 10^6/cm-sec, the measured flux of solar neutrinos is $2.89 \pm .22 \pm 0.35$, compared with the theoretical expectation of 5.69 ± 0.82 in one model and 4.4 ± 1 in another.

Both Sage and Gallex results are consistent with one another, but they are well below theoretical expectations, as if the neutrino flux from ^{7}Be capture were greatly reduced. This is interesting since the results from Homestake and Kamiokande seem to clash: they "see" only about 35% and 50% of the expected neutrino flux, respectively. According to theory, it should be the other way around since Homestake detects, in addition to ^{8}B, neutrinos from the higher ^{7}Be line, the *pep* reaction, and the CNO cycle.

This confusing experimental situation is expected to be cleared up by a new generation of experiments. A larger water Čerenkov detector, SuperKamiokande, with much improved energy resolution, is presently measuring the energy dependence of the tail of the Boron spectrum. The Solar Neutrino Observatory (SNO) at Sudbury in Canada is expected to come on line within a year. It is a *heavy* water Čerenkov detector. It has two advantages in that it will provide a more accurate determination of the energy dependence of the ^{8}B spectrum, and, thanks to its neutron detectors, will be able to measure the neutral current reaction of the solar neutrinos when they knock out a neutron from the deuterium. The neutron capture by another atom produces a gamma ray which is then observed. Unlike previous detectors, SNO will thus measure the total flux of standard model neutrinos, irrespective of flavor. Thus if the electron neutrino oscillates into one of the other two standard model neutrinos, SNO will be observing the total solar neutrino flux, as well as its flavor composition.

Finally in the Gran Sasso tunnel, BOREXINO, a water detector with scintillation counters, will measure the neutrinos produced in ^7Be capture. These new experiments have the capability of producing hard evidence for solar neutrino oscillations in the near future!

8.6 SOLAR NEUTRINO OSCILLATIONS

One explanation for the solar neutrino deficit could be neutrino vacuum oscillations among different species. This would happen if the Earth-Sun distance were just right to be in a trough in the variation of the probability with distance. Since the Earth-Sun distance is 1.5×10^{11} meters, one would need, depending on the mixing angle, $\Delta m^2 \approx 10^{-11}$ eV^2. Not all solar neutrinos would have the same oscillation length: the pp neutrinos have a much smaller oscillation length than the ^8B neutrinos. This possible explanation for the deficit is not inconsistent with the present data.

A more exciting possibility is that the deficit in solar neutrinos is due to a resonance effect that comes about when electron neutrinos traverse the sun. Emitted in the core of the sun, neutrinos must travel through most of the sun to get to earth. L. Wolfenstein (*Phys. Rev.* **D17**, 2369(1978)) remarked that the sun's interior acts as a refractive medium to neutrinos because of their coherent forward scattering. Also different flavors would have different indices of refraction. S. Mikheyev and A. Yu. Smirnov, (*Yad. Fiz.* **42**, 1441(1985); *Nuovo Cim.* **9C**, 17(1986)) then noted that, when combined with vacuum mixing, a flavor-dependent index of refraction can lead to the vanishing of the diagonal element of the effective mixing matrix, and thus produce maximum mixing even for small vacuum mixing angles. Accordingly, this phenomenon is called the MSW effect.

To see how the index of refraction comes about in quantum mechanics, consider a wave traversing a slab of matter of length R with N scattering centers per unit area. The wave function is the sum of the incoming and scattered waves,

$$\Psi(x) = \Psi_{in}(x) + \Psi_{scat}(x) \ . \tag{8.75}$$

For an incoming plane wave of momentum p, it is given in lowest order of perturbation theory by

$$\Psi(x) \approx \exp\left[ipx + \frac{2\pi i N R}{p} f_p(0)\right] , \tag{8.76}$$

where $f_p(0)$ is the forward scattering amplitude. The total wave function is just he vacuum wavefunction, multiplied by a modulating factor

$$\Psi(x) \approx e^{ip(x-R)} \; e^{ipRn(p)} , \tag{8.77}$$

where $n(p)$ is the index of refraction

$$n(p) \equiv 1 + \frac{2\pi}{p^2} N f_p(0) . \tag{8.78}$$

It depends both on the momentum of the incoming particle and on the local density of scattering centers. The imaginary part of $f_p(0)$ parametrizes the extinction (optical theorem), and the real part just alters the speed of propagation of the plane wave.

The interactions of an electron neutrino with stable matter are well known: its elastic forward scattering by electrons is caused mostly by both W and Z-mediated processes, and its interactions with neutrons and protons occur through Z-exchange. It follows that n_e, the index of refraction of the electron-neutrino weak eigenstate, is given by

$$p(n_e - 1) = \frac{2\pi}{p}\left[N_e f_p^W(0) + \sum_i N_i f_p^Z(0)\right] , \tag{8.79}$$

where the W and Z contributions to the scattering amplitudes have been singled out, and $i = 1, 2$ for protons and neutrons, respectively. On the other hand, a muon-neutrino weak eigenstate will not have any W-mediated interactions since there are no muons in the sun, although its neutral current interactions will be the same as for the electron-neutrino, leading to its own index of refraction n_μ,

$$p(n_\mu - 1) = \frac{2\pi}{p}\sum_i N_i f_p^Z(0) . \tag{8.80}$$

One might also consider a *sterile* neutrino ν_s, defined as one that has no electroweak interactions with matter; its index of refraction would be equal to one. Hence, distinct neutrino flavors propagate through matter quite differently, since for the weak eigenstates,

$$|\nu_e(x)\rangle = e^{ipxn_e}|\nu_e\rangle , \qquad |\nu_\mu(x)\rangle = e^{ipxn_\mu}|\nu_\mu\rangle . \tag{8.81}$$

These equations show that the real part of the index of refraction corresponds to a change in energy proportional to $p(n-1)$ for each flavor in the *weak eigenstate basis.*

The evolution equation of the ultrarelativistic weak eigenstates is therefore given by

$$i\frac{d}{dt}\begin{pmatrix}\nu_e\\ \nu_\mu\end{pmatrix} = H_0\begin{pmatrix}\nu_e\\ \nu_\mu\end{pmatrix} + \begin{pmatrix}p(n_e-1) & 0\\ 0 & p(n_\mu-1)\end{pmatrix}\begin{pmatrix}\nu_e\\ \nu_\mu\end{pmatrix}, \tag{8.82}$$

where

$$H_0 = \begin{pmatrix}\cos\theta_0 & \sin\theta_0\\ -\sin\theta_0 & \cos\theta_0\end{pmatrix}\begin{pmatrix}E_e & 0\\ 0 & E_\mu\end{pmatrix}\begin{pmatrix}\cos\theta_0 & -\sin\theta_0\\ \sin\theta_0 & \cos\theta_0\end{pmatrix}, \tag{8.83}$$

is the evolution operator in vacuum. After discarding all elements proportional to the unit matrix, it can be cast in the form

$$i\frac{d}{dt}\begin{pmatrix}\nu_e\\ \nu_\mu\end{pmatrix} = \frac{\pi}{L_0}\begin{pmatrix}\cos 2\theta_0+\delta(t) & \sin 2\theta_0\\ \sin 2\theta_0 & -\cos 2\theta_0-\delta(t)\end{pmatrix}\begin{pmatrix}\nu_e\\ \nu_\mu\end{pmatrix}, \tag{8.84}$$

with

$$\delta(t) \equiv \frac{L_m(t)}{2L_0}, \tag{8.85}$$

and where

$$\frac{2\pi}{L_m(t)} \equiv p[n_e(t)-n_\mu(t)]\ . \tag{8.86}$$

As the neutrino traverses matter at nearly the speed of light, its location is essentially proportional to its age, t, so that the indices of refraction develop a time-dependence through the local electron density. Furthermore, $\delta(t)$ depends on the type of neutrinos that are oscillating. For $\nu_e - \nu_\mu$ flavor oscillations, the neutral current contributions cancel out from $L_m(t)$, leaving those from the charged current interactions with electrons. A careful standard model calculation yields

$$\delta(t) = \frac{\sqrt{2}G_F p N_e(t)}{\Delta m^2}, \tag{8.87}$$

where G_F is the Fermi constant, $N_e(t)$ is the local electron density, and $\Delta m^2 = m^2_{\bar{\nu}_e} - m^2_{\bar{\nu}_\mu}$, is negative if the electron neutrino is lighter than the muon neutrino. With a heavier muon neutrino, δ is negative, and the diagonal element can vanish if

$$\delta(t_r) = -\cos 2\theta_0 \ , \tag{8.88}$$

at which point the off-diagonal elements, no matter how small, produce maximum mixing. This is the MSW effect which happens when $t = t_r$, at the critical electron density

$$N_e^{crit} = -\frac{\Delta m^2 \cos 2\theta_0}{2\sqrt{2}G_F E_\nu} \ . \tag{8.89}$$

To determine under what circumstances this resonance effect can occur, we need to introduce basic facts about the sun. Solar neutrinos are produced by nuclear reactions inside the Sun's core. There, the electron density is roughly constant, while outside it decreases exponentially

$$N_e(r) = N_c e^{-r/r_0} \ , \tag{8.90}$$

where N_c is the electron density at the core. The density is roughly one hundred Avogadro's number at one tenth the radius, and $r_0 \approx 7 \times 10^7$ meters, is about one tenth the Sun's radius. Since N_e is always less than its value N_c at the core, this gives a critical value for the neutrino energy

$$E_\nu^{min}(\text{MeV}) = -\frac{\Delta m^2 \cos 2\theta_0}{2\sqrt{2}G_F N_c} = 4 \times 10^4 \ |\Delta m^2|(\text{eV}^2) \cos 2\theta_0 \ . \tag{8.91}$$

Neutrinos with energy less than E_ν^{min} do not encounter a resonance layer. Since the detectable solar neutrinos range in energy from .23 to 14 MeV, this limits the range of fundamental parameters for which this effect is measurable.

The solution of the evolution equation that governs the mass eigenstates is complicated by the presence of the time-dependent electron density. For arbitrary time variation, it cannot be solved analytically, but for simple time dependences, such as linear and exponential variations, analytic solutions are known.

Without going into a complete analytic solution, we can offer a useful qualitative discussion of the effect, leaving a more rigorous treatment for the education of the reader.

At any given time, the *instantaneous* eigenstates of the evolution equations are given by

$$|\nu_1(t)\rangle_I = \cos\theta_m(t)|\nu_e\rangle - \sin\theta_m(t)|\nu_e\rangle \ ,$$
$$|\nu_2(t)\rangle_I = \sin\theta_m(t)|\nu_e\rangle + \cos\theta_m(t)|\nu_\mu\rangle, \tag{8.92}$$

with the instantaneous mixing angle

$$\sin^2 2\theta_m(t) = \frac{\sin^2 2\theta_0}{1 + 2\delta(t)\cos 2\theta_0 + \delta^2(t)} \ , \tag{8.93}$$

and eigenvalues

$$E_{1,2}(t) = \pm\frac{\pi}{L_0}\sqrt{1 + 2\delta(t)\cos 2\theta_0 + \delta^2(t)} \ . \tag{8.94}$$

At the critical electron density, these two eigenvalues are closest to one another, and the mixing angle is the largest.

If the time-dependence of the electron density is sufficiently slow, we can use the adiabatic approximation, according to which the true eigenkets of the evolution equation can be written as the instantaneous eigenkets modulated, in the lowest approximation, only by the evolution phase factor,

$$|\nu_{1,2}(t)\rangle_{\rm Ad} \sim e^{-i\int_{t_0}^{t} dt' E_{1,2}(t')}|\nu_{1,2}(t)\rangle_I . \tag{8.95}$$

These adiabatic states are good approximations to the neutrino wavefunctions outside of their resonance layer.

As the neutrinos enter their resonance layer, there are two lengths to consider. The first is the size of the resonance layer, over which the MSW effect is appreciable, and the second is the oscillation length of the neutrino at resonance.

Outside the Sun's core, the size of the resonance layer can be expressed in terms of the variation of the electron density by

$$\Delta r = r_0\frac{|\Delta N_e|}{N_e} \ . \tag{8.96}$$

The region of resonance is that over which the mixing angle is appreciable, say $\sin^2 2\theta_m \geq \frac{1}{2}$. From the expression for θ_m, we deduce that δ varies according to $\Delta\delta = \sin 2\theta_0$, corresponding to a variation of the density

$$\Delta N_e = \frac{\Delta m^2 \sin 2\theta_0}{2\sqrt{2}G_F E_\nu} \ . \tag{8.97}$$

It follows that the width of the resonance layer depends only on the vacuum angle and the Sun's parameter r_0

$$\Delta r = 2r_0 \tan 2\theta_0 \ . \tag{8.98}$$

The second length of interest is the oscillation length of the neutrino beam at resonance

$$L_{\text{enh}} = \frac{4\pi E_\nu}{\Delta m^2 \sin 2\theta_0} \ , \tag{8.99}$$

which depends on the neutrino energy.

We may consider two different possibilities. In the first "adiabatic" approximation, the region over which the enhancement takes place is much larger than the oscillation length. The other is the "non-adiabatic" regime where the resonance region, Δr, is much smaller than the oscillation length.

$$L_{\text{enh}} \ll \Delta r : \quad \text{Adiabatic Regime},$$

$$L_{\text{enh}} \gg \Delta r : \quad \text{Non} - \text{Adiabatic Regime} \ .$$

Which of these is relevant depends on the value of the fundamental parameters, and the neutrino energy. Qualitatively, we expect the adiabatic regime to hold if the MSW region is in the core region, and the non-adiabatic hypothesis to hold if the resonance layer is outside the core.

In the adiabatic regime, the neutrinos go through many oscillations before they leave the resonance region. Accordingly, the true solution for the neutrino eigenkets evolve in time like the adiabatic instantaneous eigenstates of the mixing matrix. An electron neutrino is born, evolves as an adiabatic state until it enters the resonance region, where it also evolves an an adiabatic state, mixing maximally with another flavor, and, through level crossing, emerges as another species to be detected on earth. In this regime there is no appreciable probability for jumping from one adiabatic state to the other.

In the non-adiabatic case, the resonance region is much smaller than the oscillation length, and only part of the wave gets lost through oscillation. Then, in the resonance layer, the true neutrino eigenket does not follow the instantaneous eigenkets of the mixing matrix: there is an appreciable jump probability between adiabatic states.

Let us follows the life of a neutrino as it traverses the Sun. An electron neutrino born at $t = t_0$ in the core of the Sun is described by the ket

$$|\nu_e\rangle = \cos\theta_b |\nu_1(t_0)\rangle_I + \sin\theta_b |\nu_2(t_0)\rangle_I \ , \tag{8.100}$$

where we have set $\theta_b = \theta_m(t_0)$. This ket evolves adiabatically until it enters its resonance layer at $t \sim t_r$

$$|\nu_e(t_r)\rangle^{\text{in}}_{\text{Ad}} = \cos\theta_b e^{-i\int_{t_0}^{t_r} E_1 dt'} |\nu_1(t_r)\rangle_I + \sin\theta_b e^{-i\int_{t_0}^{t_r} E_2 dt'} |\nu_2(t_r)\rangle_I \ . \tag{8.101}$$

To be general, let us assume that as the adiabatic states cross the resonance layer, there is a finite jump probability they be converted into one another. The neutrino beam emerges in the form

$$|\nu_e(t_r)\rangle^{\text{out}}_{\text{Ad}} = \cos\theta_b e^{-i\int_{t_0}^{t_r} E_1 dt'} [A|\nu_1(t_r)\rangle_I + B|\nu_2(t_r)\rangle_I] + \sin\theta_b e^{-i\int_{t_0}^{t_r} E_2 dt'} [-B^*|\nu_1(t_r)\rangle_I + A^*|\nu_2(t_r)\rangle_I] \ , \tag{8.102}$$

where the jump probability is

$$P_{\text{jump}} = |B|^2 \qquad ; \quad |A|^2 + |B|^2 = 1 \ . \tag{8.103}$$

In the adiabatic case, $P_{\text{jump}} = 0$. From thereon, the beam evolves adiabatically until it is detected on Earth. At the time t_d of detection, an electron-neutrino detector is sensitive to the ket

$$|\nu_e(t_d)\rangle = \cos\theta_0|\nu_1(t_d)\rangle_I + \sin\theta_0|\nu_2(t_d)\rangle_I \ . \tag{8.104}$$

The probability to detect an electron neutrino from the Sun is given by

$$P_{\nu_e\to\nu_e} = |<\nu_e(t_d)|\nu_e(t_d)\rangle^{\text{out}}_{\text{Ad}}|^2 \ . \tag{8.105}$$

This is a complicated expression which contains terms periodic in the time of detection. These disappear when we average over the location of the detector, leaving us with

$$<P_{\nu_e\to\nu_e}\rangle = \frac{1}{2} + \frac{1}{2}(|A|^2 - |B|^2)\cos 2\theta_b \cos 2\theta_0 - |AB|\sin 2\theta_b \cos 2\theta_0 \cos(2\int_{t_r}^{t_b} E_1 dt' + \alpha) \ , \tag{8.106}$$

where α is the relative phase between the jump amplitudes.

The survival probability simplifies even further when we average over the neutrino beam birthplaces

$$<< P_{\nu_e \to \nu_e} \rangle\rangle = \frac{1}{2} + \frac{1}{2}(1 - 2P_{\text{jump}}(E_\nu)) \cos 2 < \theta_b(E_\nu) \rangle \cos 2\theta_0 \ , \quad (8.107)$$

where we have indicated the dependence on the neutrino energy. The angle $< \theta_b >$ is the average mixing angle at production, assuming it does not vary violently with distance.

In the adiabatic case, we also need to compute the jump probability. In that approximation, the resonance layer is small, so we can use a linear approximation for the electron density. One finds (see problem)

$$P_{\text{jump}}(E_\nu) = \exp\left\{ -\frac{\pi}{4} \frac{\sin^2 2\theta_0}{\cos 2\theta_0} \frac{\Delta m^2}{E_\nu r_0} \right\} \ , \quad (8.108)$$

and the jump probability decreases with increasing neutrino energy. Thus in the non-adiabatic case, a neutrino will not convert if its energy is large enough. This is to be compared with the adiabatic case for which the survival probability depends on energy only through the mixing angle at birth.

These energy dependences of the survival probability make it possible to fit the results from the various solar neutrino experiments. As of now, experiments are not sensitive enough to pinpoint which of these mechanisms causes the observed deficits.

8.7 ATMOSPHERIC NEUTRINOS

In the summer of 1998, the new water Čerenkov detector in the Kamioka mine, SuperKamiokande, reported evidence (Y. Fukuda *et al*, *Phys. Rev. Lett.* **82**, 1810(1999)) for neutrino oscillations. Their underground detector is bombarded by neutrinos from the Sun (easily distinguished by their direction), as well as neutrinos and antineutrinos produced in the decay products of cosmic ray collisions with Earth's atmosphere. The composition of these *atmospheric* neutrinos is rather well understood. The pions produced in the primary interaction decay into a muon and its antneutrino. The muon proceeds to decay, producing an electron, an electron antineutrino and a muon neutrino.

$$\pi^- \to \mu^- + \overline{\nu}_\mu \ ,$$
$$\mu^- \to e^- + \overline{\nu}_e + \nu_\mu \ .$$

As the same sequence occurs for the antiparticles, one expects in the end twice as many muon neutrinos as electron neutrinos. The SuperK detector found that a significant discrepancy with theory, namely

$$\left(\frac{\nu_\mu}{\nu_e}\right)_{\text{measured}} \approx .6 \left(\frac{\nu_\mu}{\nu_e}\right)_{\text{predicted}} ,$$

in agreement with results obtained by previous detectors (IMB, Kamioka, Soudan II).

The second surprise was connected with the direction of the detected atmospheric neutrinos. A charged particle hit by a neutrino recoils in the direction of the incoming neutrino, creating its telltale Čerenkov cone. Since cosmic rays bombard the earth equally from all directions, their decay byproducts are expected to enter the detector from all angles with a specific distribution. Its measurement then allows physicists to reconstruct the isotropy of the original flux. In addition, the directions determine the distance travelled by the atmospheric neutrinos , since all are produced at the top of the atmosphere. For a certain range of parameters, this difference in travel paths makes the detector sensitive to oscillations.

SuperKamiokande reported a difference between the number of neutrinos entering the detector from above and from below. The discrepancy was correlated with the distance travelled, in a way consistent with oscillation of muon neutrinos into some other neutrino species. Their inescapable conclusion: not all muon neutrinos created in the upper atmosphere make it to the detector as muon neutrinos.

The preferred theoretical interpretation to their data is that the muon neutrinos oscillate into another neutrino, call it ν_X, such that

$$|m^2_{\nu_\mu} - m^2_{\nu_X}| = 5.9 \times 10^{-3}\ \text{eV}^2 \qquad \sin^2 2\theta_{\mu-X} \approx .95\ . \tag{8.109}$$

The flavor of ν_X has not yet been conclusively identified. It is improbable that it is an electron neutrino, as a reactor experiment (CHOOZ) has ruled out oscillation between electron and muon meutrinos with those parameters. The likely answer is that it is the τ-neutrino, but one cannot yet rule out the possibility that ν_X is an entirely new type of neutrino. Since the width of the Z-boson is already accounted for by known particles, this new neutrino must be sterile and not couple to the Z-boson.

At this time, there is tantalizing evidence for neutrino masses. In the next decade, experiments will measure the neutrino mass differences and mixings through oscillations. However, it will take sometime before an actual neutrino mass is measured, a challenge for tomorrow.

PROBLEMS

A. Derive the expression for the instantaneous neutrino mass eigenstates, and their eigenvalues, i.e. at a fixed electron density.

B. a-) Verify that the probability of detection a solar electron-neutrino, averaging over both production and detection positions, is given by

$$P = \frac{1}{2} + (\frac{1}{2} - P_{\text{jump}}) \cos 2\theta_b \cos 2\theta_0 \ .$$

b-) Show that in the appproximation that the electron density varies linearly in the transition region, the non-adiabatic jump probability is given by

$$P_{\text{jump}} = \exp\left\{ -\frac{\pi}{4} \frac{\sin^2 2\theta_0}{\cos 2\theta_0} \frac{\Delta m^2}{E_\nu r_0} \right\} \ .$$

For references see S. J. Parke, *Phys. Rev. Lett.* **10** 1275(1986), and S. P. Rosen and J.M. Gelb, *Phys. Rev.* **D34** 969(1986).

C. A possible scenario for the explanation of the solar neutrino deficit is that of vacuum oscillations. In the case of three neutrinos, work out the probability that the electron neutrino oscillates into a muon neutrino. For details, see V. Barger, K. Whisnant, S. Pakvasa, and R.J.N. Phillips, *Phys. Rev.* **D22**, 1686(1980).

D. Some time ago, evidence for a neutrino was reported, with the following parameters: a 17 KeV Majorana mass, and appreciable mixing with the electron neutrino $\sin^2\theta \sim 0.01$. Using what you have learned in this chapter and more, to give at least four reasons (two from astrophysics, two from particle physics) why it cannot exist. Fortunately, the evidence proved fallacious.

CHAPTER 9

SECOND JOURNEY:

THE AXION

9.1 QCD VACUUM ENERGY

There is no experimental evidence that the strong interactions violate either P or CP. The absence of a measurable electric dipole moment for the neutron ($d_n \leq 10^{-26}$ esu) suggests that strong CP-violation is very small, if it exists at all. Yet QCD is capable of breaking these symmetries both spontaneously and explicitly. The former through the chiral-breaking quark condensate, the latter through interactions which violate these discrete symmetries.

The strong QCD forces spontaneously break the approximate chiral symmetry of the light quarks, through a quark-antiquark condensate

$$< \mathbf{q}_L^{\dagger i} \mathbf{q}_{Rj} >_0 \equiv \Lambda^3 e^{\frac{i\gamma}{3}} \mathbf{U}^i{}_j \ , \tag{9.1}$$

where $i, j = 1, 2, 3$ denote the three light flavors. The scale parameter Λ is real. The matrix $\mathbf{U}$ can be reduced to a unitary matrix by using a $SU(3)_L \times SU(3)_R$ chiral transformation. The phase γ is normalized to the number of flavors, by setting $\det \mathbf{U} = 1$. Under P, we have

$$\mathbf{q}_L^{\dagger i} \mathbf{q}_{Rj} \to \mathbf{q}_R^{\dagger i} \mathbf{q}_{Lj} \ , \tag{9.2}$$

and under CP

$$\mathbf{q}_L^{\dagger i}\mathbf{q}_{Rj} \rightarrow \mathbf{q}_R^{\dagger j}\mathbf{q}_{Li} \ , \tag{9.3}$$

so that P-invariance of the condensate requires that

$$\mathbf{U}^\dagger\mathbf{U}^* = e^{\frac{2i\gamma}{3}}\mathbf{1} \ , \tag{9.4}$$

while CP-invariance yields

$$\mathbf{U}^\dagger\mathbf{U}^\dagger = e^{\frac{2i\gamma}{3}}\mathbf{1} \ . \tag{9.5}$$

Taking the determinant of these equations, and using $\det\mathbf{U} = 1$, we see that CP and P invariances require that $\gamma = 0$ or π, but we have (at present) no means of calculating the phase γ, since it is determined by strong QCD.

Explicit breaking of P and CP comes from the term

$$\frac{\theta}{32\pi^2}\,\mathrm{Tr}\,(\mathbf{G}_{\mu\nu}\tilde{\mathbf{G}}_{\mu\nu}) \ . \tag{9.6}$$

The dimensionless parameter θ can be absorbed in the quark masses, but the combination $\overline{\theta} = \theta -$ arg det $\mathbf{M}_q$, where $\mathbf{M}_q$ is the quark mass matrix, is physical. This term and/or the complex quark masses break the discrete symmetries explicitly. The experimental limit on the neutron's electric dipole moment requires the fundamental parameter of the standard model, $\overline{\theta}$, to be smaller than 10^{-9}.

In the chiral Lagrangian, the condensates and the quark masses produce the chirality-breaking interactions

$$\mathcal{L}_{\mathrm{break}} = \Lambda^3\sum_j m_j\mathbf{U}^j{}_j e^{i\frac{\gamma+\overline{\theta}}{3}} + \mathrm{c.c.} \ , \tag{9.7}$$

showing that the low energy physics depends on the sum of the condensate and explicit phase $\gamma + \overline{\theta}$. This expresses the lowest order shift in the vacuum energy, and it has an absolute minimum at

$$\frac{\gamma + \overline{\theta}}{3} = \pi \ , \tag{9.8}$$

which is necessary for generating successful mass relations among the pseudoscalars.

The QCD vacuum energy $E = E(\overline{\theta})$, depends on $\overline{\theta}$. In addition, the dependence must be cyclic, since the angle $\overline{\theta}$ is defined modulo 2π

$$E(\overline{\theta} + 2\pi) = E(\overline{\theta}) \ . \tag{9.9}$$

We have already seen, in the context of the chiral Lagrangian, that $E(\overline{\theta})$ acquires its minimum value at $\overline{\theta} = 0$ (mod 2π). This can also be shown in a very elegant and direct proof due to C. Vafa and E. Witten (*Phys. Rev. Lett.* **53**, 535(1984)), starting from the fundamental QCD Lagrangian. Their proof is part of a more general theorem which states that QCD does not spontaneously break any vector-like symmetry. Of course the theorem does not apply to the chiral symmetries, which are evidently broken. These authors start with the Euclidean path integral for QCD (neglecting any electroweak effects). In a volume V, it is

$$e^{-VE(\theta)} = \int \mathcal{D}\mathbf{A}\mathcal{D}\mathbf{q}\mathcal{D}\overline{\mathbf{q}} \, \exp\left(-\int d^4x \mathcal{L}_{QCD}(\theta)\right) , \tag{9.10}$$

where

$$\mathcal{L}_{QCD}(\theta) = -\frac{1}{4g^2} \, \mathrm{Tr} \, (\mathbf{G}_{\mu\nu}\mathbf{G}_{\mu\nu}) + \sum_i \overline{\mathbf{q}}_i(\not{\mathcal{D}} + m_i)\mathbf{q}_i + \frac{i\theta}{32\pi^2} \, \mathrm{Tr} \, \left(\mathbf{G}_{\mu\nu}\widetilde{\mathbf{G}}_{\mu\nu}\right) . \tag{9.11}$$

We take all the quark masses to be real, so that in the above we have chosen $\theta = \overline{\theta}$. Integrating out the fermions we obtain

$$e^{-VE(\theta)} = \int \mathcal{D}\mathbf{A} \, \det(\not{\mathcal{D}} + \mathbf{M}) e^{\int d^4x [\frac{1}{4g^2} \mathrm{Tr} \, (\mathbf{G}_{\mu\nu}\mathbf{G}_{\mu\nu}) - \frac{i\theta}{32\pi^2} \mathrm{Tr} \, (\mathbf{G}_{\mu\nu}\widetilde{\mathbf{G}}_{\mu\nu})]} . \tag{9.12}$$

It is crucial to the analysis that the imaginary unit i appears only in the last term. In the continuation from Minkowski to Euclidean space, the i in the Minkowski exponent is absorbed by the Wick rotation of dt, but an extra i appears in the Wick rotation of the Levi-Cività symbol (allowing the volume, $\epsilon_{\mu\nu\rho\sigma}dx^\mu dx^\nu dx^\rho dx^\sigma$, to remain real).

In QCD, $\det(\not{\mathcal{D}} + \mathbf{M})$ is positive and real; this is a consequence of the vector-like couplings of the quarks: for each eigenvalue λ of $\mathcal{D}$, there is another one with the opposite sign. Thus

$$\begin{aligned} \det(\not{\mathcal{D}} + \mathbf{M}) &= \prod_\lambda (i\lambda + \mathbf{M}) = \prod_{\lambda>0} (i\lambda + \mathbf{M})(-i\lambda + \mathbf{M}) , \\ &= \prod_{\lambda>0} (\lambda^2 + \mathbf{M}^2) > 0 . \end{aligned} \tag{9.13}$$

Note that if $\lambda = 0$, $\mathbf{M}$ appears by itself but we can always take $\mathbf{M} > 0$ (this is the unphysical $\theta \to \overline{\theta}$ ambiguity).

Thus if θ were zero, the integrand would be made up of purely real and positive quantities. Now, inclusion of the θ term with its i can only *reduce* the value of the path integral, which is the same as *increasing* the value of $E(\theta)$. It follows that $E(\overline{\theta})$ is minimized at $\overline{\theta} = 0$ (mod 2π).

Keeping in mind the slight *caveat* that with Yukawa couplings, the fermion determinant may no longer be positive nor real, it should now be clear that the vacuum energy, expressed either in the context of QCD or of its associated chiral model, is minimized when $\overline{\theta}$ is zero. We emphasize that this does not mean that $\overline{\theta}$ is actually zero, since in the standard model, $\overline{\theta}$ is just a parameter, not a dynamical variable, to be determined from experiment.

Thus CP-violation by the strong interactions is a *prediction* of QCD. Yet there is no experimental whiff of CP-violation by the Strong Interactions, setting a limit, $\overline{\theta} < 10^{-9}$. Why should this number be so small? This is the one of the remaining outstanding puzzle of the standard model. Theorists have come up with several possible explanations:

• The ultraviolet up-quark mass is zero. This would seem to be in conflict with chiral QCD, which measures a non-zero up-quark mass. However this is an infrared mass, not the fundamental ultraviolet quark mass: the two are known to differ by *additive* corrections generated by strong coupling QCD effects. Although the size of these corrections are presently impossible to calculate, and there is no good theoretical rationale for the u quark mass to vanish, this simple scheme would explain the lack of CP-violation by the strong interactions.

• The tree-level Lagrangian is CP-invariant: the $\overline{\theta}$ parameter is naturally zero at tree-level. This requires the tree-level determinant of the quark mass matrix to be real. The potential is arranged to break CP spontaneously, and in one class of models, the ensuing CP-violation turns out to be very small. These models simply augment the standard model chiral quarks with vector-like weak doublet with standard model hypercharge, and two vector-like weak singlets of charges 2/3 and −1/3:

$$\mathbf{V} = \begin{pmatrix} \mathbf{U} \\ \mathbf{D} \end{pmatrix}_{1/3} , \quad \overline{\mathbf{V}} = \begin{pmatrix} \mathbf{U} \\ \mathbf{D} \end{pmatrix}_{-1/3} , \quad \mathbf{U}'_{4/3} + \overline{\mathbf{U}}'_{-4/3} , \quad \mathbf{D}'_{-2/3} + \overline{\mathbf{D}}'_{2/3} \, .$$

The standard model Higgs is required to couple *only* to the chiral quarks $\mathbf{d}\overline{\mathbf{d}}$, but *not* to $\mathbf{d}\overline{\mathbf{D}}'$ $\mathbf{D}\overline{\mathbf{D}}'$ $\mathbf{D}\overline{\mathbf{d}}$, $\mathbf{D}'\overline{\mathbf{D}}$. $\Delta I_w = 0$ masses in the $\mathbf{D}\overline{\mathbf{D}}$, $\mathbf{D}'\overline{\mathbf{D}}'$ entries are allowed. CP violation comes about only through the phase carried by the $\Delta I_w = 0$ Higgs that couples to $\mathbf{d}\overline{\mathbf{D}}$ and $\mathbf{D}'\overline{\mathbf{d}}$. These rules yield the mass matrix

$$(\mathbf{d}, \mathbf{D}, \mathbf{D}') \begin{pmatrix} < H_d > & S & 0 \\ 0 & M & 0 \\ S' & 0 & M' \end{pmatrix} \begin{pmatrix} \overline{\mathbf{d}} \\ \overline{\mathbf{D}} \\ \overline{\mathbf{D}}' \end{pmatrix} , \tag{9.14}$$

shown here in block-diagonal form, and where $< H_d >, M, M'$ are real, and S, S' are complex. Its determinant is real. A similar matrix obtains in the charge 2/3 sector. There will be loop corrections to the phase of the determinant which will induce a non-zero value of $\overline{\theta}$, but these can be shown to be generically small.

• The lack of CP-violation in the strong interactions is traced to a new particle. The crucial observation, due to Peccei and Quinn (*Phys. Rev.* **D16**, 1791(1977)), is that if $\overline{\theta}$ can be made into a dynamical variable, it will naturally be driven to the value that minimizes the energy, which we have just seen is zero, thereby explaining the lack of P and CP violation by the strong interactions! This extra dynamical variable, identified by Weinberg (*Phys. Rev. Lett.* **40**, 223(1978)) and F. Wilczek (*Phys. Rev. Lett.* **40**, 279(1978)), is called it the *axion*, the (pseudo) Nambu-Goldstone boson of the global PQ symmetry.

Of these three possibilities, we will only study the last one in some detail because of its rich spectrum of experimental consequences and its possible implications for cosmology.

PROBLEMS

A. Show that, in the quark-antiquark condensate, the matrix $\mathbf{U}$ can always be chosen to be unitary.

B. Show that with N_f flavors of light quarks, the absolute minimum of the first order shift occurs when $\gamma + \overline{\theta} = N_f \pi$.

C. Show, using the Vafa-Witten theorem, that for QCD with three flavors, we have $\gamma = \pi$ and $\overline{\theta} = 0$ (see P. Sikivie and C.B. Thorn *Phys. Lett.* **B234**, 132(1990)).

D. Suppose the up-quark mass in the standard model Lagrangian is zero. Estimate the size of the effective m_u that appears in the chiral Lagrangian, and comment on the possible origin of the effect. For further details, see T. Banks, N. Seiberg and Y. Nir in *Yukawa Couplings and the Origin of Masses*, (International Press, 1994), and references therein.

E. Show that the determinant of the quark mass matrix Eq. (9.14) is real. Estimate the size of $\overline{\theta}$ generated at the one-loop level. Estimate the size of the resulting CP-violation in the weak interactions. For references, see A. Nelson *Phys. Lett.* **136B**, 387(1984), and S. Barr *Phys. Rev. Lett.* **53**, 329(1984).

9.2 AXION PROPERTIES

The classical QCD action is invariant under a constant shift of the parameter $\overline{\theta}$, since it appears multiplied by the four-divergence $\text{Tr}\,(\mathbf{G}_{\mu\nu}\widetilde{\mathbf{G}}_{\mu\nu})$. Hence if it were to be replaced by a field, the physics should be be invariant under a constant shift of that field. This describes the dynamics of the Nambu-Goldstone boson $a(x)$, which appears in the Lagrangian through the term

$$\mathcal{L}_{NG} = \frac{a(x)}{V_{PQ}}\partial_\mu J_\mu \ , \tag{9.15}$$

where V_{PQ} is the mass scale at which the global symmetry generated by the current J_μ is spontaneously broken. We use two hints to identify this symmetry. First, comparison with the QCD Lagrangian suggests that the divergence of this $U(1)$ current must be proportional to $\text{Tr}\,(\mathbf{G}_{\mu\nu}\widetilde{\mathbf{G}}_{\mu\nu})$, implying that this symmetry is anomalous, and therefore with computable strength through the usual triangle graph. Secondly, since a massless quark makes $\overline{\theta}$ unobservable through its redefinition by the extra chiral symmetry, this new symmetry must both be chiral and act on the quarks. The same conclusion follows from the anomaly diagram which involves the quark fields circulating in the triangle.

The standard model has no such global chiral U(1) symmetry; all of its global symmetries are accounted for, and none are chiral on the quarks.

To accomodate the Peccei-Quinn (PQ) symmetry, the standard model must be extended to include new degrees of freedom, assumed to be represented by local field. There must be (at least) one new complex field, which changes by a phase under the PQ chiral $U(1)$. Its phase is the axion which shifts under PQ as the quarks transform chirally, and it couples to the anomalous divergence of the current which is proportional to $\text{Tr}\,(\mathbf{G}_{\mu\nu}\widetilde{\mathbf{G}}_{\mu\nu})$. The PQ chiral symmetry is *explicitly broken* in two ways, by QCD effects (for instance instantons) and by the quark masses, since it is chiral. It follows that the axion, like the pion, is not massless, but a pseudo Nambu-Goldstone boson. Its mass must be proportional to both the strength of the strong interactions which violate the symmetry, and also to the light quark masses. This suggests that

$$M_a \sim \frac{M_\pi F_\pi}{V_{PQ}} \; ; \tag{9.16}$$

the first factor M_π is expected because as M_π tends to zero, so should the axion mass, F_π labels the strength of the strong interactions, and V_{PQ} is the scale at which the PQ symmetry is broken. When we analyze this relation in more detail, we will find it to be a little more subtle.

There are two types of axion models. In the first, the scalar field whose vacuum expectation value breaks the PQ symmetry, transforms under $SU_2 \times U_1$; in the second it is an electroweak singlet.

Assume first that the PQ order parameter belongs to scalar fields which carry electroweak quantum numbers. In order not to break electric charge, it must have one electrically-neutral component, and thus transform under the weak SU_2. The simplest choice is an SU_2 doublet, just like the Higgs doublet already present in the standard model, although one can readily envisage more complicated models, with weak triplets, etc... .

With no new quarks, the $U(1)_{PQ}$ symmetry can be simply implemented in the standard model by inventing a new Higgs field which couples to the quarks. Of several logical possibilities, simplest is to introduce two electroweak doublets, each doublet coupling to one quark charge sector only. This avoids tree-level flavor changing neutral current effects. Calling these Higgs doublets H_u and H_d, we have

$$\mathcal{L}_Y = i\widehat{\mathbf{Q}}_i\overline{\mathbf{u}}_j Y_{ij}^{(2/3)} H_u + i\widehat{\mathbf{Q}}_i\overline{\mathbf{d}}_j Y_{ij}^{(-1/3)} H_d + i\widehat{L}_i\overline{e}_j Y_{ij}^{(-1)} H_d \; . \tag{9.17}$$

The leptons couple to the same Higgs doublet as the down quarks.

This Yukawa Lagrangian has the required global PQ symmetry. The Higgs doublets, H_u and H_d, have opposite hypercharge, but the same PQ charge. The hypercharge and PQ values are summarized in the following table

	H_u	H_d	$\mathbf{Q}$	$\overline{\mathbf{u}}$	$\overline{\mathbf{d}}$	L	$\overline{e}$
Y	1	-1	$\frac{1}{3}$	$-\frac{4}{3}$	$\frac{2}{3}$	-1	2
PQ	1	1	$-\frac{1}{2}$	$-\frac{1}{2}$	$-\frac{1}{2}$	$-\frac{1}{2}$	$-\frac{1}{2}$

The PQ charges of the quarks and leptons are chosen to be chiral and flavor symmetric to ensure no possible admixture of the vector-like baryon number and the lepton numbers. With two electroweak doublets, the Higgs potential is slightly more complicated. The full potential that leaves invariant the PQ symmetry is given by

$$V(H_u, H_d) = \sum_{a=u,d} (-\mu_a^2 H_a^\dagger H_a + \lambda_{aa}(H_a^\dagger H_a)^2) \\ + \lambda_{ud} H_u^\dagger H_u H_d^\dagger H_d + \lambda'_{ud} H_u^\dagger H_d H_d^\dagger H_u \ , \tag{9.18}$$

so that with the modest addition of another Higgs doublet, the number of parameters in the potential increases dramatically from two (μ, λ) to six $(\mu_u,\ \mu_d,\ \lambda_{uu},\ \lambda_{dd},\ \lambda_{ud},\ \lambda'_{ud})$.

In this model, when both H_u and H_d acquire vacuum values, the PQ symmetry is broken at the same scale as the electroweak symmetry. This possibility makes the axion properties almost totally predictable: since V_{PQ} breaks $SU_2 \times U_1$ it cannot be larger than $G_F^{-1/2}$, the scale of weak interactions set by the Fermi coupling constant. This sets stringent bounds on the mass and coupling strength of the axion. Our empirical formula yields an axion mass in the keV range, with predicted coupling strength. This "visible" axion model has been ruled out by experiments: if the axion exists at all, it must be that $V_{PQ} \gg G_F^{-1/2}$.

Hence the PQ symmetry can only be broken by a field with no electroweak quantum numbers. The first model of this kind was proposed by J. E. Kim (*Phys. Rev. Lett.* **43**, 103(1979)), who introduced an extra quark coupled to an electroweak-singlet PQ order parameter. Since the axion couples to matter with strength inversely proportional to V_{PQ}, this type of model leads to an "invisible axion".

Another way to build such a model is to add a new electroweak singlet Higgs field $\Phi(x)$ to the previous model with two Higgs doublets. Its role is to break the PQ symmetry at a scale much larger than the electroweak scale. This field does not couple directly to quarks and leptons, but acquires a PQ quantum number by coupling to the Higgs doublets. One can envisage two couplings

$$H_u^t \tau_2 H_d \Phi \ , \qquad H_u^t \tau_2 H_d \Phi^2 \ , \tag{9.19}$$

together with their conjugates.

A model with the cubic interaction was first proposed by Zhitnitskii (*Sov. J. Nucl. Phys.* **31**, 260(1980)). The cubic interaction can be set to zero naturally, by imposing the discrete symmetry $\Phi \to -\Phi$. A model with only the quartic term is due to M. Dine, W. Fischler and M. Srednicki (*Phys. Lett.* **104B**, 199(1981)). In that case, the field Φ has PQ charge of -1, and the total potential is simply

$$V(H_u, H_d, \Phi) = V(H_u, H_d) + V_0(\Phi) + \Delta H_u^T \tau_2 H_d \Phi^2 + \text{c.c.} \ , \tag{9.20}$$

where $V_0(\Phi)$ is given by

$$V_0(\Phi) = -M^2\Phi^*\Phi + \Lambda(\Phi^*\Phi)^2 .$$

This axion model has now ten parameters in its Higgs potential!

When the three neutral scalar fields in the potential acquire vacuum values, two phase symmetries will be broken; one is the usual linear combination of hypercharge and third component of weak isospin which breaks the electroweak symmetry, the other is the PQ symmetry.

In addition to being broken spontaneously by the electroweak singlet Φ, the PQ symmetry is explicitly broken by QCD instanton-like effects, with strength determined by the anomaly diagram

$$N\delta_{AB} = \sum_a \mathrm{Tr}\,(Q_{PQ}Q^a_A Q^a_B) , \qquad (9.21)$$

where Q^a_A is the color charge of quark a, and Q_{PQ} its PQ charge. The effect of this anomalous divergence can be accounted for by adding to the Lagrangian an effective interaction proportional to the determinant of the quark mass matrix, $\det(\mathbf{q}^\dagger_L \mathbf{q}_R)$ and its conjugate. Indeed, under the PQ transformation

$$\mathbf{q} \to e^{i\alpha\gamma_5}\mathbf{q} , \qquad (9.22)$$

the determinant certainly breaks the overall continuous chiral U(1), since for N_f quark flavors

$$\det(\mathbf{q}^\dagger_L \mathbf{q}_R) \to e^{-2iN_f\alpha} \det(\mathbf{q}^\dagger_L \mathbf{q}_R) . \qquad (9.23)$$

Sikivie (*Phys. Rev. Lett.* **48**, 1156(1982)) remarked that for the special values

$$\alpha = \frac{\pi k}{N_f} , \qquad k = 1, 2, \dots , \qquad (9.24)$$

the interaction term stays invariant. Consequently the Z_{2N_f} discrete subgroup of $U(1)_{PQ}$ survives strong coupling effects. Still, this discrete subgroup is spontaneously broken by the Higgs vacuum value $< \Phi >$. For instance, in the DFS model, Φ transforms under $U(1)_{PQ}$,as

$$\Phi \to e^{i2\alpha}\Phi , \qquad (9.25)$$

leaving, for α a multiple of π, the discrete subgroup Z_2 invariant. The spontaneous breaking of $U(1)_{PQ}$ breaks the true discrete symmetry, namely

Z_{2N_f} down to Z_2. This produces N_f degenerate regions, separated by domain walls, with possibly nefarious consequences in cosmology.

At any case, the field configuration of the vacuum has a two-fold continuous degeneracy, parametrized by two angles, $\theta_0(x)$, $\theta_1(x)$, which are related to the vacuum fields by

$$
\begin{aligned}
< H_u >_0 &= \frac{v_u}{\sqrt{2}} e^{i[\theta_0(x)+\theta_1(x)]} , \\
< H_d >_0 &= \frac{v_d}{\sqrt{2}} e^{i[\theta_0(x)-\theta_1(x)]} , \\
< \Phi >_0 &= \frac{V_{PQ}}{\sqrt{2}} e^{-i\theta_0(x)} ,
\end{aligned}
\tag{9.26}
$$

where all three vacuum values are taken to be real without loss of generality. The vacuum values of the doublets are limited by the Fermi constant

$$
G_F^{-1} \equiv v^2 = v_u^2 + v_d^2 . \tag{9.27}
$$

There is no such restriction on V_{PQ}. Since Φ is a gauge singlet, quantum corrections to this potential do not detune the hierarchy $V_{PQ} \gg G_F^{-1/2}$. The kinetic terms for these fields are easily seen to be

$$
\mathcal{L}_{\text{Kin}} = \frac{1}{2}(\partial_\mu\theta_0, \partial_\mu\theta_1) \begin{pmatrix} V_{PQ}^2 + v_u^2 + v_d^2 & v_u^2 - v_d^2 \\ v_u^2 - v_d^2 & v_u^2 + v_d^2 \end{pmatrix} \begin{pmatrix} \partial_\mu\theta_0 \\ \partial_\mu\theta_1 \end{pmatrix} . \tag{9.28}
$$

Upon diagonalization to canonical form, we find

$$
\mathcal{L}_{\text{Kin}} = \frac{1}{2}\partial_\mu a(x)\partial_\mu a(x) + \frac{1}{2}\partial_\mu\theta(x)\partial_\mu\theta(x) , \tag{9.29}
$$

where the canonical fields are the axion field

$$
a(x) = \sqrt{V_{PQ}^2 + v^2 \sin^2 2\beta}\ \theta_0(x) \equiv V_{PQ}\theta_0(x) , \tag{9.30}
$$

and

$$
\theta(x) = v \cos 2\beta\theta_0(x) + v\theta_1(x) , \tag{9.31}
$$

is the field that will be eaten in the unitary gauge to become the longitudinal Z-boson. We have introduced the angle β through $\tan\beta \equiv v_u/v_d$. It is easy to read off the couplings of the axion to the quarks, by rewriting the Yukawa couplings in the form

$$i\mathbf{u}_{Ri}^{\dagger}\mathbf{u}_{Li}m_{ui}e^{i\frac{a(x)}{V_{PQ}}2\sin^2\beta}+i\mathbf{d}_{Ri}^{\dagger}\mathbf{d}_{Li}m_{di}e^{i\frac{a(x)}{V_{PQ}}2\cos^2\beta}+\text{ c.c. .} \tag{9.32}$$

Because of the relative i between the mass term and the axion term, we see that the axion couples to $\overline{\mathbf{u}}\gamma_5\mathbf{u}$ and $\overline{\mathbf{d}}\gamma_5\mathbf{d}$, the divergences of the axial currents $\overline{\mathbf{u}}\gamma_5\gamma_\mu\mathbf{u}$ and $\overline{\mathbf{d}}\gamma_5\gamma_\mu\mathbf{d}$. The coupling to the lepton is the same as that to the down quarks.

An equivalent way to arrive at this coupling is to observe that, as a Nambu-Goldstone boson, the axion couples to the divergence of the PQ current

$$J_\mu^{PQ}=V_{PQ}\partial_\mu a(x)+2\sin^2\beta\overline{\mathbf{u}}_i\gamma_\mu\gamma_5\mathbf{u}_i+2\cos^2\beta(\overline{\mathbf{d}}_i\gamma_\mu\gamma_5\mathbf{d}_i+\overline{e}_i\gamma_\mu\gamma_5 e_i)\ , \tag{9.33}$$

where i is the summed-over family index. So far the relevant part of the Lagrangian, not showing the leptons and the potential, reads

$$\begin{aligned}\mathcal{L}=&im_{ui}\mathbf{u}_{Ri}^{\dagger}\mathbf{u}_{Li}e^{i\frac{a(x)}{V_{PQ}}2\sin^2\beta}+im_{di}\mathbf{d}_{Ri}^{\dagger}\mathbf{d}_{Li}e^{i\frac{a(x)}{V_{PQ}}2\cos^2\beta}+\text{ c.c.}\\&+\frac{1}{2}\partial_\mu a(x)\partial_\mu a(x)+\theta\frac{g_3^2}{32\pi^2}\operatorname{Tr}(\mathbf{G}_{\mu\nu}\widetilde{\mathbf{G}}_{\mu\nu})+\cdots\ ,\end{aligned} \tag{9.34}$$

Under the chiral rotation

$$\mathbf{u}_{L,Ri}\to e^{\mp i\sin^2\beta\frac{a(x)}{V_{PQ}}}\mathbf{u}_{L,Ri}\ ,\qquad \mathbf{d}_{L,Ri}\to e^{\mp i\cos^2\beta\frac{a(x)}{V_{PQ}}}\mathbf{d}_{L,Ri}\ , \tag{9.35}$$

the quark kinetic term picks up the extra pieces

$$\frac{-i}{V_{PQ}}\left(\sin^2\beta\overline{\mathbf{u}}_i\gamma_\mu\gamma_5\mathbf{u}_i+\cos^2\beta\overline{\mathbf{d}}_i\gamma_\mu\gamma_5\mathbf{d}_i\right)\partial_\mu a(x)\ . \tag{9.36}$$

This transformation is anomalous so that the fermion measure in the path integral picks up a phase proportional to $\operatorname{Tr}(\mathbf{G}_{\mu\nu}\widetilde{\mathbf{G}}_{\mu\nu})$, leading to a shift in the θ parameter

$$\begin{aligned}\theta\to\theta'&=\theta+\frac{N_f}{2}(2\sin^2\beta+2\cos^2\beta)\frac{a(x)}{V_{PQ}}\ ,\\&=\theta+N_f\frac{a(x)}{V_{PQ}}\ .\end{aligned} \tag{9.37}$$

We set the parameter θ to zero by redefining the phases in the quark masses, yielding the final Lagrangian

$$\begin{aligned}\mathcal{L} =&\overline{\mathbf{u}}_i(\partial\!\!\!/ + im_{ui})\mathbf{u}_i + \overline{\mathbf{d}}_i(\partial\!\!\!/ + im_{di})\mathbf{d}_i + \frac{1}{2}\partial_\mu a(x)\partial_\mu a(x)\\ &- \frac{i}{V_{PQ}}\partial_\mu a(x)\left[\sin^2\beta\overline{\mathbf{u}}_i\gamma_\mu\gamma_5\mathbf{u}_i + \cos^2\beta\overline{\mathbf{d}}_i\gamma_\mu\gamma_5\mathbf{d}_i\right]\\ &+ N_f\frac{a(x)}{V_{PQ}}\frac{g_3^2}{32\pi^2}\,\mathrm{Tr}\,(\mathbf{G}_{\mu\nu}\tilde{\mathbf{G}}_{\mu\nu}) + \cdots .\end{aligned} \tag{9.38}$$

We are pleased to recognize, after integration by parts, that the axion couples manifestly to the divergence of J_μ^{PQ}, augmented by its anomalous part. We have achieved our purpose in replacing the θ parameter by the dynamical axion field $a(x)$. This is the fundamental Lagrangian of QCD, to which we have added the axion couplings.

PROBLEMS

A. Devise an invisible axion model which contains a new quark and only one Higgs doublet.

B. 1-) Show that the potential of Eq. (9.18) is the most general potential with PQ symmetry.
2-) Find the necessary conditions among the couplings to insure the the tree-level inequality $V \gg v_u, v_d$.
3-) Show using diagrammatic techniques that this inequality is stable against radiative corrections (technically natural).

9.3 THE AXION CHIRAL LAGRANGIAN

We now proceed to incorporate the axion into the QCD chiral Lagrangian. To that effect, we first integrate over the heavy quarks, and then make a PQ transformation on the light quarks to transfer the axion coupling into their masses. There are three light quarks for which $m \leq \Lambda_{QCD}$. In the following we assume only two light quarks, leaving it as an exercise to the reader to construct the three-light flavor axion effective low energy chiral Lagrangian. We start with

$$\mathbf{u} \rightarrow= e^{i\sin^2\beta\frac{a(x)}{V_{PQ}}\gamma_5}\mathbf{u}\ , \qquad \mathbf{d} \rightarrow e^{i\cos^2\beta\frac{a(x)}{V_{PQ}}\gamma_5}\mathbf{d}\ , \tag{9.39}$$

together with a change of the anomaly coefficient

$$N_f \frac{a(x)}{V_{PQ}} \to (N_f - 2\sin^2\beta - 2\cos^2\beta)\frac{a(x)}{V_{PQ}} = (N_f - 2)\frac{a(x)}{V_{PQ}} \ . \qquad (9.40)$$

After neglecting (integrating over) the heavy quark fields, we arrive at the relevant part of the Lagrangian

$$\begin{aligned} \mathcal{L} =& im_u \overline{\mathbf{u}} e^{i2\sin^2\beta \frac{a(x)}{V_{PQ}}\gamma_5} \mathbf{u} + im_d \overline{\mathbf{d}} e^{i2\cos^2\beta \frac{a(x)}{V_{PQ}}\gamma_5} \mathbf{d} \\ &+ (N_f - 2)\frac{a(x)}{V_{PQ}} \frac{g_3^2}{32\pi^2} \,\mathrm{Tr}\,(\mathbf{G}_{\mu\nu}\widetilde{\mathbf{G}}_{\mu\nu}) + \cdots \ . \end{aligned} \qquad (9.41)$$

Note that the presence of the heavy quarks is only reflected in the anomaly term, which is to be expected since anomalies span scales with impunity. This expression forms the basis for the formulation of the low energy chiral Lagrangian.

However, in the real world effective chiral Lagrangian with axion, there is one further complication because the axion is not the only neutral (pseudo) Nambu-Goldstone boson in the problem. With three light quark flavors, there are the pion and the η. With two light quarks, there is only the pion. Recall that the η' is not light (it is in fact heavier than the proton) because its corresponding symmetry has been explicitly broken by strong coupling effects (e.g. instantons). In the absence of the axion, the two-flavor chiral Lagrangian is given by

$$\begin{aligned} \mathcal{L} = & \frac{1}{2}\partial_\mu\eta'\partial_\mu\eta' + \frac{F_\pi^2}{16}\,\mathrm{Tr}\,(\partial^\mu\boldsymbol{\Sigma}\partial_\mu\boldsymbol{\Sigma}^{-1}) - \frac{1}{2}m_{\eta'}^2\eta'\eta' \\ & + \frac{F_\pi^2}{8}m_0\,\mathrm{Tr}\,(e^{\frac{-2i}{F_\pi}\tau^0\eta'}\boldsymbol{\Sigma}^{-1}\mathbf{M} + \text{h.c.}) \ , \end{aligned} \qquad (9.42)$$

where

$$\boldsymbol{\Sigma}(x) = e^{\frac{2i}{F_\pi}\tau^A\pi^A} \ , \qquad (9.43)$$

$\mathbf{M}$ is the light quark mass matrix, and $m_{\eta'}$ is the mass of the heavy η' meson.

What is the effect of the anomaly on this system? In the case of two flavors, the θ-term can be transmitted to the quarks by the chiral transformation

$$\mathbf{q} \to e^{i\frac{\theta}{4}\gamma_5}\mathbf{q} \ , \qquad \mathbf{q} = \mathbf{u}, \mathbf{d}, \mathbf{s} \ , \qquad (9.44)$$

which results in the shift in the phase of the condensate, or in the effective Lagrangian language,

$$\frac{\eta'}{F_\pi} \to \frac{\eta'}{F_\pi} - \frac{\theta}{4} \,. \tag{9.45}$$

To introduce the axion field in the chiral Lagrangian, we simply add its kinetic term, and rewrite the phase in terms of the axion field

$$\theta \to (N_f - 2)\frac{a(x)}{V_{PQ}} \equiv n_{lf}\frac{a(x)}{V_{PQ}} \,, \tag{9.46}$$

with the result

$$\begin{aligned}\mathcal{L}_{eff} = \frac{1}{2}\partial_\mu a \partial_\mu a + \frac{1}{2}\partial_\mu \eta' \partial_\mu \eta' + \frac{F_\pi^2}{16}\,\mathrm{Tr}\left(\partial_\mu \boldsymbol{\Sigma}^\dagger \partial_\mu \boldsymbol{\Sigma}\right) \\ -m_{\eta'}^2 (\eta'(x) + n_{lf}\frac{F_\pi}{4}\frac{a(x)}{V_{PQ}})^2\end{aligned}$$

$$+m_0\frac{F_\pi^2}{8}\,\mathrm{Tr}\left\{ e^{\frac{-2i}{F_\pi}\tau^0\eta'}\boldsymbol{\Sigma}^{-1}\begin{pmatrix} m_u e^{i2\sin^2\beta\frac{a(x)}{V_{PQ}}} & 0 \\ 0 & m_d e^{i2\cos^2\beta\frac{a(x)}{V_{PQ}}}\end{pmatrix} + \text{ h.c.}\right\} . \tag{9.47}$$

It follows that the field

$$\hat{\eta}'(x) = \frac{1}{\sqrt{1+\xi^2}}\left[\eta'(x) + \xi a(x)\right] \,, \tag{9.48}$$

where

$$\xi = \frac{n_{lf}}{4}\frac{F_\pi}{V_{PQ}} \,, \tag{9.49}$$

picks up a mass generated by instanton-like effects

$$m_{\hat{\eta}'} = m_{\eta'}\sqrt{1+\xi^2} \,, \tag{9.50}$$

while the orthogonal combination

$$\hat{a}(x) = \frac{1}{\sqrt{1+\xi^2}}\left[a(x) - \xi\eta'(x)\right] \,, \tag{9.51}$$

only picks up a "light" mass generated by the quark masses. We rewrite the quark mass terms in the hatted, yielding in the neutral sector

$$m_0\frac{F_\pi^2}{8}\,\mathrm{Tr}\,\mathbf{\Sigma}^{-1}\begin{pmatrix} m_u e^{i(\frac{\hat{a}}{V_{PQ}}g_u+\frac{\hat{\eta}'}{F_\pi}h_u)} & 0 \\ 0 & m_d e^{i(\frac{\hat{a}}{V_{PQ}}g_d+i\frac{\hat{\eta}'}{F_\pi}h_d)} \end{pmatrix} + \text{h.c.}\,, \quad (9.52)$$

where

$$g_u = \frac{1}{\sqrt{1+\xi^2}}[2\sin^2\beta + \frac{n_{lf}}{2}]\,, \qquad g_d = \frac{1}{\sqrt{1+\xi^2}}[2\cos^2\beta + \frac{n_{lf}}{2}]\,, \quad (9.53)$$

and

$$\begin{aligned} h_u &= \frac{1}{\sqrt{1+\xi^2}}(-2+\sin^2\beta\frac{n_{lf}}{2}\left(\frac{F_\pi}{V_{PQ}}\right)^2)\,, \\ h_d &= \frac{1}{\sqrt{1+\xi^2}}(-2+\cos^2\beta\frac{n_{lf}}{2}\left(\frac{F_\pi}{V_{PQ}}\right)^2)\,, \end{aligned} \quad (9.54)$$

Computations are made simpler by concentrating solely on the neutral sector. It is easy to see that the potential is given by

$$V = -m_0\frac{F_\pi^2}{4}(m_u\cos\theta_u + m_d\cos\theta_d)\,, \quad (9.55)$$

with

$$\theta_u = \frac{\hat{a}(x)}{V_{PQ}}g_u + \hat{\eta}'(x)\frac{h_u}{F_\pi} + \frac{2}{F_\pi}\pi_3(x)\,, \qquad \theta_d = \frac{\hat{a}(x)}{V_{PQ}}g_d + \hat{\eta}'(x)\frac{h_d}{F_\pi} - \frac{2}{F_\pi}\pi_3(x)\,,$$

In the above, $m_{u,d}$ are positive and real, so that the minimum of the potential occurs when

$$\theta_u = 2n_u\pi\,, \qquad \theta_d = 2n_d\pi\,, \quad (9.56)$$

where $n_{u,d} = 0, 1, 2, \ldots$. We have already seen that there is a mass term for the $\hat{\eta}'$ field, which fixes it to zero at minimum. Hence, neglecting ξ^2, the minimum conditions reduce to

$$\frac{\hat{a}(x)}{V_{PQ}}N_f = 2(n_u+n_d)\pi\,, \quad (9.57)$$

which shows that the PQ solution does indeed solve the problem, by allowing

$$\overline{\theta} = \frac{N_f \hat{a}}{V_{PQ}} , \tag{9.58}$$

to relax to zero. This analysis also shows that the vacuum is degenerate *mod* 2π, reproducing our earlier observation.

The masses of the light bosons (π and $\hat{a}$) obtained by expanding the potential, and setting $\hat{\eta}' = 0$,

$$V \approx m_0 \frac{F_\pi^2}{4}[m_u(\frac{\hat{a}(x)}{V_{PQ}}g_u + \frac{2}{F_\pi}\pi_3(x))^2 + m_d(\frac{\hat{a}(x)}{V_{PQ}}g_d - \frac{2}{F_\pi}\pi_3(x))^2] + \cdots , \tag{9.59}$$

leading to the mass matrix

$$m_0(\hat{a} \quad \pi^3)\begin{pmatrix} \frac{F_\pi^2}{4V_{PQ}^2}(m_u g_u^2 + m_d g_d^2) & \frac{F_\pi}{2V_{PQ}}(m_u g_u - m_d g_d) \\ \frac{F_\pi}{2V_{PQ}}(m_u g_u - m_d g_d) & (m_u + m_d) \end{pmatrix}\begin{pmatrix} \hat{a} \\ \pi^3 \end{pmatrix} . \tag{9.60}$$

Since we are dealing with invisible axion models, we can simplify life and expand in $\frac{F_\pi}{V_{PQ}}$. The analysis yields the neutral pion mass,

$$m_{\pi^0}^2 = m_0(m_u + m_d) + \mathcal{O}((\frac{F_\pi}{V_{PQ}})^2) ,$$

and the axion mass

$$m_a^2 = m_{\pi^0}^2 N_f^2 (\frac{F_\pi}{2V_{PQ}})^2 \frac{m_u m_d}{(m_u + m_d)^2} + \mathcal{O}(\frac{F_\pi^4}{V_a^4}) . \tag{9.61}$$

The eigenstates are

$$\begin{aligned} \pi^0(x) &= \pi_3(x) + \frac{F_\pi}{2V_{PQ}}\left\{\frac{m_u g_u - m_d g_d}{m_u + m_d}\right\}\hat{a}(x) + \mathcal{O}(\frac{F_\pi^2}{V_{PQ}^2}) , \\ a_{phys}(x) &= \hat{a}(x) - \frac{F_\pi}{2V_{PQ}}\left\{\frac{m_u g_u - m_d g_d}{m_u + m_d}\right\}\pi_3(x) + \mathcal{O}(\frac{F_\pi^2}{V_{PQ}^2}). \end{aligned} \tag{9.62}$$

In the invisible axion model, it is clear that the mixing is minimal since $F_\pi \ll V_{PQ}$. It is important to realize that the properties of the invisible axion are determined by very few parameters, namely V_{PQ} which fixes both the overall strength of its couplings to matter and its mass, as well as model-dependent coupling constants to specific quarks and leptons. In the next section, we address phenomenological constraints on V_{PQ} and possible modes of detection. These rely on the electromagnetic interactions of the axion, which we have hitherto neglected.

Axion detection

In the original axion model where the breaking of the PQ symmetry is at electroweak scale, all of the axion properties are determined in terms of one parameter, the angle β. Its mass is given by

$$m_a = \frac{N_f}{\sin 2\beta} \frac{\sqrt{m_u m_d}}{m_u + m_d} \frac{F_\pi m_{\pi^0}}{v} \approx \frac{147.2}{\sin 2\beta}\ \text{keV} \ . \tag{9.63}$$

Its couplings to matter are through the term

$$\mathcal{L}_{int} = \frac{1}{2v} a(x) \partial_\mu J^{PQ}_\mu \ , \tag{9.64}$$

where, neglecting the lepton contributions,

$$\begin{aligned} J^{PQ}_\mu = v\partial_\mu a - \frac{1}{2}\sum_{i=1}^{3}(\cot\beta \overline{\mathbf{u}}_i\gamma_\mu\gamma_5\mathbf{u}_i - \tan\beta\overline{\mathbf{d}}_i\gamma_\mu\gamma_5\mathbf{d}_i) \\ + \frac{3}{(m_u + m_d)\sin 2\beta}(m_d\overline{\mathbf{u}}_1\gamma_\mu\gamma_5\mathbf{u}_1 + m_u\overline{\mathbf{d}}_1\gamma_\mu\gamma_5\mathbf{d}_1) \ . \end{aligned} \tag{9.65}$$

This current is the canonical PQ current minus a contribution from the two light quarks to make it anomaly-free, and orthogonal to the neutral pion current. This original axion model is in contradiction with experiment.

In invisible axion models, V_{PQ} is no longer constrained by the electroweak scale. This has two consequences; it reduces its couplings to matter and makes it extremely light. Thus, the possibility of its detection through laboratory experiments seems remote, hence the name invisible.

But there are serious restrictions on V_{PQ} : a lower bound set by astrophysical considerations, and an upper bound based on the axion's contribution to the universe's energy density. Remarkably the two still leave a "small" (we are conforming to the standard accuracy of pre-nucleosynthesis cosmology by displaying the error in the exponent) window

$$10^{9\pm1}\ \text{GeV} \le f_a = \frac{V_{PQ}}{N} \le 10^{12\pm1}\ \text{GeV}\ . \tag{9.66}$$

Any value outside this range leads to severe contradictions with our present understanding of stellar and cosmological evolutions.

In the following we do not detail these bounds but refer the reader to the many excellent texts on the subject, notably *The Early Universe, po. cit.*.

The lower bound is set by noting that the axion provides a pathway for energy loss in stars. Once created inside a star, the wee-weakly interacting axion can escape the star, causing greater energy loss and faster aging. The lower bound is derived from the existence of old stars, such as red giants.

The tracing of the cosmological evolution of the axion provides the upper bound . The axion stays in thermal equilibrium with the primal heat bath until it decouples at a temperature $T \approx V_{PQ}$. Relativistic axions are simply red-shifted away. As the universe cools to a temperature of the order of Λ_{QCD}, the axion develops a mass, and therefore a parabolic potential. At later times, the axion is at the bottom of this parabola, but earlier it might be anywhere. Thus it starts oscillating about the minimum, with a classical field of the form

$$< a >= A(t) \cos m_a t\ , \tag{9.67}$$

where $A(t)$ is of order of V_{PQ}. The axion hardly decays and so the oscillations are not damped. The result of this coherent oscillation is to produce an energy density in the axion field which evolves like matter. Thus it grows faster than radiation. Since it is proportional to V_{PQ}, it can close the universe for too large a value of V_{PQ}. The complicated detailed calculations, performed by many groups, all lead to the same upper bound.

Although this range of allowed V_{PQ} leads to a very tiny coupling, Sikivie (*Phys. Rev. Lett.* **51**, 1415(1983)) pointed out that the coupling to electromagnetism can still lead to an effect observable in the laboratory: the conversion of an axion into a photon inside an electromagnetic cavity.

At the level of the fundamental quark-lepton Lagrangian, electromagnetic interactions are introduced by replacing the derivative by the Maxwell covariant derivative. However when quarks and leptons undergo chiral rotations, the anomalous divergence induces a term along $F_{\mu\nu}\widetilde{F}_{\mu\nu}$ term as well, leading to an induced interaction for the axion of the form

$$\mathcal{L}_{int} = N_e \frac{1}{32\pi^2} \frac{a(x)}{V_{PQ}} F_{\mu\nu}\widetilde{F}_{\mu\nu}\ , \tag{9.68}$$

where in the DFS model, with three chiral families, $N_e = 4$.

To obtain the full coupling in the effective low energy chiral Lagrangian, we have to shift to account for the chiral transformation of the light quarks, and also for the diagonalization of the axion-η'- pion system. Subtracting out the anomaly of the **u** and **d** quarks means that the anomaly coefficient is now

$$N'_e = N_e - \frac{8}{3}\sin^2\beta - \frac{2}{3}\cos^2\beta \ . \tag{9.69}$$

This equation forms the basis for the possible detection of the *invisible* axion. The more axions in the cavity, the more conversions will be observed. If axions are the main constituents of dark matter, it is conceivable their presence could be detected in such cavities. An ongoing experiment has so far been unable to "tune" on the axion (C. Hagmann *et al*, *Phys. Rev. Lett.* **80**, 2043(1998)).

PROBLEMS

A. Show, using symmetry arguments, that the axion mass vanishes if any quark mass is zero.

B. Show that in the original "visible" axion model (see W. Bardeen and H. Tye, *Phys. Lett.* **B74**, 229(1978)), the axion mass is given by

$$m_a = \frac{N_f}{\sin 2\beta}\frac{\sqrt{m_u m_d}}{m_u + m_d}\frac{F_\pi m_{\pi^0}}{v}$$

C. Start from the results of chapter 5 to derive the axion mass for three flavors. Verify that in the limit of $m_{u,d}/m_s \to 0$ it reduces to the expression derived in the text.

CHAPTER **10**

THIRD JOURNEY:

SUPERSYMMETRY

The generalization of the standard model to $N = 1$ supersymmetry is remarkably easy. Quarks and leptons are taken to be the fermionic components of chiral Wess-Zumino multiplets, adding to the standard model their spin-zero superpartners, the squarks and sleptons. The gauge bosons are themselves the spin one component of vector supermultiplets, introducing new spinors superpartners, the gluinos, the winos and the bino. The sole standard model Higgs needs to be replaced by two scalar Higgs bosons of opposite hypercharge, both forming chiral supermultiplets, creating a vector-like pair of spinor doublets, the Higgsinos. An odd number of Weyl fermions cannot be implemented if there is more than one supersymmetry, which is the reason we only focus on $N = 1$ supersymmetry. In the following, familiarity with the basic constructions of $N = 1$ supersymmetry is necessary. The neophyte is urged to work through the Appendix, where the necessary notations and technology are developed.

10.1 THE $N = 1$ STANDARD MODEL

There are several ways to generalize the standard model to satisfy $N = 1$ supersymmetry. The most "economical" introduces only those fields necessary to reproduce the ($N = 0$) standard model and satisfy theoretical consistency. When their couplings are chosen to reproduce just the global symmetries of the standard model, it is called the minimal supersymmetric standard model (MSSM). Even with the minimal set of fields, there can be new couplings which respect the gauge symmetries. They all involve superpartners, and are thus absent from the $N = 0$ standard model. Before discussing these non-minimal extensions, we start with the construction of the simplest $N = 1$ standard model.

The Minimal $N = 1$ Standard Model (MSSM)

All gauge bosons of the standard model are now part of a gauge multiplet, described (in the Wess-Zumino gauge) by a chiral vector multiplet $\mathcal{W}$. As a result, all have their spin one-half gaugino superpartners, distinguished from their spin one partners by a twiddle. We have then, eight gluinos, $\widetilde{\mathbf{g}}$, one Bino, $\widetilde{B}^0$, one charged and one neutral Wino, $\widetilde{W}^{\pm}$, and $\widetilde{W}^0$.

Every quark and lepton of the standard model is viewed as the spinor component of a Wess-Zumino chiral superfield, generically denoted by Φ,

$$L \to \Phi_L\ , \quad \mathbf{Q} \to \Phi_{\mathbf{Q}}, \quad \overline{\mathbf{u}} \to \Phi_{\overline{\mathbf{u}}}, \quad \overline{\mathbf{d}} \to \Phi_{\overline{\mathbf{d}}}\ , \quad \overline{e} \to \Phi_{\overline{e}}\ ;$$

it follows that each fermion helicity has a scalar superpartner with the same electroweak quantum number

$$\begin{aligned}
\Phi_L &: \begin{pmatrix} \nu_L \\ e_L \end{pmatrix} \text{ and } \begin{pmatrix} \widetilde{\nu}_L \\ \widetilde{e}_L \end{pmatrix} \quad \text{(leftslepton)}\ , \\
\Phi_{\overline{e}} &: \quad \overline{e}_L \quad \text{and} \quad \widetilde{e}^*_R \quad \text{(rightantislepton)}\ , \\
\Phi_{\mathbf{Q}} &: \begin{pmatrix} \mathbf{u}_L \\ \mathbf{d}_L \end{pmatrix} \text{ and } \begin{pmatrix} \widetilde{\mathbf{u}}_L \\ \widetilde{\mathbf{d}}_L \end{pmatrix} \quad \text{(leftsquark)}\ , \\
\Phi_{\overline{\mathbf{u}}} &: \quad \overline{\mathbf{u}}_L \quad \text{and} \quad \widetilde{\mathbf{u}}^*_R \quad \text{(rightantisquark)}\ , \\
\Phi_{\overline{\mathbf{d}}} &: \quad \overline{\mathbf{d}}_L \quad \text{and} \quad \widetilde{\mathbf{d}}^*_R \quad \text{(rightantisquark)}\ .
\end{aligned} \tag{10.1}$$

Squarks and sleptons are also given twiddles: for example $\widetilde{\mathbf{u}}^*_R$ is an electroweak-singlet, color-antitriplet, complex scalar field representing the right up-like squark. The subscript R indicates it is the superpartner of the right-handed up quark, (not its helicity–it has no spin). Family indices have been suppressed, but this association is repeated for each of the three chiral families.

It may seem uneconomical to double the numbers of degrees of freedom, but this is done here in the context of enlarging a space-time symmetry. There is a historical precedent for such a dramatic increase in the number of elementary particles: the requirement of Lorentz invariance, through the Dirac equation, forced the introduction of the positron, and physicists of the 1930's saw their world double with the addition of antimatter.

In $N = 1$ supersymmetry, the standard model Higgs doublet must be viewed as the scalar component of a chiral superfield, which contains one weak doublet of Weyl fermions, the Higgsinos. These chiral fermions have the same electroweak quantum numbers as the Higgs (hypercharged electroweak doublets), which generate contributions to two different anomalies. The first is the usual ABJ triangle anomaly associated with hypercharge. The second is Witten's global anomaly, according to which, any theory with an odd number of chiral fermions which transform as $SU(2)$ doublets, path-integrates to zero. The simplest way to remedy these two potential disasters is to introduce another Higgsino doublet with opposite hypercharge, to act as the vector-like completion of the first, and cancel both anomalies. By the reverse argument, this introduces a new Higgs spin zero doublet with opposite hypercharge. The minimal $N = 1$ model therefore must contain two Higgs chiral superfields:

$$\begin{aligned} \Phi_{H_d} &: \begin{pmatrix} H_d^0 \\ H_d^- \end{pmatrix}_{-1} \quad \text{and} \quad \begin{pmatrix} \widetilde{H}_d^0 \\ \widetilde{H}_d^- \end{pmatrix}_{-1} \quad \text{(Higgsino)} , \\ \Phi_{H_u} &: \begin{pmatrix} H_u^+ \\ H_u^0 \end{pmatrix}_{1} \quad \text{and} \quad \begin{pmatrix} \widetilde{H}_u^+ \\ \widetilde{H}_u^0 \end{pmatrix}_{1} \quad \text{(Higgsino)} . \end{aligned} \tag{10.2}$$

The Yukawa interactions of the MSSM are the same as in the $N = 0$ standard model, but of course written in a manifestly supersymmetric-invariant form. Hence they appear in the superpotential

$$W_{MSSM} = \mathbf{Y}_{ij}^{u} \Phi_{\mathbf{Q}}^{i} \Phi_{\overline{\mathbf{u}}}^{j} \Phi_{H_u} + \mathbf{Y}_{ij}^{d} \Phi_{\mathbf{Q}}^{i} \Phi_{\overline{\mathbf{d}}}^{j} \Phi_{H_d} + \mathbf{Y}_{ij}^{\ell} \Phi_{L}^{i} \Phi_{\overline{e}}^{j} \Phi_{H_d} , \tag{10.3}$$

where $i, j = 1, 2, 3$ are the family indices of the three chiral families.

This superpotential reproduces the Yukawa couplings of the standard model, but with one important difference. We recall that the one Higgs field of the standard model does double duty, coupling to charge $2/3$

quarks, and its conjugate to charge -1/3, -1 quarks and leptons, and giving all fermions their masses. Supersymmetry does not allow conjugates to appear in the superpotential which has to be holomorphic in the superfields. As a result, hypercharge conservation forbids the same Higgs superfield from coupling analytically to both sectors. With only one Higgs field, some quarks and/or leptons would stay massless.

It follows that a second Higgs supermultiplet is needed, not only to satisfy theoretical consistency, but also to reproduce the fact that all quarks and charged leptons are massive.

The reduction of the flavor Yukawa matrices $\mathbf{Y}_{ij}$ proceeds as in the $N = 0$ model. Without loss of generality, we can bring the lepton Yukawa to diagonal form,

$$\mathbf{Y}^{\ell}_{ij} \to \mathbf{Y}^{\ell}_{ii} \ . \tag{10.4}$$

We then diagonalize the down-quark Yukawa sector, by setting

$$\mathbf{Y}^{d} = \mathbf{U}^{T}_{d}\mathbf{M}_{d}\mathbf{V}_{d} \ , \qquad (\mathbf{M}_{d} \text{ diagonal}) \ , \tag{10.5}$$

and by rewriting the Lagrangian in terms of the superfields

$$\Phi'^{i}_{\overline{\mathbf{d}}} = (\mathbf{V}_{d}\Phi_{\overline{\mathbf{d}}})^{i}, \quad \Phi'^{i}_{\mathbf{Q}} = (\mathbf{U}_{d}\Phi_{\mathbf{Q}})^{i} \ .$$

The same reduction of the up-quark Yukawa matrix yields

$$\mathbf{Y}^{u} = \mathbf{U}^{T}_{u}\mathbf{M}_{u}\mathbf{V}_{u} \ , \qquad (\mathbf{M}_{u} \text{ diagonal}) \ , \tag{10.6}$$

while redefining

$$\Phi'^{i}_{\overline{\mathbf{u}}} = (\mathbf{V}_{u}\Phi_{\overline{\mathbf{u}}})^{i} \ . \tag{10.7}$$

The $\mathbf{V}_u$ matrix disappears from the Lagrangian, and we have no further freedom for $\Phi_{\mathbf{Q}}$. The cubic superpotential (after dropping the primes) becomes, suppressing all color and electroweak indices,

$$\mathbf{Y}^{\ell}_{ii}\Phi^{i}_{L}\Phi^{i}_{\overline{e}}\Phi_{H_d} + \mathbf{M}^{ii}_{d}\Phi^{i}_{\mathbf{Q}}\Phi^{i}_{\overline{\mathbf{d}}}\Phi_{H_d} + \Phi^{i}_{\mathbf{Q}}\widehat{\mathcal{U}}^{ji}\mathbf{M}^{jj}_{u}\Phi^{j}_{\overline{u}}\Phi_{H_u} \ , \tag{10.8}$$

with the flavor-mixing matrix

$$\widehat{\mathcal{U}} = \mathbf{U}^{t}_{u}\mathbf{U}_{d} \ ; \tag{10.9}$$

it reduces to the CKM matrix, after the usual Iwasawa decomposition. The Yukawa couplings of the quarks and leptons in the $N = 0$ and $N = 1$ models are the same, except that we now have two Higgs of opposite hypercharge.

Unfortunately, this simple cubic superpotential cannot by itself reproduce the real world for two different reasons: it is invariant under unphysical global symmetries, and it generates a potential which does not have a physically-acceptable ground state.

This superpotential contains the global symmetries of the $N = 0$ standard model. In the lepton sector, it conserves the three lepton numbers; in the quark sector, no distinction between families is allowed, and only the total quark (baryon) number is conserved. Global transformations on the superfields are of the form

$$\Phi_s \to e^{i\eta n_s} \Phi_s \ , \tag{10.10}$$

where s denotes the species: L, $\overline{e}$, $\overline{\mathbf{u}}$, $\overline{\mathbf{d}}$, or $\mathbf{Q}$. The global transformations which preserve supersymmetry obey the relations

$$\begin{aligned} n_{L_i} + n_{\overline{e}_i} + n_{H_d} &= 0 \ , \qquad i = 1, 2, 3 \ , \\ n_{\mathbf{Q}} + n_{\overline{\mathbf{u}}} + n_{H_u} &= 0 \ , \qquad \text{any flavor} \ , \\ n_{\mathbf{Q}} + n_{\overline{\mathbf{d}}} + n_{H_d} &= 0 \ , \qquad \text{any flavor} \ . \end{aligned} \tag{10.11}$$

Assume for a moment only one chiral family. With seven superfields, there are seven phases. Constrained by the three couplings in the superpotential, this leaves four independent symmetries, identified with

	n_L	$n_{\overline{e}}$	n_Q	$n_{\overline{u}}$	$n_{\overline{\mathbf{d}}}$	n_{H_u}	n_{H_d}
L	1	-1	0	0	0	0	0
B	0	0	1/3	-1/3	-1/3	0	0
Y	-1	2	1/3	-4/3	2/3	1	-1
PQ	-1/2	-1/2	-1/2	-1/2	-1/2	1	1

We recognize the total lepton number (L), the vectorial baryon number (B), the chiral Peccei-Quinn (PQ) symmetry, and of course the local hypercharge symmetry (Y). When the full complement of three families is added, there are two additional global symmetries, the relative lepton numbers, $L_e - L_\mu$ and $L_\mu - L_\tau$. Except for the Peccei-Quinn symmetry, all these symmetries appear in the standard model.

Supersymmetric theories can have a special global $U(1)$ symmetry, called R-symmetry. Invariance under R is achieved if, under a phase change of the Grassmann variables, $\theta \to e^{i\beta}\theta$, the superpotential transforms as $W \to e^{2i\beta}W$. Since our superpotential contains only cubic superfield interactions, the theory is (so far) also R-invariant with

$$\Phi_s \to e^{i2\beta/3}\Phi_s \ , \tag{10.12}$$

for all chiral superfields. Under this symmetry, the chiral spinor superfields that contain the gauge bosons transform as

$$\mathcal{W}^a(x,\theta) \to e^{i\beta}\mathcal{W}^a(x, e^{i\beta}\theta) \ , \tag{10.13}$$

(shown here for $SU(2)^W$ only). The gauginos transform the same way

$$\lambda^a(x) \to e^{i\beta}\lambda^a(x) \ . \tag{10.14}$$

We see that invariance under R-symmetry requires the gauginos to be massless, since their Majorana masses, $\widehat{\lambda}^a\lambda^a$, transform under R.

To have escaped detection, gauginos must be massive, so that R-symmetry is broken in the real world. We note in passing that R-symmetry is an anomalous chiral symmetry.

The chiral PQ symmetry also causes a problem. As it is carried only by fields that transform as weak isospinors, the PQ symmetry is necessarily broken at electroweak scales. This leads to a weakly coupled axion with a mass in the keV range, a possibility that has been experimentally ruled out. Hence this symmetry must be broken as well.

This embarassment of undesirable symmetries is somewhat alleviated by introducing another term in the superpotential, the so-called μ-term

$$\mu\Phi^t_{H_u}\tau_2\Phi_{H_d} \ . \tag{10.15}$$

It is a mass term, which violates neither supersymmetry nor any electroweak symmetry, and gives a common mass μ, to the Higgs and Higgsinos. It also breaks both PQ and R symmetries, but one linear combination

$$R' = R + \frac{1}{3}PQ \ , \tag{10.16}$$

is left invariant, since this term has $\Delta PQ = 2$, and $\Delta R = -2/3$. However, this left-over R' symmetry (henceforth we drop the prime) is still nefarious enough to cause phenomenological havoc, as it implies massless gauginos.

However, we have not yet discussed the breaking of supersymmetry, as a result of which we expect the gauginos to acquire masses. This will explicitly break R-symmetry, and with the μ-term, PQ symmetry as well. One obvious conflict with phenomenology will at least be avoided. Clearly the μ term is necessary, and the superpotential of the *minimal* $N = 1$ standard model (MSSM) contains both the cubic terms of Eq (8.3), and the quadratic μ term. We note that the mass scale of the μ-term is an embarassment for supersymmetry since it can have any value, as it is unrestricted by symmetry considerations. Irrespective of its genesis, the μ-term must be present in the effective low energy theory, to resolve conflicts with phenomenology.

To conclude, if this $N = 1$ supersymmetric model is to describe the real world, one must add to it interactions which break supersymmetry, in order to generate a mass gap between the known particles and their superpartners, break the electroweak and R symmetries. These are not all unconnected since breaking of supersymmetry generates a mass for the gauginos, which as we have seen, automatically breaks R-symmetry. However, the gaugino mass terms are of the Majorana type, that is, quadratic in the gaugino fields; as such they are invariant under a discrete remnant of R symmetry for which the gaugino fields change sign. This discrete symmetry is called R-parity.

Naively, one might have thought that, even after adding, through supersymmetry breaking, gaugino mass terms, the minimal $N = 1$ standard model would still have an apparent discrete Z_4 symmetry, under which the fields transform as

$$\lambda \to -\lambda \; ; \qquad \Phi_s \to i\Phi_s \; ; \qquad \Phi_{H_{u,d}} \to -\Phi_{H_{u,d}} \; , \tag{10.17}$$

where λ represents any of the gauginos. Since it is an R-symmetry, particles and their superpartners do not share the same multiplicative quantum number.

Since the Higgs scalar doublets are odd under this symmetry, we might have expected that electroweak breaking would break it down to Z_2, creating potential domain wall problems, with bad cosmological consequences. Fortunately it can be shown that the "broken" part of this symmetry is in fact expressible in terms of hypercharge, baryon number and lepton number, and thus no discrete symmetry is broken at the electroweak scale, and the Z_4 symmetry is in fact a Z_2 symmetry. Identified as R-parity, it is an *exact* symmetry of the minimal $N = 1$ standard model. It is easy to see that all quarks, leptons, and Higgs bosons are even under R-parity; all their superpartners are odd.

The existence of this extra symmetry has far-reaching experimental consequences. Superpartners, which have odd R-parity, must be pair-produced in the laboratory, from ordinary matter which is R-even. In addition, R-parity conservation requires that they decay into an odd number of superpartners. At the end of a chain of sequential decays, there will remains one odd R-parity particle, the lightest superpartner (LSP). Kinematically forbidden to further decay, this particle is *absolutely stable.*

This imposes several restrictions on the nature of this particle. Consistency with cosmology implies that at some point particles and their superpartners were in thermal equilibrium at temperatures above the mass of the superpartners. As the Universe cooled, the superpartners decayed away, leaving behind the stable lightest supersymmetric particle (LSP). These particles should be present everywhere in our present Universe as well. If its present cosmological abundance is significant, the LSP has to be neutral. Its present abundance depends on the circumstances of its production and its mass. Even if neutral, there are severe bounds since they could overclose the universe, if too massive and/or abundant. An exciting possibility is that the LSP makes up most if not all of the "dark matter" which pervades our universe.

The Non-Minimal $N = 1$ Model

Even with the chiral superfields of the standard model, we can add to the superpotential other renormalizable terms which are invariant under both supersymmetry and the gauge groups $SU(3)^c \times SU(2)^W \times U(1)$. They are, suppressing all family indices and isospin coupling matrices,

$$\Phi_{\overline{\mathrm{d}}}\Phi_{\overline{\mathrm{d}}}\Phi_{\overline{\mathrm{u}}}\ ; \qquad \Phi_{\mathbf{Q}}\Phi_{\overline{\mathrm{d}}}\Phi_{L}\ ; \qquad \Phi_{L}\Phi_{L}\Phi_{\overline{e}}\ , \qquad \Phi_{L}\Phi_{H_u}\ . \tag{10.18}$$

All violate some global symmetry. The first violates quark number by three units, so that baryon number is broken (*mod* 3), leaving behind the discrete group Z_3. The three other terms violate lepton number by one unit. While these new terms are allowed by supersymmetry, the stringent experimental limits on both baryon and lepton number conservation indicate that they are restricted to appear with tiny coupling coefficients, if at all.

They all violate the R-symmetry (*mod* 4), since they have $R = 3/2$, which does not leave any discrete remnant. There is also a parity under which all weak doublets are odd, all singlets even, but this is a consequence of invariance under the weak $SU(2)$.

In this non-minimal model, it is easy to see that there is no R-parity, and therefore no stable superparticle, although in some models the LSP could be long-lived, since the non-minimal interactions have to be small.

PROBLEMS

A. The left-handed lepton doublets and the Higgs doublet have the same gauged electroweak quantum numbers, but different lepton number. List the objections to building a model where the Higgs is the superpartner of a lepton doublet.

B. Since the top quark mass is much larger than any other quark and lepton, we might want to build a model with only one Higgs doublet coupling to the charge 2/3 sector, and cancel its anomalies by adding new fields. Build a model of this type.

C. Show that the baryon number violating coupling in the non-minimal extension cannot by itself cause proton decay, but can cause baryon-antibaryon oscillations.

D. 1-) Find the transformations of all the fields of the MSSM under the Z_4 in the text. Show that no discrete symmetry is broken at the electroweak scale.
2-) Show that all particles of the MSSM can be assigned an R-parity. Discuss the consequences of this assignment when the non-minimal interactions are considered?

10.2 SUSY BREAKING IN THE MSSM

It is obvious that to account for data, supersymmetry must be broken. We discount the so-called hard breaking, which would affect the ultraviolet structure of the theory, and destroy one of the rationales for supersymmetry: perturbative control in the ultraviolet. This suggests that we only consider breaking generated by the vacuum: spontaneous breaking. We therefore seek a theory whose vacuum state which breaks $N = 1$ supersymmetry.

Like turtles which carry their own houses, supersymmetric theories generate their own potential. We start by asking if the MSSM is capable of the heroic deed of breaking any symmetries. The full tree-level potential of the minimal $N = 1$ standard model is the sum of the squares of its F- and D-terms,

$$V = \sum_{i=H_u,H_d,L,\mathbf{Q},\overline{e},\overline{\mathbf{u}},\overline{\mathbf{d}}} F_i^\dagger F_i + \frac{1}{2}(D^2 + D^a D^a + D^A D^A) \ . \tag{10.19}$$

There is one D-term for each of the three gauge groups

$$\begin{aligned} U(1) : D &= \frac{1}{2}g_1[-\widetilde{L}^\dagger_{Li}\widetilde{L}_{Li} + 2\widetilde{e}^*_{Ri}\widetilde{e}_{Ri} + \frac{1}{3}\widetilde{\mathbf{Q}}^\dagger_{Li}\widetilde{\mathbf{Q}}_{Li} + \frac{4}{3}\widetilde{\mathbf{u}}^\dagger_{Ri}\widetilde{\mathbf{u}}_{Ri} \\ &\quad - \frac{2}{3}\widetilde{\mathbf{d}}^\dagger_{Ri}\widetilde{\mathbf{d}}_{Ri} + H^\dagger_u H_u - H^\dagger_d H_d] , \\ SU(2)^W : D^a &= g_2[\widetilde{L}^\dagger_{Li}\frac{\tau^a}{2}\widetilde{L}_{Li} + \widetilde{\mathbf{Q}}^\dagger_{Li}\frac{\tau^a}{2}\widetilde{\mathbf{Q}}_{Li} + H^\dagger_u\frac{\tau^a}{2}H_u + H^\dagger_d\frac{\tau^a}{2}H_d] , \\ SU(3)^c : D^A &= g_3[\widetilde{\mathbf{Q}}^\dagger_{Li}\frac{\lambda^A}{2}\widetilde{\mathbf{Q}}_{Li} + \widetilde{\mathbf{u}}^\dagger_{Ri}\frac{\lambda^A}{2}\widetilde{\mathbf{u}}_{Ri} + \widetilde{\mathbf{d}}^\dagger_{Ri}\frac{\lambda^A}{2}\widetilde{\mathbf{d}}_{Ri}] . \end{aligned} \tag{10.20}$$

The F terms are obtained from the MSSM superpotential

$$\begin{aligned} F_{H_u} &= \mu H_d - \widetilde{\mathbf{u}}^*_R\mathbf{M}_u\mathcal{U}\widetilde{\mathbf{Q}}_L , \\ F_{H_d} &= -\mu H_u - \widetilde{e}^*_R\mathbf{M}_e\widetilde{L}_L - \widetilde{\mathbf{d}}^*_R\mathbf{M}_d\widetilde{\mathbf{Q}}_L , \\ F_{\mathbf{Q}} &= \mathbf{M}_d\widetilde{\mathbf{d}}_R H_d + \mathcal{U}^t\mathbf{M}_u\widetilde{\mathbf{u}}^*_R H_u , \end{aligned} \tag{10.21}$$

$$F_L = \mathbf{M}_e\widetilde{e}^*_R H_d , \qquad F_{\overline{e}} = \widetilde{L}^t\mathbf{M}_e H_d , \tag{10.22}$$

$$F_{\overline{\mathbf{u}}} = \widetilde{\mathbf{Q}}^t_L\mathcal{U}^t\mathbf{M}_u\tau_2 H_u , \qquad F_{\overline{\mathbf{d}}} = \widetilde{\mathbf{Q}}^t\mathbf{M}_d\tau_2 H_d . \tag{10.23}$$

We see that, except for the μ term, this potential is purely quartic in the fields. The μ term gives an equal mass to the Higgs and the Higgsinos, and also generates cubic couplings among the Higgs and sleptons and squarks.

To determine if the MSSM potential breaks either supersymmetry and/or electroweak symmetry, we note that, as a sum of squares, its minimum occurs when all the auxiliary fields are set to zero. There are many field configurations for which this is true.

The simplest example is that for which all the fields as well as the potential vanish. We recall (from the Appendix) that spontaneous breaking of supersymmetry requires that the action of a supersymmetry transformation on the vacuum not vanish. From the supersymmetry algebra, it follows that the vacuum energy must be positive. Hence the MSSM potential does not by itself break supersymmetry.

To see how far this potential is from the real world, let us work out its value at the electroweak minimum where

$$H_u = \frac{v_u}{\sqrt{2}}\begin{pmatrix}0\\1\end{pmatrix} ; \qquad H_d = \frac{v_d}{\sqrt{2}}\begin{pmatrix}1\\0\end{pmatrix} , \tag{10.24}$$

and set all the other fields to zero (unless we want L-violation through the only charge conserving possibility $\widetilde{\nu}_L \neq 0$ which violates L spontaneously). Using

$$H_u^\dagger H_u = -H_u^\dagger \frac{\tau^3}{2} H_u = \frac{|v_u|^2}{4} \,, \qquad H_d^\dagger H_d = H_d^\dagger \frac{\tau^3}{2} H_d = \frac{|v_d|^2}{4} \,, \quad (10.25)$$

we find that, in the electroweak vacuum, some D and F terms do not vanish, namely

$$D = \frac{1}{4} g_1 [|v_u|^2 - |v_d|^2] \,, \qquad D^3 = -\frac{1}{4} g_2 [|v_u|^2 - |v_d|^2] \,, \qquad (10.26)$$

and

$$F_{H_u} = \mu \frac{v_d}{\sqrt{2}} \begin{pmatrix} 0 \\ i \end{pmatrix} \,, \qquad F_{H_d} = \mu \frac{v_u}{\sqrt{2}} \begin{pmatrix} i \\ 0 \end{pmatrix} \,. \qquad (10.27)$$

The electroweak vacuum configuration is not a minimum of the MSSM potential, and breaks supersymmetry.

Without a μ term, all F terms would vanish in the electroweak vacuum. Also, there would be an infinite number of degenerate vacua where the D terms all vanish as well, leaving supersymmetry unbroken, with $|v_u| = |v_d|$. This leaves the vacuum values of the Higgs doublet undetermined, at least at the classical level. This is an example of a *flat direction*, a ubiquitous feature of supersymmetric models. Alas, phenomenology requires a μ term, which "lifts" that vacuum degeneracy, by setting the minimum at $|v_u| = |v_d| = 0$. Since this potential breaks neither supersymmetry nor electroweak symmetry, we conclude that the MSSM must be augmented by mechanisms that generate both supersymmetry and/or electroweak symmetry breakings.

There are many renormalizable models of spontaneous supersymmetry breaking. Models with F-breaking were first investigated by L. O'Raifeartaigh (*Nucl. Phys.* **B96**, 331(1975)), and are discussed in the Appendix. Their general feature is a global R-like symmetry, and they satisfy the sum rule

$$\mathrm{Str}\mathcal{M}^2 \equiv \sum_{J=0,1/2} (-1)^{2J}(2J+1)m_J^2 = 0, \qquad (10.28)$$

relating the masses of the particles and their superpartners.

Other models, first proposed by P. Fayet and J. Iliopoulos (*Phys. Lett.* **B51**, 461(1974)), generate D-breaking. They require a local $U(1)$ symmetry, to allow for a gauge singlet term linear in its associated D field. In these models, the same sum rule is modified to read

$$\mathrm{Str}\mathcal{M}^2 = gD \sum_i q_i \,, \qquad (10.29)$$

where q_i are the charges of the Weyl fermions. If the anomaly of this $U(1)$ is cancelled in a vector-like way, the right-hand side is zero.

In both cases, these sum rules cause phenomenological problems, although they can be modified by quantum corrections. For that reason, these models have not proved easy to implement. In the context of global supersymmetry, susy breaking produces a massless Nambu-Goldstone fermion, called the Goldstino. In supergravity, the Goldstino becomes part of the gravitino, the massive superpartner of the graviton, and the right-hand-side of these sum rules acquires new contributions, which may alleviate phenomenological concerns.

Fortunately, we can describe the impact of supersymmetry-breaking on the MSSM without committing ourselves to a specific theory for the breaking mechanism. Indeed, in the context of an effective theory, it is natural to parametrize the spontaneous breaking of a symmetry by adding soft terms to its Lagrangian. These are mass terms of dimension two and cubic interactions of scalars with dimension three. Intuitively these terms cannot affect the theory in the ultraviolet, since they all have prefactors with positive mass dimensions, which vanish in the limit where all masses are taken to zero, relative to the scale of interest. Soft breaking is not to be viewed as fundamental, but as an effective manisfestation of a more fundamental theory of symmetry breaking. We have already used soft-breaking terms in Chapter 5, to describe chiral symmetry breaking in the context of the effective low energy chiral Lagrangian that describes strong interactions. We now know that at the fundamental level, these soft-breaking terms are generated by the quark mass terms in QCD, the real fundamental theory of the strong interactions. As of this writing, we have no equivalently compelling picture of the fundamental theory that generates susy breaking. Its formulation remains one of the deepest mysteries of particle physics, since it impacts directly on the cosmological constant problem.

The useful strategy we are going to pursue is to add to the MSSM soft terms that break supersymmetry but not electroweak symmetry. We can think of several types of masses that explicitly break supersymmetry: mass terms for the superpartners of the massless chiral fermions; mass terms for each Higgs doublet but not for the Higgsinos; and finally gaugino mass terms. These clearly split the mass degeneracy between particles and their superpartners, thus breaking supersymmetry. There are also terms of dimension three that describe interactions among the scalar components of chiral superfields. In order to preserve electroweak symmetry, they are taken to be of the same form as the superpotential, with the superfields replaced by their scalar components. At the end, it takes more than sixty parameters with mass dimension (and as many phases) to describe the soft-breaking of the MSSM!

It is clear that, to make further progress, we need to make some simplifying assumptions to reduce the number of soft parameters.

The general soft symmetry breaking potential for the MSSM contains three types of mass terms. The Higgs mass terms

$$m_{H_u}^2 H_u^\dagger H_u + m_{H_d}^2 H_d^\dagger H_d + \hat{\mu}(H_u^t \tau_2 H_d + \text{c.c.}) \;, \tag{10.30}$$

where $\hat{\mu}$ is often written as $B\mu$, where B is a dimensionless parameter. The squark and slepton mass terms are

$$\sum_{ij}\Big(m_{\widetilde{\mathbf{Q}}ij}^2 \widetilde{\mathbf{Q}}_i^\dagger \widetilde{\mathbf{Q}}_j + m_{\widetilde{\mathbf{u}}ij}^2 \widetilde{\mathbf{u}}_{Ri}^\dagger \widetilde{\mathbf{u}}_{Rj} + m_{\widetilde{\mathbf{d}}ij}^2 \widetilde{\mathbf{d}}_{Ri}^\dagger \widetilde{\mathbf{d}}_{Rj} + m_{\widetilde{L}ij}^2 \widetilde{L}_i^\dagger \widetilde{L}_j + m_{\widetilde{e}ij}^2 \widetilde{e}_{Ri}^\dagger \widetilde{e}_{Rj} \Big) \;. \tag{10.31}$$

If these terms violate only supersymmetry, the slepton mass matrices must be diagonal to preserve the relative lepton number symmetries. Note that in view of the recent experimental findings on flavor mixing among neutrinos, this restriction no longer applies. No such symmetry restriction exists for the squark mass matrices, but in order to avoid flavor changing neutral processes, they must have numerically small off-diagonal components. This requirement has led many physicists to assume that all the squark mass matrices are diagonal. A more extreme attitude is to assume that *all* the soft squark and slepton masses are the same, implying some sort of universality of the susy-breaking mechanism.

There are also three gaugino mass terms

$$\frac{1}{2}\sum_{l=1}^{3} M_l \widehat{\lambda}_l \lambda_l + \text{c.c.} \;; \tag{10.32}$$

they break R-symmetry, but preserve R-parity. These mass terms are supplemented by the cubic interactions

$$\sum_{i,j}\Big(\mathcal{A}_u^{ij} \widetilde{\mathbf{u}}_{Ri} H_u \widetilde{\mathbf{Q}}_j + \mathcal{A}_d^{ij} \widetilde{\mathbf{d}}_{Ri} H_d \widetilde{\mathbf{Q}}_j + \mathcal{A}_e^{ij} \widetilde{e}_{Ri} H_d \widetilde{L}_j + \text{c.c.} \Big) \;. \tag{10.33}$$

The form of the coefficient matrices $\mathcal{A}_{\mathbf{u},\mathbf{d},e}^{ij}$ should depend on the mechanism that generates the Susy-breaking.

PROBLEMS

A. Find the exact number of all the possible soft terms that break supersymmetry in the MSSM, and preserve only the *gauged* electroweak symmetries.

B. Show that the sum rules for F and D breaking of global supersymmetry applied to the MSSM lead to contradictions with experiment. Discuss possible ways to avoid this conflict.

C. Determine all possible higher order operators made out of the MSSM superfields. Hint: organize the catalog of these operators in terms of their baryon and lepton number properties. For details, see T. Gerghetta, Chris Kolda and S. Martin, *Nucl. Phys.* **B468**, 37(1996).

10.3 THE TREE LEVEL MSSM

The addition of the soft terms to the MSSM superpotential opens the way for the analysis of the mass eigenstates of the MSSM. The low energy spectrum of the quarks and leptons remains much the same, but the model implies serious modifications in the Higgs sector. Also, it enables us to discuss the spectrum of the extra spin one-half fermions in the models, dubbed *charginos* and *neutralinos*, as well as the scalar *squarks* and *sleptons*.

The Higgs Sector

The Higgs sector of the MSSM contains two complex doublets, $H_{u,d}$, that describe two charged fields and four neutral real fields. Electroweak breaking causes one charged combination to be "eaten" by the charged W boson, and one neutral by the Z-boson. This leaves one charged Higgs field H^+ and its conjugate, one CP-odd neutral scalar A, and two CP-even neutrals the *little* Higgs field h, and the *big* Higgs field H. The MSSM Higgs sector is much richer than that of the standard model which has only one scalar Higgs field. With the soft-breaking terms, the scalar potential of the MSSM is given by

$$\begin{aligned} V = B\mu(H_u^+ H_d^- - H_u^0 H_d^0) + \text{ c.c.} + \sum_{i=u,d} (|\mu|^2 + m_{H_i}^2) H_i^\dagger H_i + \\ + \frac{1}{8} g_1^2 [H_u^\dagger H_u - H_d^\dagger H_d]^2 + \frac{1}{8} g_2^2 [H_u^\dagger \tau^a H_u - H_d^\dagger \tau^a H_d]^2 \ . \end{aligned} \tag{10.34}$$

The quadratic terms all come from the D-terms, so that all quartic couplings are squares of gauge couplings.

In this tree-level form of the potential, all of its parameters depend on an arbitrary renormalization scale. We can recover (one-loop) renormalization group-independence by adding the (one-loop) effective potential. We may think that we can always choose the renormalization scale so as to make these corrections small, thereby justifying our neglect of the quantum corrections in the analysis that follows. However, the reader must be warned that the *values* of the parameters in the vacuum can be very sensitive to that choice of scale. Therefore, while a complete analysis requires that these corrections be included, they are not necessary for the discussion of the potential near its minimum provided its parameters are directly expressed in terms of physical masses.

We have not included in this potential the squark and slepton fields which must not play a role in electroweak symmetry breaking. This can be justified if some inequalities among the parameters are satisfied.

The potential contains, however, the charged fields $H^{\pm}_{u,d}$, through for example,

$$H_u^\dagger H_u = |H_u^0|^2 + |H_u^+|^2 \; . \tag{10.35}$$

Fortunately, they do not acquire a *vev*. To see this, we first note that by a weak isospin rotation we can set without loss of generality $< H_u^+ >= 0$. Then it is easy to see that the vanishing of the derivative at minimum requires that $< H_d^+ >= 0$ as well. Hence charge is preserved and we can simplify our discussion by setting these fields to zero, yielding the potential

$$\begin{aligned} V = \sum_{i=u,d} (|\mu|^2 + m^2_{H_i})|H_i^0|^2 - B\mu H_u^0 H_d^0 + \text{ c.c.} \\ +\frac{1}{8}(g_1^2 + g_2^2)(|H_u^0|^2 - |H_d^0|^2)^2 \; . \end{aligned} \tag{10.36}$$

A careful analysis of the phases show that this potential cannot violate CP. As a result all the physical Higgs particles are CP eigenstates.

We see that the quartic terms vanish along the *flat direction* $|H_u^0| = |H_d^0|$. A positive quadratic term requires

$$2|\mu|^2 + m^2_{H_u} + m^2_{H_d} > 2B\mu \; , \tag{10.37}$$

to insure that the potential is bounded from below. For this potential to break electroweak symmetry, the determinant of the mass matrix must be negative, so that one linear combination of the fields have a negative mass. This means that

$$B^2\mu^2 > (|\mu|^2 + m_{H_u}^2)(|\mu|^2 + m_{H_d}^2) \ . \qquad (10.38)$$

Interestingly, these two conditions are not compatible if $m_{H_u}^2 = m_{H_d}^2$, a fact that is in accord with their renormalization group evolution.

Let us assume that all these conditions are satisfied and electroweak breaking takes place. At minimum, we set $< H_{u,d}^0 >= v_{u,d}$. While the combination $v_u^2 + v_d^2$ is fixed by the Fermi coupling constant, the MSSM introduces a new angle β, defined by the ratio of the vacuum values

$$\tan\beta \equiv \frac{v_u}{v_d} \ . \qquad (10.39)$$

Minimization with respect to H_u^0 and H_d^0 respectively, yield

$$|\mu|^2 + m_{H_u}^2 - B\mu\cot\beta = \frac{1}{2}m_Z^2\cos 2\beta \ , \qquad (10.40)$$

and

$$|\mu|^2 + m_{H_d}^2 - B\mu\tan\beta = -\frac{1}{2}m_Z^2\cos 2\beta \ , \qquad (10.41)$$

where m_Z is the tree-level Z mass: $m_Z^2 = v^2(g_1^2 + g_2^2)/8$. Note that the left-hand-side of these equations contains the μ parameter while the right-hand-side can be no larger than half the Z boson mass. This shows that the μ parameter must be, barring unlikely cancellations, of the same order of magnitude as electroweak breaking. This is puzzling since its value in the MSSM is not restricted by any symmetry considerations. This points to new symmetries beyond those of the MSSM, through which its value is linked to the susy-breaking mechanism.

One can evaluate this potential around this minimum. In the unitary gauge, we write the neutral Higgs fields as

$$H_u^0 = \frac{1}{\sqrt{2}}e^{iA/v_u}(v_u + h_u(x)) \ , \qquad H_d^0 = \frac{1}{\sqrt{2}}e^{iA/v_d}(v_d + h_d(x)) \ . \ (10.42)$$

Plugging back in the potential, we see that the neutral CP-odd A field gets a mass only from the $B\mu$ term. Easy algebra yields

$$m_A^2 = \frac{2B\mu}{\sin 2\beta} \ . \qquad (10.43)$$

The two neutral fields are mixed both by the $B\mu$ and the D-terms. This mixing is described by a new angle α, in terms of which the CP-even mass eigenstates, $h(x)$ and $H(x)$ are written

$$\begin{pmatrix} h \\ H \end{pmatrix} \equiv \begin{pmatrix} \cos\alpha & -\sin\alpha \\ \sin\alpha & \cos\alpha \end{pmatrix} \begin{pmatrix} h_u \\ h_d \end{pmatrix} . \tag{10.44}$$

The mixing angle is best expressed in terms of the masses

$$\frac{\sin 2\alpha}{\sin 2\beta} = \frac{m_A^2 + m_Z^2}{m_H^2 - m_h^2} , \qquad \frac{\cos 2\alpha}{\cos 2\beta} = \frac{m_A^2 - m_Z^2}{m_H^2 - m_h^2} , \tag{10.45}$$

where the heavier H mass is

$$m_H^2 = \frac{1}{2}(m_A^2 + m_Z^2) + \sqrt{(m_A^2 + m_Z^2)^2 - (2m_Z m_A \cos 2\beta)^2} , \tag{10.46}$$

and the lighter one weighs in at

$$m_h^2 = \frac{1}{2}(m_A^2 + m_Z^2) - \sqrt{(m_A^2 + m_Z^2)^2 - (2m_Z m_A \cos 2\beta)^2} , \tag{10.47}$$

The minus sign shows that the *little* Higgs cannot be arbitrarily heavy. It replaces the Higgs field of the standard model. Its Yukawa couplings to quarks and leptons are the same as the Higgs field, except for multiplicative factors:

$$\frac{\cos\alpha}{\sin\beta} \text{ for up quarks} , \qquad -\frac{\sin\alpha}{\cos\beta} \text{ for down quarks and leptons} . \tag{10.48}$$

Similarly the *big* Higgs couples with factors

$$\frac{\sin\alpha}{\cos\beta} \text{ for up quarks} , \qquad \frac{\cos\alpha}{\sin\beta} \text{ for down quarks and leptons} . \tag{10.49}$$

The CP-odd A field couples to up quarks according to $\cot\beta$, and to the down quarks and charged leptons as $\tan\beta$. However the masses of A and H can be made large, simply by cranking up the value of $B\mu$. Not so for the *little* Higgs. Expanding in inverse powers of m_A^2 yields

$$m_h^2 = m_Z^2 \cos^2 2\beta - m_A^2 (\frac{m_Z}{m_A})^4 (2 - \sin^2 2\beta) + \cdots , \tag{10.50}$$

so that it is bounded *at tree-level* to be lighter than the Z boson. The current (LEP-1999) bound on the standard model Higgs ($\geq$ 95 GeV) rules out the tree-level potential, but we shall see later that the large top quark Yukawa couplings engenders large positive contributions to its mass, saving the MSSM potential from phenomenological disaster.

There exists a limit where the little Higgs h has exactly the same properties as those of the standard model Higgs. Note from (10.45) that

$$\tan 2\alpha = \frac{m_A^2 + m_Z^2}{m_A^2 - m_Z^2} \tan 2\beta \ , \tag{10.51}$$

which shows that if m_A is sufficiently large, $\beta = \alpha - \pi/2$. Hence, $\cos\beta = -\sin\alpha$, $\sin\beta = \cos\alpha$, and the little Higgs couplings reduce to those of the standard model Higgs.

Finally the MSSM contains one charged Higgs field, with mass

$$m_{H^+}^2 = m_A^2 + m_W^2 \ , \tag{10.52}$$

which is also heavy. Although charged, it may not be detectable until a high energy linear collider is built.

To conclude, the Higgs sector of the MSSM contains one light Higgs boson, h, with standard model like properties and mass bounded from above; it is accompanied by three other states, A, H, and H^+ , expected to be heavier than the little Higgs, with masses that become nearly degenerate the heavier they are. In terms of fundamental input parameters, the physical properties of the Higgs potential are determined by $|\mu|$, and the soft breaking parameters $m_{H_{u,d}}^2$ and $B\mu$.

Charginos and Neutralinos

The MSSM contains two new charged fermions, the Higgsinos, $\widetilde{H}_u^+$, $\widetilde{H}_d^-$, and the Winos $\widetilde{W}^{\pm}$. These states will be mixed by electroweak breaking, to yield two *charginos*, $\widetilde{C}_{1,2}$ mass eigenstates. The universal gaugino interaction of the form

$$g_2 \widetilde{W}^{-T} \tau_2 \widetilde{H}_u^+ H_u + g_2 \widetilde{W}^{+T} \tau_2 \widetilde{H}_d^- H_d \ , \tag{10.53}$$

result in mixing terms when the Higgs doublets get their *vev*'s. Adding the direct mass terms for the Higgsinos and Winos, we obtain the mass matrix

$$\begin{pmatrix} M_2 & \sqrt{2}\sin\beta m_W \\ \sqrt{2}\cos\beta m_W & \mu \end{pmatrix} \ , \tag{10.54}$$

where m_W is the W-mass. Its diagonalization involves separate unitary matrices acting on its left and right, and therefore two mixing angles. This yields the charginos mass eigenstates

$$\begin{aligned} \widetilde{C}_1^{\pm} &= \cos\phi_{\pm}\widetilde{W}^{\pm} - \sin\phi_{\pm}\widetilde{H}_u^{\pm} , \\ \widetilde{C}_2^{\pm} &= \sin\phi_{\pm}\widetilde{W}^{\pm} + \cos\phi_{\pm}\widetilde{H}_u^{\pm} , \end{aligned} \tag{10.55}$$

with masses

$$m^2_{\widetilde{C}_{1,2}} = \frac{1}{2}(M_2^2+\mu^2+2m_W^2)^2 \mp \sqrt{4(M_2^2+\mu^2+2m_W^2)^2 - (\mu M_2 - m_W^2 \sin 2\beta)^2)} . \tag{10.56}$$

Perhaps a more useful characterisation of the masses is through the relations

$$m^2_{\widetilde{C}_1} + m^2_{\widetilde{C}_2} = (M_2^2+\mu^2+2m_W^2)^2 , \tag{10.57}$$

which has no β dependence, and

$$m^2_{\widetilde{C}_1} m^2_{\widetilde{C}_2} = \mu M_2 - m_W^2 \sin 2\beta . \tag{10.58}$$

The MSSM also contains four extra neutral spin one-half particles: two neutral Higgsinos, $\widetilde{H}_u^0$, $\widetilde{H}_d^0$, and two neutral gauginos $\widetilde{B}$, $\widetilde{W}^0$. Electroweak breaking causes them to mix into four mass eigenstates called the *neutralinos*, which we denote by $\widetilde{N}_i$, $i = 1, \ldots, 4$, $\widetilde{N}_1$ being the lightest. As for the charginos, their mixing is due to the universal gaugino interactions. These mass eigenstates are determined from the 4×4 Majorana matrix

$$\begin{pmatrix} M_1 & 0 & -\cos\beta\sin\theta_{\mathrm{w}} m_Z & \sin\beta\sin\theta_{\mathrm{w}} m_Z \\ 0 & M_2 & \cos\beta\cos\theta_{\mathrm{w}} m_Z & -\sin\beta\cos\theta_{\mathrm{w}} m_Z \\ -\cos\beta\sin\theta_{\mathrm{w}} m_Z & \cos\beta\cos\theta_{\mathrm{w}} m_Z & 0 & -\mu \\ \sin\beta\sin\theta_{\mathrm{w}} m_Z & -\sin\beta\cos\theta_{\mathrm{w}} m_Z & -\mu & 0 \end{pmatrix} . \tag{10.59}$$

Here θ_{w} is the Weinberg angle. Since it is a Majorana mass matrix, it is diagonalized by one 4×4 unitary matrix. It is parametrized by six mixing angles and ten phases. The analytical expression of the masses and mixing angles is very messy and not very illuminationg, except that the product of the neutralino masses obeys a nice identity

$$\widetilde{N}_1\widetilde{N}_2\widetilde{N}_3\widetilde{N}_4 = -\mu^2 M_1 M_2 + \mu m_W^2 (M_1 + M_2 \tan\theta_{\mathrm{w}}) \sin 2\beta . \tag{10.60}$$

The right-hand-side is negative if $\mu < 0$ and $M_1 M_2 > 0$. If $\mu > 0$, the mass determinant may still be positive, but this implies a chargino lighter than the W boson. Hence the mass determinant is negative, and it is easy to show that one mass eigenvalue is negative. We have the further relations

$$|\widetilde{N}_1| + |\widetilde{N}_2| + |\widetilde{N}_3| - |\widetilde{N}_4| = M_1 + M_2 \ , \tag{10.61}$$

as well as

$$m^2_{\widetilde{N}_1} + m^2_{\widetilde{N}_2} + m^2_{\widetilde{N}_3} + m^2_{\widetilde{N}_4} = M_1^2 + M_2^2 + 2\mu^2 + 2m_Z^2 \ . \tag{10.62}$$

Comparison with (10.57) shows that the average mass squared of the neutralinos is less than that of the charginos.

It is instructive to examine the chargino and neutralino masses in the limit of small electroweak breaking, where we can set

$$m_z \ll |\mu \pm M_{1,2}| \ . \tag{10.63}$$

In that limit, the lightest neutralino is very nearly the Bino, and the lightest chargino is the Wino. The chargino masses are to be

$$\begin{aligned} m_{\widetilde{C}_1} &= M_2 - \frac{m_W^2 \theta_{\mathrm{w}} (M_2 + \mu \sin 2\beta)}{\mu^2 - M_2^2} + \cdots \ , \\ m_{\widetilde{C}_2} &= \mu + \frac{m_W^2 (\mu + M_2 \sin 2\beta)}{\mu^2 - M_2^2} + \cdots \ . \end{aligned} \tag{10.64}$$

Similarly, the two lightest neutralino masses are

$$\begin{aligned} m_{\widetilde{N}_1} &= M_1 - \frac{m_Z^2 \sin^2\theta_{\mathrm{w}} (M_1 + \mu \sin 2\beta)}{\mu^2 - M_1^2} + \cdots \ , \\ m_{\widetilde{N}_2} &= M_2 - \frac{m_W^2 (M_2 + \mu \sin 2\beta)}{\mu^2 - M_2^2} + \cdots \ . \end{aligned} \tag{10.65}$$

It is interesting that the masses of the lightest chargino $\widetilde{C}_1$ and the second neutralino $\widetilde{N}_2$ are nearly degenerate. We leave it as an exercise to the reader to work out $m_{\widetilde{N}_{3,4}}$ in the same approximation. We note that since the MSSM is invariant under R parity, the lighest superpartner (LSP) will be absolutely stable. A prime candidate is the lighest neutralino $\widetilde{N}_1$, which could account for the dark matter that pervades the universe.

To conclude, the chargino and neutralino sectors of the MSSM are closely linked. Their masses depend on the two electroweak gaugino masses and on μ. In that sector, the lightest particle is a neutralino, with probably a degenerate chargino-neutralino at the next level, and two higher mass neutralinos.

Squarks and Sleptons

There are severe phenomenological restrictions on the flavor structure of the input squark and slepton mass matrices. The standard model breaks explicitly, albeit controllably, the quark flavor symmetries. With the recent result from SuperKamiokande, we know that the relative lepton numbers are broken as well. For the soft-breaking terms to preserve these delicate patterns, the squark and slepton matrices must be nearly flavor-diagonal; otherwise the exchange of virtual squarks and sleptons will produce large flavor-changing effects, unless of course the sparticles are very heavy. In the following analysis, we assume that squarks and sleptons mass matrices are diagonal.

The masses of the squarks and sleptons depend primarily on the soft-breaking parameters. Left and right squarks and sleptons are mixed by electroweak breaking, in an amount that depends on their Yukawa couplings, on the value of μ, and on the value of $\tan\beta$. Hence the spectrum of squark and slepton masses of the third chiral family will stand out because of their large Yukawa couplings.

The two top squarks, $\widetilde{t}_{L,R}$, are mixed by the mass matrix

$$\begin{pmatrix} m^2_{\widetilde{Q}_3} + m_t^2 + \Delta_{\widetilde{Q}_3} & m_t(A_t + \mu\cot\beta) \\ m_t(A_t + \mu\cot\beta) & m^2_{\widetilde{t}_R} + m_t^2 + \Delta_{\widetilde{t}_R} \end{pmatrix} , \tag{10.66}$$

where for each squark (and slepton) with third component of isospin I_{3L} and weak hypercharge Y, the D-terms contribute

$$\Delta_{\widetilde{q}} = m_Z^2 \cos 2\beta (I_{3L}\cos^2\theta_{\rm w} - Y\sin^2\theta_{\rm w}) . \tag{10.67}$$

We conclude that appreciable mixing is expected as a result of the large top mass. As a result, one mass eigenstate will be much lower than the central mass. Diagonalization yields two top squark mass eigenstates, $\widetilde{t}_{1,2}$, characterized by a mixing angle $\theta_{\widetilde{t}}$. Similarly, the sbottom mass matrix is

$$\begin{pmatrix} m^2_{\widetilde{Q}_3} + m_b^2 + \Delta_{\widetilde{Q}_3} & m_b(A_b + \mu\tan\beta) \\ m_b(A_b + \mu\tan\beta) & m^2_{\widetilde{b}_R} + m_b^2 + \Delta_{\widetilde{b}_R} \end{pmatrix} , \tag{10.68}$$

yielding two mass eigenstates $\widetilde{b}_{1,2}$, with mixing angle $\theta_{\widetilde{b}}$ not as large as for the top squarks. Finally the two physical stau leptons, $\widetilde{\tau}_{1,2}$ are the eigenstates of the matrix

$$\begin{pmatrix} m^2_{\widetilde{L}_3} + m^2_\tau + \Delta_{\widetilde{L}_3} & m_\tau(A_\tau + \mu \tan\beta) \\ m_\tau(A_\tau + \mu \tan\beta) & m^2_{\widetilde{\tau}_R} + m^2_\tau + \Delta_{\widetilde{\tau}_R} \end{pmatrix} . \tag{10.69}$$

These matrices are to be compared to the mass matrices for squarks, and sleptons of the first two families which are nearly diagonal

$$m^2_{\widetilde{u}_1} = m^2_{\widetilde{Q}_1} + \Delta_{\widetilde{u}_L} , \qquad m^2_{\widetilde{d}_1} = m^2_{\widetilde{Q}_1} + \Delta_{\widetilde{d}_L} , \tag{10.70}$$

so that the left squarks of the first family are split only by electroweak breaking

$$m^2_{\widetilde{u}_1} - m^2_{\widetilde{d}_1} = -\cos 2\beta m^2_W . \tag{10.71}$$

A similar relation obtains for the left selectron and its sneutrino. The right squarks satisfy

$$m^2_{\widetilde{u}_2} = m^2_{\widetilde{u}_R} + \Delta_{\widetilde{u}_R} , \qquad m^2_{\widetilde{d}_2} = m^2_{\widetilde{d}_R} + \Delta_{\widetilde{d}_R} , \tag{10.72}$$

so that

$$m^2_{\widetilde{u}_2} - m^2_{\widetilde{d}_2} = m^2_Z \cos 2\beta \sin^2\theta_{\rm w} . \tag{10.73}$$

Of course the central masses of these particles depend on the values of the soft-breaking parameters and on their renormalization group evolutions.

In the extreme case where *all* input squark and slepton masses are taken to be the same, we expect the lighest squark to be the light stop $\widetilde{t}_1$ because of mixing effects. To conclude, this sector of the MSSM depends on yet another set of parameters, the input squark and slepton masses.

Finally, we note the last fermion of the MSSM, the gluino. As the only fermion with color octet properties, it does not mix, and its mass is solely determined by its input mass M_3, properly continued to lower scales.

PROBLEMS

A. Show that the Higgs potential (10.34) preserves both electric charge and CP.

B. Work out the mass, CP values, and couplings to the leptons and quarks of the Higgs sector fields H^+, A, h and H.

C. In the case where μ is positive and the determinant of the neutralino mass matrix is negative, derive a bound for the lowest chargino mass. Compare with the current LEP bound, when M_1 is less than M_2.

D. Find the expressions for the two heaviest neutralino masses in the limit of Eq. (10.63), and show that $\widetilde{C}_2$ and $\widetilde{N}_3$ are nearly degenerate.

10.4 THE ONE-LOOP MSSM

With the addition of supersymmetry, the low energy quantum corrections to the standard model are not significantly altered. At higher energies, however, there are some dramatic changes. For one, the perky Higgs self-coupling of the standard model is replaced by the square of gauge couplings, and the perturbative evolution of the gauge and Yukawa couplings can proceed to much shorter distances without encountering non-perturbative effects.

In general, local field theories with supersymmetry have more tractable ultraviolet behavior than generic field theories: there are fewer infinities because of the cancellations between boson and fermion loops, resulting in simpler radiative corrections to the tree-level MSSM.

To be compared to experiment, the MSSM must include a mechanism that breaks supersymmetry. We continue to assume that it enters the MSSM through soft-breaking terms that do not affect its ultraviolet structure. There will also be, in addition to the usual quantum corrections of any local field theory, many interesting effects, depressed by inverse powers of the supersymmetry-breaking scale.

Non-Renormalization Theorem

In $N = 1$ theories, a theorem states that the superpotential (F-term) is not renormalized in perturbation theory, but it is not valid beyond perturbation theory.

Applied to a field theory with Wess-Zumino multiplets, it means that only their kinetic term (which is a D-term) can be renormalized. Indeed, direct calculations show these models to have only wave function renormalization. When interactions with vector superfields are added, one finds that only the D-terms generate infinities, which include the gauge couplings.

This theorem can be proved order by order in perturbation theory, using the elegant techniques of supergraphs in superspace. There is however a more direct proof which relies on the holomorphicity of the superpotential as a function of the chiral superfields.

Consider a renormalizable theory that contains only chiral superfields, Φ_i, with a generic superpotential

$$W = m^{ij}\Phi_i\Phi_j + \lambda^{ijk}\Phi_i\Phi_j\Phi_k\ . \tag{10.74}$$

Following N. Seiberg (*Phys. Lett.* **B318**, 469(1993)), we consider all the parameters which appear in this superpotential as the vacuum values of chiral superfields. These chiral superfields have no kinetic terms, and act as background fields. This technique can be used just as well for field theories without supersymmetry, but it does not lead to interesting results. With supersymmetry, it becomes very powerful, because the superpotential must be holomorphic in the superfields, including the background superfields. Hence the superpotential must be analytic in the parameters as well.

In theories with gauge couplings, these remarks can be combined with the asymptotic freedom of the gauge couplings to construct the full non-perturbative effective superpotential. Here we only concern ourselves with the non-renormalization theorem in the perturbative realm.

In the absence of a superpotential, the theory with N chiral superfields has a large global symmetry: $SU(N)\times U(1)\times U(1)_R$, where the last $U(1)$ is an R symmetry. The N superfields Φ_i transform as the fundamental of $SU(N)$ and have unit phases under the two $U(1)$'s.

We can think of the parameters in the superpotential in terms of one chiral superfield Θ_{ij} which transforms as the second rank symmetric $SU(N)$ tensor, with charges $(-2,0)$ under the two $U(1)$'s, and another Θ_{ijk} which transforms as a third rank symmetric tensor of $SU(N)$ with charges $(-3,-1)$. Their *vev*'s yield the parameters m_{ij} and λ_{ijk}, respectively. Since their vacuum values break the global symmetry spontaneously, the superpotential, including quantum corrections, should be invariant under the global group of the kinetic terms, and holomorphic in both the Φ_i and in the fields whose *vev* give the couplings in W.

The most general superpotential consistent with these symmetries, analytic in the superfields Φ_i, is structurally given by

$$W_{\text{eff}} = m^{ij}\Phi_k\Phi_l f(\{z\})\ . \tag{10.75}$$

Here f is an arbitrary analytic function of the set $\{z\}$, which stands for combinations linear in the superfields and invariant under the phase symmetries

$$\{z\} = (m^{-1})_{ij}\lambda^{klm}\Phi_n\ . \tag{10.76}$$

The contraction on the $SU(N)$ indices is assumed for each term in f so as to give an $SU(N)$ invariant. This expression clearly contains higher powers in the chiral superfields than originally present at the classical level. However, the crucial point is made by noting that all the higher powers can be generated from one-particle reducible tree-level diagrams. For instance, suppressing all indices, the terms proportional to Φ^M will behave as

$$\frac{\lambda^{M-2}}{m^{M-3}}\Phi^M \ . \tag{10.77}$$

One can check that these interactions can all be generated by tree graphs, which are reducible. Hence these will not appear in the one-particle irreducible effective action.

This proves that in this theory there are no renormalizations of the superpotential (F-terms). It also means that if a given term is initially absent from the superpotential, it will not be generated by radiative corrections. In this sense, absence of certain interaction terms in the superpotential are technically natural in the absence of any other symmetry considerations. In this simple theory without gauge terms, the result applies beyond perturbation theory, although the theory is not asymptotically free.

With the addition of gauge fields, the above analysis goes through, but with strong coupling effects in the infrared, the superpotential acquires new non-perturbative terms, thus breaking the non-renormalization theorem.

Finally, it must be emphasized that the lack of F-terms in the quantum corrections does not imply that the couplings in the superpotential are not renormalized. In fact, as we see in the next section, both Yukawa couplings and mass terms run with scale, but only because of the wavefunction renormalizations associated with the various fields.

The one-loop Renormalization Group Equations

In $N = 1$ supersymmetric gauge theories, there are two types of infinities, associated with the gauge couplings, and the wave-function renormalizations of the matter chiral superfields, which give rise to anomalous dimensions. In $N = 1$ supersymmetry, the familiar renormalization group equation can be written *to all orders* (see D. R. T. Jones and L. Mezincescu, *Phys. Lett.* **136B**, 242(1984), and M. T. Grisaru and P. C. West, *Nucl. Phys.* **B254**, 249(1985)) in the form

$$\frac{dg}{dt} = -\frac{\alpha^2}{2\pi}\frac{3c_A - \sum_i c_i(1-\gamma_i)}{1-\alpha c_A/2\pi} \ . \tag{10.78}$$

In this formula, c_A the Dynkin index of the adjoint representation, accounts for the contributions from the gauge bosons and gauginos. The sum is over the chiral matter supermultiplets, each transforming as a representation with Dynkin index c_i. We see that the contributions from the gauge supermultiplet can be summed to all orders. However, with matter superfields, we need to consider γ_i, the anomalous dimension of the ith matter chiral superfield. Its exact expression is not known, but has to be evaluated order by order in perturbation theory.

Thus at the one-loop MSSM, we have

$$\beta_a = -\frac{1}{16\pi^2} b_a g_a^3 + \cdots , \tag{10.79}$$

where $a = 1,\ 2,\ 3$, for the three gauge groups $SU(3) \times SU(2) \times U(1)$ of the standard model. For $SU(N)$, the Dynkin index of the fundamental is 1/2 and that of the adjoint is N, so that in the MSSM with three chiral families, and two Higgs superfields, we find that $b_1 = -\frac{33}{5}$; $b_2 = -1$, $b_3 = 3$: both the hypercharge and weak $SU(2)$ couplings are heading to Landau poles in the deep ultraviolet. Since their values are small at experimental scales, such disasters are not encountered until well beyond Planck energies, as in QED.

The parameters of the superpotential also run with scale. The coefficient of the μ-term satisfies

$$\frac{d\mu}{dt} = \frac{\mu}{16\pi^2}[\ \mathrm{Tr}\{3\mathbf{Y}_u^\dagger \mathbf{Y}_u + 3\mathbf{Y}_d^\dagger \mathbf{Y}_d + \mathbf{Y}_e^\dagger \mathbf{Y}_e\} - 3(\frac{1}{5}g_1^2 + g_2^2)] , \tag{10.80}$$

while the Yukawa coupling matrices run according to

$$\frac{d\mathbf{Y}_{u,d,e}}{dt} = \frac{1}{16\pi^2}\mathbf{Y}_{u,d,e}\boldsymbol{\beta}_{u,d,e} . \tag{10.81}$$

To one-loop, direct calculations yield

$$\begin{aligned}
\boldsymbol{\beta}_u &= 3\mathbf{Y}_u^\dagger \mathbf{Y}_u + \mathbf{Y}_d^\dagger \mathbf{Y}_d + 3\mathrm{Tr}\{\mathbf{Y}_u^\dagger \mathbf{Y}_u\} - (\frac{13}{15}g_1^2 + 3g_2^2 + \frac{16}{3}g_3^2) , \\
\boldsymbol{\beta}_d &= 3\mathbf{Y}_d^\dagger \mathbf{Y}_d + \mathbf{Y}_u^\dagger \mathbf{Y}_u + \mathrm{Tr}\{3\mathbf{Y}_d^\dagger \mathbf{Y}_d + \mathbf{Y}_e^\dagger \mathbf{Y}_e\} - (\frac{7}{15}g_1^2 + 3g_2^2 + \frac{16}{3}g_3^2) , \\
\boldsymbol{\beta}_e &= 3\mathbf{Y}_e^\dagger \mathbf{Y}_e + \mathrm{Tr}\{3\mathbf{Y}_d^\dagger \mathbf{Y}_d + \mathbf{Y}_e^\dagger \mathbf{Y}_e\} - (\frac{9}{5}g_1^2 + 3g_2^2) .
\end{aligned} \tag{10.82}$$

We note that the μ term and the Yukawa coupling matrices are stable under the renormalization group in the sense that if any is set to zero, this implies that its derivative with respect to scale is also zero. This is in accord with both chiral symmetry and the non-renormalization theorem.

As we can see from the above, the running of the Yukawa couplings is determined by two contributions of opposite signs. This is also true in the standard model as well, but with different relative coefficients. This opens the possibility of a fixed point at which both contributions cancel.

The evolution of the vacuum expectation values of the Higgs bosons can be written in the form

$$\frac{d\ln v_{u,d}}{dt} = \frac{1}{16\pi^2}\gamma_{u,d} \; , \tag{10.83}$$

where, to one-loop, the anomalous dimensions are given by

$$\begin{aligned} \gamma_u &= \frac{3}{4}(\frac{1}{5}g_1^2 + g_2^2) - 3\mathrm{Tr}\{\mathbf{Y}_u^\dagger \mathbf{Y}_u\} \; , \\ \gamma_d &= \frac{3}{4}(\frac{1}{5}g_1^2 + g_2^2) - 3\mathrm{Tr}\{\mathbf{Y}_d^\dagger \mathbf{Y}_d\} - \mathrm{Tr}\{\mathbf{Y}_e^\dagger \mathbf{Y}_e\} \; . \end{aligned} \tag{10.84}$$

We see that at one-loop, the running of the gauge couplings is unaffected by the other sectors of the theory. If we assume the average mass of the superpartners to be M_{SUSY}, we can run the standard model gauge couplings from their experimental values to M_{SUSY}, using the standard model β-functions, and then run them from M_{SUSY} to the deep ultraviolet, using the β-functions of the MSSM.

The result is quite suggestive. The values of the gauge couplings converge in the ultraviolet to a common value. In addition, the scale at which they get close is well below the Planck scale. The convergence of the gauge couplings of the standard model is matched by the well-known fact that the standard model fermions have the right quantum numbers to fall in representations of the gauge groups $SU(5)$, the smallest group to contain the standard model, or $SO(10)$, if a right-handed neutrino is included.

Another intriguing fact is that the Yukawa couplings of the b and τ also converge *at the same scale*, indicating perhaps that with supersymmetry, the standard model is the broken remnant of a more unified gauge theory.

Higgs Loop Corrections

We have already mentioned that the low energy Higgs potential needs to be corrected by loop effects. One particular correction has proven to be crucial for comparing the MSSM to experiment. The big mass of the top quark implies a large Yukawa coupling, which can generate important corrections to tree-level results. Of particular interest are the corrections to the potential generated by top and stop quark loops; they occur with opposite sign and strength, when supersymmetry is unbroken. However with different stop and top masses, the cancellation is not perfect, resulting in a significant correction since the top Yukawa coupling is large. The minimization of the effective potential shows that the *little* Higgs mass gets a significant positive correction. This effect was worked out by many groups, but below we describe an elegant derivation, following Y. Okada, N. Yamaguchi, and T. Yanagida, (*Phys. Lett.* **B262**, 54(1991), and references therein).

We have remarked that, in the limit $m_A \gg m_Z$, the *little* Higgs behaves much as its standard model namesake. Thus, below the Susy scale, we can write its dynamics in terms of an effective potential with a quartic coupling constant, $\lambda_{\rm eff}$, just as in the standard model. The only difference is that at the Susy scale, $M_{\rm susy}$, a typical superpartner mass, this coupling must satisfy the boundary condition

$$\lambda_{\rm eff}(M_{\rm susy}) = \frac{1}{4}(g_1^2 + g_2^2)\cos^2 2\beta. \tag{10.85}$$

Below that scale, the coupling evolve according to its renormalization group equation derived by including only those particles lighter than $M_{\rm susy}$. For the large y_t required by the top mass, the y_t^4 term dominates the RG equation

$$16\pi^2 \frac{d\lambda_{\rm eff}}{dt} = -12 y_t^4 + \cdots \ , \tag{10.86}$$

where the dots refer to gauge and self-coupling contributions which are numerically small. The resulting solution is simply

$$\lambda_{\rm eff}(t) = \frac{1}{4}(g_1^2 + g_2^2)\cos 2\beta + \frac{3}{4\pi^2} y_t^2 \ln \frac{M_{\rm susy}}{t} \ . \tag{10.87}$$

In this expression, $M_{\rm susy}$ stands for the stop quark masses, as the effect comes from stop and top quark loops. The minimization of the potential proceeds as before, with the result

$$m_h^2 = m_Z^2 \cos^2 2\beta + \frac{3\sqrt{2}}{4\pi^2} G_F m_t^4 \ln \frac{m_{\tilde{t}_1} m_{\tilde{t}_2}}{m_t^2} \ . \tag{10.88}$$

It relaxes the upper bound found at tree-level, but only by a limited amount, since this is a radiative correction in a perturbative theory. A hallmark of the MSSM is that the *little* Higgs cannot be arbitrarily massive.

Running the Softies

As we have seen, a successful comparison of the MSSM with low energy data requires the addition of soft terms that break supersymmetry. These soft breaking parameters enter the MSSM at some scale Λ_{MSSM}, thus providing boundary conditions to the renormalization group equations. Their solutions yield the parameters in the infrared (at experimental energies), to be used to analyze the mass spectrum of the MSSM. The value of the scale Λ_{MSSM} is not known, although some theoretical prejudices place it in the deep ultraviolet.

The soft-breaking parameters at cut-off should be derivable from a more fundamental theory of Susy-breaking, but while many theories of supersymmetry-breaking have been proposed, none is so compelling to warrant its inclusion in this chapter. Eventually, the experimental discovery of superparticles, and the measurements of their masses and mixings, should determine the soft parameters.

In lieu of a theory and inputs from experiments, it behooves us to make assumptions as to the form of the soft parameters. Assuming that supersymmetry exists at low energy, there are already stringent experimental constraints. These come from the absence of significant flavor-changing neutral current processes. As we have already seen, this has a natural explanation in the $N = 0$ standard model through the GIM mechanism. Also, no direct processes that violate lepton number, such as the decay $\mu \rightarrow e + \gamma$, have been observed. Large violations could come from quark and lepton flavor violation by means of large off-diagonal terms in the squark and slepton masses, and through scalar cubic interactions. Through virtual exchanges of squarks and sleptons, these would yield large flavor-changing effects among ordinary quarks and leptons.

These considerations suggest that the input squark and slepton soft-breaking masses should be (nearly) flavor-diagonal, and that the flavor-changing scalar cubic interactions be no worse than those in the Yukawa coupling. The nature of the mechanism that produces these terms is a matter of debate, but it points at some sort of flavor universality: Susy-breaking has to be generated through interactions that are flavor-blind. A prime candidate for the transmitter of Susy-breaking to the MSSM is (super)gravity. This suggests a possible set of simplifying assumptions:

♠ *Squark and Slepton masses are flavor-diagonal.*
or its more restrictive version

◇ *squark and slepton masses are equal at cut-off*

$$m_{\tilde{q}}^2 = m_{\tilde{e}}^2 \equiv m_0^2 \ \text{at} \ \Lambda_{MSSM} \ . \tag{10.89}$$

♠ *The cubic scalar coupling matrices $\mathcal{A}_{\mathbf{u},\mathbf{d},e}$ are proportional to the Yukawa matrices*

$$\mathcal{A}_{\mathbf{u},\mathbf{d},e}^{ij} = A_{\mathbf{u},\mathbf{d},e} Y_{\mathbf{u},\mathbf{d},e}^{ij} \ \text{at} \ \Lambda_{MSSM} \ , \tag{10.90}$$

where $A_{\mathbf{u},\mathbf{d},e}$ are three complex mass parameters.

♠ *All three gaugino masses are equal at cut-off*

$$M_1 = M_2 = M_3 \equiv m_{1/2} \ \text{at} \ \Lambda_{MSSM} \ . \tag{10.91}$$

The gaugino mass unification is natural in Grand Unified Theories where all three gauge groups of the standard model are unified into one structure at cut-off.

Ultimately, these, or any other assumptions must be tested by experiments. In the chiral approximation of QCD, the soft breaking parameters were "measured" as, for example, in the Gell-Mann Okubo sum rule.

The one-loop evolution of the gaugino masses with scale is given by

$$\frac{d \ln M_a}{dt} = -\frac{1}{8\pi^2} b_a g_a^2 \ , \tag{10.92}$$

resulting in the (one-loop) scale-independence of the ratios

$$\frac{M_a(t)}{g_a^2(t)} = \frac{M_a(t_\Lambda)}{g_a^2(t_\Lambda)} \ . \tag{10.93}$$

At one-loop, the Higgs masses evolve according to

$$\begin{aligned} 8\pi^2 \frac{dm_{H_u}^2}{dt} &= \sum_{i,j} (3|Y_u^{ji}|^2 (m_{H_u}^2 + m_{\tilde{Q}_{Li}}^2 + m_{\tilde{u}_{Rj}}^2) + |\mathcal{A}_{\mathbf{u}}^{ij}|^2) \\ &\quad + \frac{3}{10} g_1^2 \mathrm{Tr}\{Y m^2\} - \frac{3}{5} g_1^2 M_1^2 - 3 g_2^2 M_2^2 \ . \end{aligned} \tag{10.94}$$

$$
\begin{aligned}
8\pi^2 \frac{dm^2_{H_d}}{dt} = \sum_{i,j} \Big(& |Y_e^{ji}|^2 (m^2_{H_d} + m^2_{\widetilde{L}_{Li}} + m^2_{\widetilde{e}_{Rj}}) + |\mathcal{A}_e^{ij}|^2 \\
& + 3|Y_d^{ji}|^2 (m^2_{H_d} + m^2_{\widetilde{Q}_{Li}} + m^2_{\widetilde{d}_{Rj}}) + |\mathcal{A}_{\mathbf{d}}^{ij}|^2 \Big) \\
& - \frac{3}{10} g_1^2 \mathrm{Tr}\{Y m^2\} - \frac{3}{5} g_1^2 M_1^2 - 3 g_2^2 M_2^2 \ .
\end{aligned}
\qquad (10.95)
$$

Squark and slepton masses evolve as well

$$
\begin{aligned}
8\pi^2 \frac{dm^2_{\widetilde{Q}_{Li}}}{dt} = \sum_{j} \Big(& |Y_u^{ji}|^2 (m^2_{H_u} + m^2_{\widetilde{Q}_{Li}} + m^2_{\widetilde{u}_{Rj}}) + |\mathcal{A}_{\mathbf{u}}^{ij}|^2 \\
& + |Y_d^{ji}|^2 (m^2_{H_d} + m^2_{\widetilde{Q}_{Li}} + m^2_{\widetilde{d}_{Rj}}) + |\mathcal{A}_{\mathbf{d}}^{ij}|^2 \Big) \\
& + \frac{1}{10} g_1^2 \mathrm{Tr}\{Y m^2\} - \frac{1}{15} g_1^2 M_1^2 - 3 g_2^2 M_2^2 - \frac{16}{3} g_3^2 M_3^2 \ .
\end{aligned}
\qquad (10.96)
$$

$$
\begin{aligned}
8\pi^2 \frac{dm^2_{\widetilde{u}_{Ri}}}{dt} = \sum_{j} \Big(& 2|Y_u^{ij}|^2 (m^2_{H_u} + m^2_{\widetilde{u}_{Ri}} + m^2_{\widetilde{Q}_{Lj}}) + |\mathcal{A}_{\mathbf{u}}^{ij}|^2 \Big) \\
& - \frac{2}{5} g_1^2 \mathrm{Tr}\{Y m^2\} - \frac{16}{15} g_1^2 M_1^2 - \frac{16}{3} g_3^2 M_3^2 \ ,
\end{aligned}
\qquad (10.97)
$$

$$
\begin{aligned}
8\pi^2 \frac{dm^2_{\widetilde{d}_{Ri}}}{dt} = \sum_{j} \Big(& 2|Y_d^{ij}|^2 (m^2_{H_d} + m^2_{\widetilde{d}_{Ri}} + m^2_{\widetilde{Q}_{Lj}}) + |\mathcal{A}_{\mathbf{d}}^{ij}|^2 \Big) \\
& + \frac{1}{5} g_1^2 \mathrm{Tr}\{Y m^2\} - \frac{4}{15} g_1^2 M_1^2 - \frac{16}{3} g_3^2 M_3^2 \ ,
\end{aligned}
\qquad (10.98)
$$

$$
8\pi^2 \frac{dB}{dt} = \sum_{i,j} (3\mathcal{A}_{\mathbf{u}}^{ij} Y_u^{\dagger ij} + 3\mathcal{A}_{\mathbf{d}}^{ij} Y_d^{ij} + \mathcal{A}_{\mathbf{e}}^{ij} Y_e^{ij}) - \frac{3}{5} g_1^2 M_1 - 3 g_2^2 M_2 \ . \qquad (10.99)
$$

The equations for the cubic couplings are much more complicated. We include them in their full glory, although for practical purposes only the third family's contribution need to be taken into account

$$\frac{1}{16\pi^2}\frac{d\mathcal{A}_e^{ij}}{dt} =$$
$$4(\mathbf{Y}_e\mathbf{Y}_e^\dagger)^{ik}\mathcal{A}_e^{kj} + 5\mathcal{A}_e^{ik}(\mathbf{Y}_e^\dagger\mathbf{Y}_e)^{kj} + 2(\mathcal{A}_e^{km}Y_e^{*km} + 3\mathcal{A}_d^{km}Y_d^{*km})Y_e^{ij}$$
$$-6(\frac{3}{5}g_1^2M_1 + g_2^2M_2)Y_e^{ij} + \mathrm{Tr}\{3\mathbf{Y}_d^\dagger\mathbf{Y}_d + \mathbf{Y}_e^\dagger\mathbf{Y}_e\}\mathcal{A}_e^{ij} - (\frac{9}{5}g_1^2 + 3g_2^2)\mathcal{A}_e^{ij}\ ,$$

$$\frac{1}{16\pi^2}\frac{d\mathcal{A}_d^{ij}}{dt} =$$
$$4(\mathbf{Y}_d\mathbf{Y}_d^\dagger)^{ik}\mathcal{A}_d^{kj} + 5\mathcal{A}_d^{ik}(\mathbf{Y}_d^\dagger\mathbf{Y}_d)^{kj} + \mathcal{A}_d^{ik}(\mathbf{Y}_u^\dagger\mathbf{Y}_u)^{kj} + 2(\mathbf{Y}_d\mathbf{Y}_u^\dagger)^{ik}\mathcal{A}_u^{kj}$$
$$+2(\mathcal{A}_e^{km}Y_e^{*km} + 3\mathcal{A}_d^{km}Y_d^{*km})Y_d^{ij} - \left(\frac{14}{15}g_1^2M_1 + 6g_2^2M_2 + \frac{32}{3}g_3^2M_3\right)Y_d^{ij}$$
$$+ \mathrm{Tr}\{3\mathbf{Y}_d^\dagger\mathbf{Y}_d + \mathbf{Y}_e^\dagger\mathbf{Y}_e\}\mathcal{A}_d^{ij} - (\frac{7}{15}g_1^2 + 3g_2^2 + \frac{16}{3}g_3^2)\mathcal{A}_d^{ij}\ ,$$

$$\frac{1}{16\pi^2}\frac{d\mathcal{A}_u^{ij}}{dt} =$$
$$(\mathbf{Y}_u\mathbf{Y}_u^\dagger)^{ik}\mathcal{A}_u^{kj} + 5\mathcal{A}_u^{ik}(\mathbf{Y}_u^\dagger\mathbf{Y}_u)^{kj} + \mathcal{A}_u^{ik}(\mathbf{Y}_d^\dagger\mathbf{Y}_d)^{kj} + 2(\mathbf{Y}_u\mathbf{Y}_d^\dagger)^{ik}\mathcal{A}_d^{kj}$$
$$+ 6\mathcal{A}_u^{km}Y_u^{*km} - \left(\frac{26}{15}g_1^2M_1 + 6g_2^2M_2 + \frac{32}{3}g_3^2M_3\right)Y_u^{ij}$$
$$+ 3\mathcal{A}_u^{ij}\mathrm{Tr}\{\mathbf{Y}_u^\dagger\mathbf{Y}_u\} - (\frac{13}{15}g_1^2 + 3g_2^2 + \frac{16}{3}g_3^2)\mathcal{A}_u^{ij}\ .$$

Summation over the repeated indices k, m is implied. For more details, we refer the interested reader to N.K. Falck, *Z. Phys. C* **30**, 247(1986). In these formulae,

$$\mathrm{Tr}\{Ym^2\} = m_{H_u}^2 - m_{H_d}^2 + \sum_{\text{families}} (m_{\widetilde{Q}_L}^2 - 2m_{\widetilde{u}_R}^2 + m_{\widetilde{d}_R}^2 - m_{\widetilde{L}_L}^2 + m_{\widetilde{e}_R}^2)\ . \tag{10.100}$$

Note that in an anomaly free theory, this term is zero if all the masses are equal at some scale. That it is zero at such a scale can be seen by using the fact that there is no mixed gravitational anomaly. In such a case, $\mathrm{Tr}\{Ym^2\} = m^2\mathrm{Tr}\{Y\} = 0$. To show that it remains true at all scales, one also needs the cancellation of the other triangle anomalies. For example, in the evolution equations for the Higgs, squark and slepton masses, the $g_1^2M_1^2$ terms come from one-loop mass corrections involving two bino-particle-sparticle vertices and are therefore proportional to Y^2. Thus one needs to have the $U(1)$ anomaly cancellation condition $\text{Tr}\{Y^3\} = 0$ in order to show $d\text{Tr}\{Ym^2\}/dt = 0$.

These equations bring down the parameters from Λ_{MSS} to experimental scales. Conversely, they can be used to determine the class of soft breaking parameters at the cut-off that are in accord with experiment. In practice, it is sufficient to keep only the Yukawa couplings of the third family, which considerably simplifies the evolution equations.

Although we do not know the values of the soft Susy-breaking parameters at Λ_{MSSM}, the eventual measurements of the masses of the superpartners will impose constraints on the basic assumptions, once the evolution equations are taken into account.

As an example, let us assume the simplest possible pattern for the input parameters: a common mass m_0 for the squark and sleptons, and a common gaugino mass $m_{1/2}$. One can easily solve the one-loop equations, and obtain the slepton and scalar masses of the two lightest families. As we have said, the masses of the third family superpartners are more complicated due to the large Yukawa couplings. It is convenient to introduce the scale-dependent quantities

$$C_i(t) = \begin{pmatrix} 3/5 \\ 3/4 \\ 4/3 \end{pmatrix} \times \frac{1}{2\pi^2} \int_t^{\Lambda_{MSSM}} dt g_i(t)^2 M_i(t)^2 \ , \tag{10.101}$$

associated with the Y, $SU(2)$ and $SU(3)$ couplings, respectively. Their actual values depend on the input gaugino masses. We can then evaluate the seven physical squark and slepton masses of a light family in terms of m_0^2, and the three C_i. We obtain

$$m^2_{\widetilde{d}_L} = m_0^2 + C_3 + C_2 + \frac{1}{36} C_1 + \Delta_d \ , \tag{10.102}$$

$$m^2_{\widetilde{u}_L} = m_0^2 + C_3 + C_2 + \frac{1}{36} C_1 + \Delta_u \ , \tag{10.103}$$

$$m^2_{\widetilde{u}_R} = m_0^2 + C_3 + \frac{4}{9} C_1 + \Delta_{\overline{u}} \ , \tag{10.104}$$

$$m^2_{\widetilde{d}_R} = m_0^2 + C_3 + \frac{1}{9} C_1 + \Delta_{\overline{d}} \ , \tag{10.105}$$

$$m^2_{\widetilde{e}_L} = m_0^2 + C_2 + \frac{1}{4} C_1 + \Delta_e \ , \tag{10.106}$$

$$m^2_{\widetilde{\nu}_L} = m_0^2 + C_2 + \frac{1}{36} C_1 + \Delta_\nu \ , \tag{10.107}$$

$$m^2_{\widetilde{e}_R} = m_0^2 + C_1 + \Delta_{\overline{e}} \ . \tag{10.108}$$

One can immediately infer several sum rules. Some which relate members of the same weak isospin multiplet, like

$$m^2_{\tilde{d}_L} - m^2_{\tilde{u}_L} = m^2_{\tilde{e}_L} - m^2_{\tilde{\nu}_L} = -\cos(2\beta)M_W^2 \ , \tag{10.109}$$

do not depend on the assumptions of universal m_0. Others, such as

$$2(m^2_{\tilde{u}_R} - m^2_{\tilde{d}_R}) + (m^2_{\tilde{d}_R} - m^2_{\tilde{d}_L}) + (m^2_{\tilde{e}_L} - m^2_{\tilde{e}_R}) = \frac{10}{3}\sin^2\theta_{\rm w} M_Z^2 \cos 2\beta \ . \tag{10.110}$$

directly test the universality of m_0, and does not depend on the gaugino mass inputs. On the other hand, gaugino mass universality can be tested through the sum rule

$$m^2_{\tilde{e}_L} - m^2_{\tilde{e}_R} - (20)^2\cos 2\beta = C_0(m^2_{\tilde{d}_R} - m^2_{\tilde{e}_R} - (36)^2\cos 2\beta) \ , \tag{10.111}$$

where

$$C_0 = \frac{C_2 - \frac{3}{4}C_1}{C_3 - \frac{8}{9}C_1} \ . \tag{10.112}$$

One can also obtain sum rules for the third family squark and sleptons. We note

$$m^2_{\tilde{t}_1} + m^2_{\tilde{t}_2} - 3m^2_{\tilde{b}_L} - 2m_t^2 = m^2_{\tilde{u}_L} + m^2_{\tilde{u}_R} - 3m^2_{\tilde{d}_L} \ , \tag{10.113}$$

which relates masses of squarks and quarks of different families. For more details and other sum rules, see S. Martin and P. Ramond *Phys. Rev.* **D48**, 5365(1993).

Electroweak Breaking

Another tantalizing pattern emerges by using simple boundary conditions in the ultraviolet and continuing them to the infrared. It is simply that Supersymmetry breaking can trigger electroweak breaking, as discovered by L.E.Ibáñez and G.G.Ross *Phys. Lett.* **B110**, 215(1982), L. Alvarez-Gaumé, M. Claudson and M. Wise, *Nucl. Phys.* **B207**, 96(1982), and K. Inoue, A. Kakuto, H. Komatsu and S. Takeshita *Prog. Theor. Phys.* **68**, 927(1982). This does not prove supersymmetry, but rather provides yet another piece of circumstantial evidence for its existence.

The basic reason is that a large Yukawa coupling can start with a positive scalar mass *squared* in the ultraviolet, and drive it negative in the infrared. Nature provides us with such a large coupling since the top quark is so massive. To see this effect, we can neglect all the Yukawa couplings except that of the top quark. With all the soft-breaking masses equal, we have

$$\mathrm{Tr}\{Ym^2\} = m^2\mathrm{Tr}\{Y\} = 0 \ , \tag{10.114}$$

at the initial scale, and it remains zero at all scales. The renormalization group equations become

$$8\pi^2 \frac{d}{dt}\begin{pmatrix} m^2_{H_u} \\ m^2_{\mathbf{t}_R} \\ m^2_{\mathbf{t}_L} \end{pmatrix} = y_t^2 \begin{pmatrix} 3 & 3 & 3 \\ 2 & 2 & 2 \\ 1 & 1 & 1 \end{pmatrix}\begin{pmatrix} m^2_{H_u} \\ m^2_{\mathbf{t}_R} \\ m^2_{\mathbf{t}_L} \end{pmatrix} - \begin{pmatrix} \frac{3}{5} & 3 & 0 \\ \frac{16}{15} & 0 & \frac{16}{3} \\ \frac{1}{15} & 3 & \frac{16}{3} \end{pmatrix}\begin{pmatrix} g_1^2 M_1^2 \\ g_2^2 M_2^2 \\ g_3^2 M_3^2 \end{pmatrix} . \tag{10.115}$$

The Yukawa coupling drives the mass squared negative in the infrared, counterbalanced by the effect of the gaugino masses.

Neglecting the effect of the gaugino masses, we see that all three masses decrease in the infrared. fortunately, the Higgs mass squared decreases the most, and it turn negative at the largest scale. This triggers breakdown of electroweak symmetry, exactly along the lines observed experimentally, by a weak isodoublet. It is remarkable that the numerical factors work out just right to ensure that it is the neutral Higgs mass squared that turns negative before any of the squark masses, thus avoiding the spontaneous breakdown of color and/or electric charge.

PROBLEMS

A. Starting from the one-loop formula for the gauge β functions of non-supersymmetric theories, derive the one-loop Eq. (10.78). Also verify the numerical values of the b_a for the MSSM.

B. 1-) In a simplified model with only one family of quarks and leptons, derive the one-loop RG equation of the τ lepton Yukawa coupling.
2-) Derive the one-loop anomalous dimension for the lepton fields. Given the non-renormalization theorem, discuss in detail why the Yukawa couplings still run with scale.

C. 1-) Revisit the issue of an infrared fixed point for the top and bottom Yukawa couplings in the context of the MSSM. Compare your results with the standard model (see problem in chapter 6).
2-) Solve the one-loop equations analytically, and discuss the possible infrared behavior of the couplings.

D. 1-) Use the latest values of the three gauge couplings as measured or extrapolated at M_Z as boundary conditions to the gauge coupling one-loop evolution equations. Assume that all superpartner masses are at some scale M_{SUSY}. Evaluate the possible values of M_{SUSY} and the scale M_X at which all three can meet and still be perturbative.
2-) Repeat the above for y_b/y_τ, and discuss your results in terms of the top quark mass.

E. Starting, from Eq (10.78), without matter, discuss the possibility of an infrared fixed point and its domain of validity. For reference, see T. Banks and A. Zaks, *Nucl. Phys.* **B196**, 189(1982).

F. Derive the one-loop renormalization equation for gaugino masses.

G. Show in the one-loop approximation, that if $\mathrm{Tr}\{Ym^2\}$ vanishes at some scale, it stays zero at all scales.

Appendix : Supersymmetry Toolbox

This Appendix introduces the basic features of theories with global $N = 1$ supersymmetry using the language of local field theory. In the absence of gravity, $N = 1$ supersymmetry employs two collections of fields, each connected by supersymmetry transformations. The first is the chiral or Wess-Zumino supermultiplet, which consists of one left-handed Weyl spinor and one complex scalar field. The second is the gauge supermultiplet, containing the Yang-Mills gauge bosons, and the gauginos, their spin 1/2 supersymmetric partners. Since these supermultiplets already contains fields readily identified with those in the standard model, the $N = 1$ supersymmetric standard model is naturally described in terms of these multiplets and their interactions.

A.1 THE CHIRAL SUPERMULTIPLET

The simplest set of fields on which $N = 1$ supersymmetry is realized is the chiral or Wess-Zumino multiplet which contains the three fields,

$$\begin{aligned} &\varphi(x)\ ,\ \text{a complex scalar}\ , \\ &\psi(x)\ ,\ \text{a Weyl spinor}\ , \\ &F(x),\ \ \text{a complex auxiliary field}\ . \end{aligned} \tag{A.1}$$

The free Lagrangian is given by

$$\mathcal{L}_0^{WS} \equiv \partial_\mu\varphi^*\partial^\mu\varphi + \psi^\dagger\sigma^\mu\partial_\mu\psi + F^*F\ . \tag{A.2}$$

Up to a surface term, it is invariant under the transformations

$$\begin{aligned} \delta\varphi &= \widehat{\alpha}\psi\ , \\ \delta\psi &= \alpha F - \overline{\sigma}^\mu\widehat{\alpha}^\dagger\partial_\mu\varphi\ , \\ \delta F &= -\alpha^\dagger\sigma^\mu\partial_\mu\psi\ . \end{aligned} \tag{A.3}$$

Here α, a Weyl spinor with Grassmann components, is the global parameter of the supersymmetry transformation. Grassmann variables are just anticommuting numbers, so that for any two

$$(\widehat{\zeta}\chi)^* = \zeta^\dagger\sigma_2^*\chi^* = -\zeta^\dagger\widehat{\chi}^\dagger\ , \tag{A.4}$$

which explains the absence of an i in front of the fermion kinetic term. The fields φ and ψ, with respective canonical dimensions, -1 and $-3/2$, are physical, while F is not, with the non-canonical dimension of -2. The supersymmetry parameter has dimension $1/2$. Note that F transforms as a total divergence.

Under two supersymmetry transformations, labelled δ_1 and δ_2, with parameters α_1 and α_2, we find that

$$[\delta_1, \delta_2]\begin{pmatrix}\varphi \\ \psi \\ F\end{pmatrix} = (\alpha_1^\dagger\sigma^\mu\alpha_2 - \alpha_2^\dagger\sigma^\mu\alpha_1)\partial_\mu\begin{pmatrix}\varphi \\ \psi \\ F\end{pmatrix}\ . \tag{A.5}$$

This equation shows that the result of two supersymmetry transformations is nothing but a space-time translation by the amount

$$\delta x^\mu = (\alpha_1^\dagger\sigma^\mu\alpha_2 - \alpha_2^\dagger\sigma^\mu\alpha_1)\ , \tag{A.6}$$

recalling that $P_\mu = -i\partial_\mu$ is the generator of translations. Supersymmetry transformations act as square roots of translations, and generalize the Poincaré group. We can readily verify this algebra on one of the fields

$$\begin{aligned}\delta_1\delta_2 F &= -\alpha_2^\dagger\sigma^\mu\partial_\mu\delta_1\psi\ ,\\ &= -\alpha_2^\dagger\sigma^\mu\alpha_1\partial_\mu F - \alpha_2^\dagger\sigma^\mu\overline{\sigma}^\rho\widehat{\alpha}_1^\dagger\partial_\mu\partial_\rho\varphi\ .\end{aligned} \tag{A.7}$$

The symmetry of $\partial_\mu\partial_\rho\varphi$ allows us to set

$$\sigma^\mu\overline{\sigma}^\rho = \frac{1}{2}(\sigma^\mu\overline{\sigma}^\rho + \sigma^\rho\overline{\sigma}^\mu) = g^{\mu\rho}\ , \tag{A.8}$$

leading to

$$\delta_1\delta_2 F = -\alpha_2^\dagger\sigma^\mu\alpha_1\partial_\mu F - \alpha_2^\dagger\widehat{\alpha}_1^\dagger g^{\mu\nu}\partial_\mu\partial_\nu\varphi\ . \tag{A.9}$$

Now $\alpha_2^\dagger\widehat{\alpha}_1^\dagger$ is symmetric under the $(1 \leftrightarrow 2)$ interchange, and drops out from the commutator, giving the desired result

$$[\delta_1, \delta_2]F = (\alpha_1^\dagger\sigma^\mu\alpha_2 - \alpha_2^\dagger\sigma^\mu\alpha_1)\partial_\mu F\ . \tag{A.10}$$

The other two expressions for φ and ψ work out in a similar way, making use of Fierz identities applied to ψ.

All these results can be neatly summarized by introducing a two-component Weyl spinor Grassmann variable θ. We construct the *chiral superfield* $\Phi(x, \theta)$ which depends only on θ. Since θ is Grassmann, its cube vanishes, resulting, without loss of generality, in the expansion

$$\Phi(x, \theta) = \varphi(x) + \widehat{\theta}\psi(x) + \frac{1}{2}\widehat{\theta}\theta F(x)\ . \tag{A.11}$$

We then express the supersymmetry transformation as acting on the fields,

$$\delta\Phi = \delta\varphi + \widehat{\theta}\delta\psi + \frac{1}{2}\widehat{\theta}\theta\delta F\ , \tag{A.12}$$

and rewrite it as an operator acting on the coordinates

$$\delta\Phi = \left[\widehat{\alpha}\frac{\partial}{\partial\theta} + \alpha^\dagger\sigma^\mu\theta\partial_\mu\right]\Phi\ , \tag{A.13}$$

where the Grassmann derivative is defined through

$$\frac{\partial}{\partial\theta}\,\widehat{\theta} = 1\ . \tag{A.14}$$

Expressing the supersymmetry transformations in this way enables us to directly derive the commutator formula. It also clearly shows why the change in the coefficient of $\widehat{\theta}\theta$ is a total divergence: it can only come from the term linear in θ in the supersymmetry generator, which contains the space-time derivative operator.

We can write the effect of a supersymmetry transformation on the chiral superfield in another way, namely as

$$\Phi(x^\mu, \theta) \to \Phi(x^\mu + \alpha^\dagger\sigma^\mu\theta, \theta + \alpha) \ , \tag{A.15}$$

but it is somewhat ackward since the change in x^μ is not real. If we write the change in terms of its real plus imaginary parts,

$$\alpha^\dagger\sigma^\mu\theta = \frac{1}{2}(\alpha^\dagger\sigma^\mu\theta - \theta^\dagger\sigma^\mu\alpha) + \frac{1}{2}(\alpha^\dagger\sigma^\mu\theta + \theta^\dagger\sigma^\mu\alpha) \ , \tag{A.16}$$

we find that the imaginary part of the shift can be written as the change of $\theta^\dagger\sigma_\mu\theta/2$. This suggests that we replace x^μ by

$$y^\mu = x^\mu + \frac{1}{2}\theta^\dagger\sigma^\mu\theta \ , \tag{A.17}$$

and consider the chiral superfield as a function of y_μ. It is of course no longer chiral (a function of θ alone), its expansion being given by

$$\begin{aligned}\Phi(y^\mu, \theta) =& \varphi(x) + \widehat{\theta}\psi(x) + \frac{1}{2}\widehat{\theta}\theta F(x) + \frac{1}{2}\theta^\dagger\sigma^\mu\theta\partial_\mu\varphi(x) \\ & - \frac{1}{4}\widehat{\theta}\theta\theta^\dagger\sigma^\mu\partial_\mu\psi(x) + \frac{1}{16}|\widehat{\theta}\theta|^2\partial^\mu\partial_\mu\varphi(x) \ ,\end{aligned} \tag{A.18}$$

using some Fierzing and the identity

$$\theta^\dagger\sigma^\mu\theta\theta^\dagger\sigma^\nu\theta = \frac{1}{2}g^{\mu\nu}|\widehat{\theta}\theta|^2 \ . \tag{A.19}$$

The manifestly real superfield

$$V(x^\mu, \theta, \theta^*) = \Phi^*(y^\mu, \theta)\Phi(y^\mu, \theta) \ , \tag{A.20}$$

depends on both θ and θ^*,

$$V^*(x^\mu, \theta, \theta^*) = V(x^\mu, \theta, \theta^*) \ , \tag{A.21}$$

and transforms under supersymmetry in a more pleasing way, namely with a real change in the coordinate x^μ

$$V(x^\mu, \theta, \theta^*) \to V(x^\mu + \frac{1}{2}[\alpha^\dagger\sigma^\mu\theta - \theta^\dagger\sigma^\mu\alpha], \theta + \alpha, \theta^* + \alpha^*) \ . \qquad (A.22)$$

Its expansion is given by

$$\begin{aligned} V(x, \theta, \theta^*) = &\ \varphi^*(x)\varphi(x) + [\widehat{\theta}\psi\varphi^* - \theta^\dagger\widehat{\psi}^\dagger\varphi] \\ &+ \frac{1}{2}[\widehat{\theta}\theta\varphi^* F - \theta^\dagger\widehat{\theta}^\dagger\varphi F^* + \theta^\dagger\sigma^\mu\theta(\varphi^*\partial_\mu\varphi - \partial_\mu\varphi^*\varphi - \psi^\dagger\sigma_\mu\psi)] \\ &- \frac{1}{4}\theta^\dagger\widehat{\theta}^\dagger(2F^*\widehat{\psi} + \varphi\partial_\mu\psi^\dagger\sigma^\mu - \partial_\mu\varphi\psi^\dagger\sigma^\mu)\theta \\ &- \frac{1}{4}\widehat{\theta}\theta\theta^\dagger(2\widehat{\psi}^\dagger F + \varphi^*\sigma^\mu\partial_\mu\psi - \sigma^\mu\psi\partial_\mu\varphi^*) \\ &+ \frac{1}{8}|\widehat{\theta}\theta|^2\Big(2F^*F - \partial_\mu\varphi^*\partial^\mu\varphi + \frac{1}{2}(\varphi^*\partial^\mu\partial_\mu\varphi + \varphi\partial^\mu\partial_\mu\varphi^*) \\ &\qquad + \psi^\dagger\sigma^\mu\partial_\mu\psi - \partial_\mu\psi^\dagger\sigma^\mu\psi\Big) \ . \end{aligned} \qquad (A.23)$$

The alert student will recognize the last term as the Lagrange density, plus an overall divergence.

It is a bit tedious to verify the transformation laws of all the components of the real superfield. Here we illustrate it only for terms of the form $\partial_\mu\varphi\varphi^*$. On the one hand, we get from the shifts in coordinates

$$\begin{aligned} &V(x^\mu + \frac{1}{2}(\alpha^\dagger\sigma^\mu\theta - \theta^\dagger\sigma^\mu\alpha), \theta + \alpha, \theta^* + \alpha^*) \\ &\quad = \frac{1}{2}\varphi^*(\alpha^\dagger\sigma^\mu\theta - \theta^\dagger\sigma^\mu\alpha)\partial_\mu\varphi + \frac{1}{2}(\alpha^\dagger\sigma^\mu\theta + \theta^\dagger\sigma^\mu\alpha)\varphi^*\partial_\mu\varphi + \cdots \\ &\quad = \alpha^\dagger\sigma^\mu\theta\varphi^*\partial_\mu\varphi + \cdots \ . \end{aligned}$$

On the other hand, by varying the fields directly, we obtain the very same term

$$\begin{aligned} \widehat{\theta}\delta\psi\varphi^* &= -\widehat{\theta}\overline{\sigma}^\mu\sigma_2\alpha^*\partial_\mu\varphi\varphi^* + \cdots \ , \\ &= \alpha^\dagger\sigma^\mu\theta\partial_\mu\varphi\varphi^* + \cdots \ . \end{aligned}$$

We can also interpret the supersymmetric change on the real superfield in terms of the action of differential operators, namely

$$\delta V = \left\{ \widehat{\alpha}\left(\frac{\partial}{\partial\theta} + \frac{1}{2}\overline{\sigma}^{\mu}\widehat{\theta}^{\dagger}\partial_{\mu}\right) - \alpha^{\dagger}\left(\widehat{\left(\frac{\partial}{\partial\theta}\right)}^{\dagger} - \frac{1}{2}\sigma^{\mu}\theta\partial_{\mu}\right)\right\} V(x,\theta,\theta^*) \ , \quad (A.24)$$

where we have used the identity

$$\theta^{\dagger}\sigma^{\mu}\alpha = -\widehat{\alpha}\overline{\sigma}^{\mu}\widehat{\theta}^{\dagger} \ .$$

These equations suggest we introduce the generators of supersymmetry

$$\begin{aligned} Q &= \frac{\partial}{\partial\theta} + \frac{1}{2}\overline{\sigma}^{\mu}\widehat{\theta}^{\dagger}\partial_{\mu} \ , \\ \widehat{Q}^{\dagger} &= \widehat{\left(\frac{\partial}{\partial\theta}\right)}^{\dagger} - \frac{1}{2}\sigma^{\mu}\theta\partial_{\mu} \ , \end{aligned} \quad (A.25)$$

to write the change in the real superfield

$$\delta V = (\widehat{\alpha}Q - \alpha^{\dagger}\widehat{Q}^{\dagger})V(x,\theta,\theta^*) \ , \quad (A.26)$$

The supersymmetry generators satisfy the anticommutation relations

$$\begin{aligned} \{Q, Q\} &= \{Q^*, Q^*\} = 0 \ , \\ \{Q, Q^*\} &= \overline{\sigma}^{\mu}\frac{\partial}{\partial x_{\mu}} \ . \end{aligned} \quad (A.27)$$

Added to the generators of the Poincaré group, these generators form the super-Poincaré group. It follows that the particles described by supersymmetry must form irreducible representations of this supergroup.

Any representation of the super Poincaré group can be organized in terms of representations of its subgroup, the Poincaré group. It is easy to see that the supersymmetry generators commute with the translations,

$$[Q, P_{\mu}] = 0 \ . \quad (A.28)$$

Hence they also commute with $P_{\mu}P^{\mu}$, the mass squared Casimir operator of the Poincaré subgroup. We can therefore split our analysis in terms of the mass.

It is simplest to start with massless representations. The massless representations of the Poincaré group are labelled by the helicity λ which runs over positive and negative integer and half-integer values. In local field theory, each helicity state $|\lambda >$ is accompanied by its CPT conjugate $|-\lambda >$; for example, the left polarized photon $|\lambda = +1 >$ and its CPT conjugate the right polarized photon $|\lambda = -1 >$.

In the infinite momentum frame, with only $P_0 = P_3 \neq 0$, the supersymmetry algebra reduces to the Clifford algebra

$$\{Q_1, Q_1^*\} = iP_0 \ ,$$

all other anticommutators being zero. There is only one supersymmetry operator (that is why it is called $N = 1$). Together with its conjugate, they act like the creation and annihilation operators of a one-dimensional fermionic harmonic oscillator. Starting from any state $| \lambda >$, we can generate only one other state $Q^* | \lambda >$, which has helicity $\lambda + 1/2$. A second application of the raising operator yield zero since $Q_1^2 = 0$. This construction yields the only massless irreducible representation of $N = 1$ supersymmetry: two states, differing by half a unit of helicity. The lowest representations are

– The Wess-Zumino multiplet corresponds to the representation $| 0 > \oplus | 1/2 >$, together with its CPT conjugate $| 0 > \oplus | -1/2 >$. It describes one Weyl fermion and two scalar degrees of freedom.

– The gauge multiplet contains the states $| 1 > \oplus | 1/2 >$, together with their CPT conjugates $| -1 > \oplus | -1/2 >$. These describe a vector particle and a Weyl fermion.

There is an infinite of representations with higher helicities. We should note the Rarita-Schwinger representation $| 3/2 > \oplus | 1 >$ and its conjugate, as well as the graviton-gravitino combination, made up of $| 2 > \oplus | 3/2 >$, plus conjugate. The latter appears in supergravity, the local generalization of supersymmetry.

It is obvious that the number of bosonic and fermionic degrees of freedom match exactly in these representations. This can be seen in the field theory multiplets as well: using the equations of motion, the chiral multiplet has two fermionic degrees of freedom, exactly matched by the complex scalar field. If the equations of motion are not used, the number of fermions doubles, but the excess in fermions is exactly matched by adding two boson fields, the complex auxiliary field F.

The massive representations of the super Poincaré group can be obtained by assembling massless multiplets, using the group-theoretical equivalent of the Higgs mechanism. We leave to the reader the construction of the lowest lying supermultiplets.

Clearly, the real supermultiplet is highly reducible. It can be checked that the covariant derivative operator

$$\mathcal{D} \equiv \frac{\partial}{\partial \theta} - \frac{1}{2}\overline{\sigma}^{\mu}\widehat{\theta}^{\dagger}\partial_{\mu} \ , \tag{A.29}$$

and its complex conjugate anticommute with the generators of supersymmetry. By requiring that they vanish on the real superfield, we obtain the chiral superfield.

The construction of supersymmetric invariants is facilitated by the use of Grassmann variables. We have already noted that the highest component of a superfield transforms as a four-divergence, so that its integral over space-time is supersymmetric invariant. We can extract this component through the operation of Grassmann integration, defined by

$$\int d\theta \equiv 0 \ , \ \int d\theta\theta \equiv 1 \ ; \tag{A.30}$$

note that since θ has dimension 1/2, $d\theta$ has the opposite dimension, -1/2. Integration enables us to rewrite the invariant in the form

$$\int d^4x \int d^2\theta \ \Phi(x,\theta) = \int d^4x F \ . \tag{A.31}$$

It follows that the integral of a chiral superfield over superspace (x, θ), is a supersymmetric invariant.

However, any product of $\Phi(x,\theta)$ is itself a chiral superfield, since $\widehat{\theta}\theta\theta = 0$. For any number of chiral superfields $\Phi_a, \ a = 1, \ldots, N$, all the quantities

$$\int d^4x \int d^2\theta \ \Phi_{a_i} \cdots \Phi_{a_n} \quad \text{for all } a_i \text{ and } n \ , \tag{A.32}$$

are supersymmetric invariants.

For a real superfield, transforming under supersymmetry like $V(x, \theta, \theta^*)$, it is easy to show that its component along $|\widehat{\theta}\theta|^2$, called the D-term, also transforms as a four-divergence. Its space-time integral is therefore a supersymmetric invariant. By integrating over both θ and θ^*, we can extract the D-term. Indeed we have already seen that the kinetic part of the Lagrangian is a D-term

$$\mathcal{L}_{kin} = \int d^2\theta \int d^2\overline{\theta} \ |\Phi(y_{\mu}, \theta)|^2 \ . \tag{A.33}$$

It has the right dimension: the superfield has dimension one, and the four Grassmann integral bring dimension two.

The potential part of the Lagrangian is given by

$$V = \int d^2\theta W(\Phi) + c.c. \ , \tag{A.34}$$

where the function W is called the superpotential; it depends *only* on the chiral superfields, *not* their conjugates: it is a holomorphic function of the superfields. In renormalizable theories, it is at most cubic in the chiral superfields

$$W = m_{ij}\Phi_i\Phi_j + \lambda_{ijk}\Phi_i\Phi_j\Phi_k \ . \tag{A.35}$$

It is straightforward to see that the physical potential is simply expressed in terms of the superpotential

$$V(\varphi) = \sum_i F_i^* F_i = \sum_i |\frac{\partial W(\varphi)}{\partial \varphi_i}|^2 \ ; \tag{A.36}$$

it is obviously positive definite, a general feature of global supersymmetry. The components of the quadratic polynomial are given by

$$\int d^2\theta \Phi_1 \Phi_2 = (\varphi_1 F_2 + \varphi_2 F_1 - \widehat{\psi}_1 \psi_2) \ , \tag{A.37}$$

With this term alone in the superpotential, the equations of motions for the auxiliary fields are

$$F_1^* = -m\varphi_2 \ , \qquad F_2^* = -m\varphi_1 \ . \tag{A.38}$$

Substituting their solutions, we find

$$-m^2|\varphi_1|^2 - m^2|\varphi_2|^2 - m\widehat{\psi}_1\psi_2 \ . \tag{A.39}$$

These are the mass terms for four real scalars and one Dirac fermion of mass m, with the mass sum rule

$$\sum_{J=0} m^2 = 2 \sum_{J=1/2} m^2 \ , \tag{A.40}$$

where we count one Dirac = 2 Weyl fermions. We can rewrite this equation in the form

$$\mathcal{S}trM^2 \equiv \sum_{J=0,1/2} (2J+1)(-1)^{2J} m_J^2 = m^2 + m^2 - 2m^2 = 0 \ . \tag{A.41}$$

A cubic superpotential

$$\int d^2\theta \Phi_1\Phi_2\Phi_3 = (\varphi_1\varphi_2 F_3 + \varphi_1 F_2 \varphi_3 + F_1 \varphi_2\varphi_3 \\ - \varphi_1 \widehat{\psi}_2 \psi_3 - \varphi_2 \widehat{\psi}_1 \psi_3 - \varphi_3 \widehat{\psi}_1 \psi_2) \ ,$$

contains the renormalizable Yukawa interactions, and, after using the equations of motion of the auxiliary fields, quartic renormalizable self-interactions. Clearly, higher order polynomials yield non-renormalizable interactions.

The kinetic term of the chiral multiplet has a special global symmetry, called R-symmetry, not found in non-supersymmetric models. It is not an internal symmetry since it does not commute with supersymmetry. R-symmetry is a global phase symmetry on the Grassmann variables

$$\theta \to e^{i\beta}\theta \ , \qquad \theta^* \to e^{-i\beta}\theta^* \ . \tag{A.42}$$

The Grassmann measures transform in the opposite way

$$d\theta \to e^{-i\beta}d\theta \ , \qquad d\theta^* \to e^{i\beta}d\theta^* \ . \tag{A.43}$$

Clearly, the Grassmann integration measure for the kinetic term is invariant, so that the most general R-type transformation that leaves the kinetic integrand invariant is

$$\Phi_i(y_\mu, \theta) \to e^{in_i\beta}\Phi_i(y_\mu, e^{i\beta}\theta) \ . \tag{A.44}$$

In terms of components, this means that

$$\varphi_i \to e^{in_i\beta}\varphi_i \ , \qquad \psi_i \to e^{i(n_i-1)\beta}\psi_i \ . \tag{A.45}$$

This symmetry is not necessarily shared by the superpotential, unless it transforms under R as

$$W \to e^{2i\beta}W \ , \tag{A.46}$$

to match the transformation of the Grassmann measure, further restricting the form of the superpotential.

To implement internal symmetries, we simply assume that superfields transform as representation of the internal symmetry group. If the invariance is global, the kinetic part is automatically invariant, as it sums over all the internal degrees of freedom. The superpotential may or may not be invariant, depending on its form.

We close this section by working out certain functions of superfields which arise in discussing non-perturbative aspects of supersymmetric theories. As we have seen, products of chiral superfields are themselves chiral superfields, so that any special function of a chiral superfield is defined through its series expansion.

Logarithm

Given a chiral superfield

$$\Phi = \varphi(x) + \widehat{\theta}\psi(x) + \frac{1}{2}\widehat{\theta}\theta F(x) \ ,$$

we have

$$\ln\Phi = \ln\varphi + \ln[1 + \widehat{\theta}\psi'(x) + \frac{1}{2}\widehat{\theta}\theta F'(x)] \ , \tag{A.47}$$

where

$$\psi' = \frac{\psi}{\varphi} \ , \qquad F' = \frac{F}{\varphi} \ . \tag{A.48}$$

We then use the series expansion of the logarithm to obtain

$$\begin{aligned}\ln\Phi &= \ln\varphi + (\widehat{\theta}\psi' + \frac{1}{2}\widehat{\theta}\theta F') - \frac{1}{2}(\widehat{\theta}\psi' + \frac{1}{2}\widehat{\theta}\theta F')^2, \\ &= \ln\varphi + \widehat{\theta}\psi' + \frac{1}{2}\widehat{\theta}\theta(F' + \frac{1}{2}\widehat{\psi}'\psi') \ ,\end{aligned} \tag{A.49}$$

using the Fierz identities.

Power

The arbitrary power of a chiral superfield is given by its series expansion, since

$$\begin{aligned}\Phi^a &= \varphi^a\{1 + \widehat{\theta}\psi' + \frac{1}{2}\widehat{\theta}\theta F'\}^a, \\ &= \varphi^a\{1 + a\widehat{\theta}\psi' + \frac{1}{2}a\widehat{\theta}\theta F' + \frac{1}{2}a(a-1)(\widehat{\theta}\psi')^2\} \ ,\end{aligned}$$

which, after a Fierz, yields the exact result

$$\Phi^a = \varphi^a[1 + a\widehat{\theta}\psi' + \frac{1}{2}\widehat{\theta}\theta(aF' - \frac{a(a-1)}{2}\widehat{\psi}'\psi')] \ . \tag{A.50}$$

PROBLEMS

A. Using Fierz transformations, prove that $\theta\widehat{\theta}\theta = 0$.

B. Verify explicitly that the commutator of two supersymmetry transformations on the Weyl fermion component of a chiral superfield is indeed a translation.

C. Verify that the covariant derivative operator $\mathcal{D}$ defined in Eq. $(A.29)$ anticommutes with the supersymmetry generators.

D. Show that a chiral superfield is a real superfield that obeys the constraint $\mathcal{D}V = 0$.

E. Show that the D-term of a real superfield transforms as a four-divergence.

A.2 THE REAL SUPERFIELD

Although we have already constructed a real superfield out of a chiral superfield, we should be able to build one directly in terms of the four real Grassmann variables contained in the Weyl spinor θ. An elegant way to proceed is to rewrite the two-component Weyl into a four component Majorana spinor. In the Majorana representation, all four components of a Majorana spinor are real anticommuting degrees of freedom. The real superfield is then the most general function of the Majorana spinor

$$\Theta \equiv \begin{pmatrix} \theta \\ -\sigma_2\theta^* \end{pmatrix} , \qquad (A.51)$$

shown here in the Weyl representation. Because all these components anticommute, the expansion will stop at the fourth order. We can use naive counting to determine the number of each component at each order. There are four components to the first order in Θ, $4\cdot 3/2 = 6$ components at the second order, $4\cdot 3\cdot 2/(1\cdot 2\cdot 3) = 4$ at the third, and finally $4\cdot 3\cdot 2\cdot 1/(1\cdot 2\cdot 3\cdot 4) = 1$ component at the fourth. Hence a real superfield contains $(1,4,6,4,1)$ degrees of freedom, half of them commuting, half anti-commuting. We can form the six quadratic covariants

$$\overline{\Theta}\Theta, \quad \overline{\Theta}\gamma_5\Theta, \quad \overline{\Theta}\gamma_5\gamma_\mu\Theta , \qquad (A.52)$$

where the bar denotes the usual Pauli adjoint

$$\overline{\Theta} \equiv \Theta^\dagger\gamma^0 . \qquad (A.53)$$

It is easy to check the reality conditions

$$(\overline{\Theta}\Theta)^* = -\overline{\Theta}\Theta , \qquad (\overline{\Theta}\gamma_5\Theta)^* = \overline{\Theta}\gamma_5\Theta , \qquad (\overline{\Theta}\gamma_5\gamma_\mu\Theta)^* = -\overline{\Theta}\gamma_5\gamma_\mu\Theta . \qquad (A.54)$$

Finally, by using the further identities

$$\overline{\Theta}\Theta\overline{\Theta} = -\overline{\Theta}\gamma_5\Theta\overline{\Theta}\gamma_5 = \frac{1}{4}\overline{\Theta}\gamma_5\gamma_\mu\Theta\overline{\Theta}\gamma_5\gamma^\mu , \qquad (A.55)$$

$$\overline{\Theta}\gamma_5\gamma_\mu\Theta\overline{\Theta} = -\overline{\Theta}\Theta\overline{\Theta}\gamma_5\gamma_\mu , \qquad \overline{\Theta}\gamma_5\gamma^\mu\Theta\overline{\Theta}\gamma_5\gamma^\mu\Theta = g^{\mu\nu}(\overline{\Theta}\Theta)^2 , \quad (A.56)$$

we are able to write the most general Lorentz covariant expansion of a real superfield

$$\begin{aligned} V(x^\mu, \Theta) = A(x) + i\overline{\Theta}\Psi(x) + i\overline{\Theta}\Theta M(x) + \overline{\Theta}\gamma_5\Theta N(x) \\ + i\overline{\Theta}\gamma_5\gamma^\mu\Theta A_\mu(x) + \overline{\Theta}\Theta\overline{\Theta}\Lambda(x) + (\overline{\Theta}\Theta)^2 D(x) . \end{aligned} \qquad (A.57)$$

It can be shown that in Weyl notation the same real superfield reads

$$\begin{aligned} V(x^\mu, \theta, \theta^*) = A(x) - i(\widehat{\theta}\psi + \theta^\dagger\widehat{\psi}^\dagger) \\ - i\widehat{\theta}\theta C - i\theta^\dagger\widehat{\theta}^\dagger C^* + i\theta^\dagger\sigma^\mu\theta A_\mu \\ + \widehat{\theta}\theta\theta^\dagger\widehat{\lambda}^\dagger + \theta^\dagger\widehat{\theta}^\dagger\widehat{\theta}\lambda + |\widehat{\theta}\theta|^2 D , \end{aligned} \qquad (A.58)$$

where

$$C(x) = M(x) - iN(x) , \qquad (A.59)$$

and

$$\Psi(x) = \begin{pmatrix} \psi(x) \\ -\sigma_2\psi^*(x) \end{pmatrix} . \qquad (A.60)$$

It is evident from this equation that the real superfield contains a chiral superfield and its conjugate, made up of the non-canonical fields A, ψ, and C. We can therefore always write it in the form

$$V(x, \theta, \theta^*) = -i(\Phi(x, \theta) - \Phi^*(x, \theta)) + V'(x, \theta, \theta^*) , \qquad (A.61)$$

where

$$\Phi(x, \theta) = \frac{1}{2}\Big(B(x) + iA(x)\Big) + \widehat{\theta}\psi(x) + \widehat{\theta}\theta C(x) . \qquad (A.62)$$

If the real superfield is taken to be dimensionless, the vector field A_μ and the Weyl spinor λ have the right canonical dimension to represent a gauge field and a spinor field. The real superfield then describes the vector supermultiplet we have encountered in classifying the representations of the super Poincaré group, but with many extra degrees of freedom, which happen to fall neatly in chiral multiplets. This is no accident, since these extra fields in fact turn out to be gauge artifacts.

To conclude this section, let us work out the power of a real superfield, which turns out to be useful in several contexts. Consider

$$V^a = [A(1+X)]^a \ , \tag{A.63}$$

where X is a real superfield with all its components normalized by A.

$$X = i\overline{\Theta}\Psi' + i\overline{\Theta}\Theta M' + \overline{\Theta}\gamma_5\Theta N' + i\overline{\Theta}\gamma_5\gamma^\mu\Theta A'_\mu + \overline{\Theta}\Theta\overline{\Theta}\Lambda' + (\overline{\Theta}\Theta)^2 D' \ , \tag{A.64}$$

where the prime denotes division by A. Then, noting that $X^5 = 0$, a little bit of algebra gives

$$\begin{aligned} V^a = A^a[1 + aX + a(a-1)\frac{X^2}{2!} \\ + a(a-1)(a-2)\frac{X^3}{3!} + a(a-1)(a-2)(a-3)\frac{X^4}{4!}] \ . \end{aligned} \tag{A.65}$$

The Fierz identity shown here for any two Dirac four component spinors

$$\Psi\overline{\Lambda} = -\frac{1}{4}\overline{\Lambda}\Psi - \frac{1}{4}\gamma_5\overline{\Psi}\gamma_5\Lambda + \frac{1}{4}\gamma_5\gamma^\rho\overline{\Psi}\gamma_5\gamma^\rho\Lambda - \frac{1}{4}\gamma^\rho\overline{\Psi}\gamma_\rho\Lambda + \frac{1}{2}\sigma_{\mu\nu}\overline{\Psi}\sigma^{\mu\nu}\Lambda \ , \tag{A.66}$$

is used repeatedly to rewrite the powers of X in terms of the standard expansion for a real superfield. We leave it as an exercise in fierce Fierzing to work out the general formula. Here we just concentrate on the D-term. The contributions to the D-term are as follows:

$$
\begin{aligned}
X \ &: (\overline{\Theta}\Theta)^2 D' ; \\
X^2 \ &: 2i\overline{\Theta}\Psi'\overline{\Theta}\Theta\overline{\Theta}\Lambda' + (i\overline{\Theta}\Theta M' + \overline{\Theta}\gamma_5\Theta N' + i\overline{\Theta}\gamma_5\gamma^\mu\Theta A'_\mu)^2 \ , \\
&= (\overline{\Theta}\Theta)^2 \left\{ -\frac{i}{2}\hat{\overline{\Lambda}}\Psi' - M'^2 + N'^2 - A'_\mu A'^\mu \right\} \ ; \\
X^3 \ &: -3(i\overline{\Theta}\Theta M' + \overline{\Theta}\gamma_5\Theta N' + i\overline{\Theta}\gamma_5\gamma^\rho\Theta A'_\rho)(\overline{\Theta}\Psi')^2 \ , \\
&= \frac{3}{4}(\overline{\Theta}\Theta)^2 (iM'\overline{\Psi}'\Psi' - N'\overline{\Psi}'\gamma_5\Psi' - A'^\rho\overline{\Psi}'\gamma_5\gamma_\rho\Psi') \ ; \\
X^4 \ &: (i\Theta\Psi')^4 \ , \\
&= \frac{1}{16}(\overline{\Theta}\Theta)^2(\overline{\Psi}'\Psi')^4[1 + 1 + g^\mu_\mu] \ .
\end{aligned}
$$

Putting it all together, we obtain for the D-term

$$
\begin{aligned}
(V^a)_D = &A^a[aD' + \frac{a(a-1)}{2}(-\frac{i}{2}\overline{\Lambda}'\Psi' - M'^2 + N'^2 - A'_\mu A'^\mu) \\
&+ \frac{1}{8}a(a-1)(a-2)(iM'\overline{\Psi}'\Psi' - N'\overline{\Psi}'\gamma_5\Psi' - A^{\rho'}\overline{\Psi}'\gamma_5\gamma_\rho\Psi')] \ .
\end{aligned}
\tag{A.67}
$$

One can use these formulae to show that the real superfield, expunged of its chiral components, satisfies $\hat{V}^3 = 0$.

PROBLEMS

A. Verify the identities in Eqs. (A.55) and (A.56).

B. Verify the form of the expansion of a real superfield given in Eq. (A.58).

C. Show that a dimensionless real superfield that satisfies $V^3 = 0$ contains only a gauge field, a Weyl fermion and a real auxiliary field.

D. Starting from the transformation law of a real superfield, derive the transformation of the three fields, A_μ, λ, and D.

E. Derive the expression of the exponential of a real superfield.

A.3 THE VECTOR SUPERMULTIPLET

One massless representation of the super-Poincaré is the vector supermultiplet, containing a gauge potential and the gaugino, its associated Weyl fermion. The analysis of the previous section shows that they come accompanied by a real auxiliary field. Taking the Abelian case for simplicity, we are led to consider the three fields

$$\begin{aligned} A_\mu(x) &: \text{ a gauge field} \\ \lambda(x) &: \text{ a Weyl spinor (the gaugino)}, \\ D(x) &: \text{ an auxiliary field }. \end{aligned} \tag{A.68}$$

The auxiliary field is here to provide the right count between bosonic and fermionic degreees of freedom. Without using the massless Dirac equation, the spinor is described by four degrees of freedom, and the gauge field is described by three degrees of freedom, leaving D to make up the balance. With the use of the equations of motion, both the Weyl field and the massless gauge field have two degrees of freedom, and the auxiliary field disappears. Sometimes the gaugino is called a Majorana fermion, but there should be no confusion between a Weyl fermion and a Majorana fermion: in two-component notation they look exactly the same. The Action

$$S = \int d^4x[-\frac{1}{4}F_{\mu\nu}F^{\mu\nu} + \lambda^\dagger\sigma^\mu\partial_\mu\lambda + \frac{1}{2}D^2] , \tag{A.69}$$

where $F_{\mu\nu} = \partial_\mu A_\nu - \partial_\nu A_\mu$, is invariant under the following supersymmetry transformations

$$\begin{aligned} \delta A_\mu &= \frac{-i}{\sqrt{2}}(\lambda^\dagger\sigma_\mu\alpha + \alpha^\dagger\sigma_\mu\lambda) , \\ \delta\lambda &= \frac{1}{\sqrt{2}}(D + \frac{i}{2}\sigma^{\mu\nu}F_{\mu\nu})\alpha , \\ \delta D &= \frac{1}{\sqrt{2}}(\partial_\mu\lambda^\dagger\sigma^\mu\alpha - \alpha^\dagger\sigma^\mu\partial_\mu\lambda) , \end{aligned} \tag{A.70}$$

where

$$\sigma^{\mu\nu} = \frac{1}{2}(\overline{\sigma}^\mu\sigma^\nu - \overline{\sigma}^\nu\sigma^\mu) . \tag{A.71}$$

We see that D transforms as a four-divergence, making its space-time integral is a supersymmetric invariant. Let us check the commutation relations of the algebra on the fields:

$$\begin{aligned}\delta_1\delta_2 D &= \frac{1}{\sqrt{2}}(\partial_\mu \delta_1\lambda^\dagger \sigma^\mu \alpha_2 - \alpha_2^\dagger \sigma^\mu \partial_\mu \delta_1 \lambda) \ , \\ &= \frac{1}{2}(\alpha_1^\dagger \sigma^\mu \alpha_2 - \alpha_2^\dagger \sigma^\mu \alpha_1)\partial_\mu D - \\ &\quad - \frac{i}{4}\left((\sigma^{\rho\sigma}\alpha_1)^\dagger \sigma^\mu \alpha_2 - \alpha_2^\dagger \sigma^\mu \sigma^{\rho\sigma}\alpha_1\right)\partial_\mu F_{\rho\sigma} \ .\end{aligned} \tag{A.72}$$

Since we have

$$\sigma^{\rho\sigma\dagger} = -\overline{\sigma}^{\rho\sigma} = -\frac{1}{2}(\sigma^\rho \overline{\sigma}^\sigma - \sigma\overline{\sigma}^\rho) \ , \tag{A.73}$$

the identity

$$\sigma^\mu \sigma^{\rho\tau} = -i\epsilon^{\mu\rho\tau\delta}\sigma_\delta + g^{\mu\rho}\sigma^\tau - g^{\mu\tau}\sigma^\rho \ , \tag{A.74}$$

allows us to rewrite this equation as

$$[\delta_1, \delta_2]D = \alpha_{[1}^\dagger \sigma^\mu \alpha_{2]}\partial_\mu D + \frac{i}{4}\alpha_{[1}^\dagger(\overline{\sigma}^{\rho\tau}\sigma^\mu + \sigma^\mu \sigma^{\rho\tau})\alpha_{2]}\partial_\mu F_{\rho\tau} \ , \tag{A.75}$$

leading us to the simpler form

$$\begin{aligned}[\delta_1, \delta_2]D =& (\alpha_1^\dagger \sigma^\mu \alpha_2 - \alpha_2^\dagger \sigma^\mu \alpha_1)\partial_\mu D \\ &+ \frac{1}{2}(\alpha_1^\dagger \sigma_\lambda \alpha_2 - \alpha_2^\dagger \sigma_\lambda \alpha_1)\epsilon^{\mu\rho\tau\lambda}\partial_\mu F_{\rho\tau} \ .\end{aligned} \tag{A.76}$$

The last term vanishes because of the Bianchi identity. (What if it did not? Any implications for the monopole?) Similarly, we compute

$$\begin{aligned}[\delta_1, \delta_2]A_\mu &= \frac{-i}{\sqrt{2}}(\delta_1\lambda^\dagger \sigma_\mu \alpha_2 + \alpha_2^\dagger \sigma_\mu \delta_1 \lambda) - (1 \leftrightarrow 2) \ , \\ &= \frac{1}{4}\alpha_1^\dagger(\overline{\sigma}^{\rho\tau}\sigma_\mu - \sigma_\mu \sigma^{\rho\tau})\alpha_2 F_{\rho\tau} - (1 \leftrightarrow 2) \ , \\ &= (\alpha_1^\dagger \sigma^\rho \alpha_2 - \alpha_2^\dagger \sigma^\rho \alpha_1)F_{\rho\mu} \ ,\end{aligned} \tag{A.77}$$

skipping over several algebraic steps. The right hand side contains the desired term, namely $\partial_\rho A_\mu$, but it also contains $-\partial_\mu A_\rho$; clearly it could not be otherwise from the transformation laws: their right-hand side is manifestly gauge invariant, whereas δA_μ certainly is not. Indeed our result can be rewritten in the suggestive form

$$[\delta_1, \delta_2] A_\mu = (\alpha_1^\dagger \sigma^\rho \alpha_2 - \alpha_2^\dagger \sigma^\rho \alpha_1) \partial_\rho A_\mu - \partial_\mu \Sigma \ , \tag{A.78}$$

where the last term is a gauge transformation, with a field dependent gauge function given by

$$\Sigma = (\alpha_1^\dagger \sigma^\rho \alpha_2 - \alpha_2^\dagger \sigma^\rho \alpha_1) A_\rho \ . \tag{A.79}$$

This equation shows clearly that a supersymmetry transformation (in this form) is accompanied by a gauge transformation. It also means that this description of the gauge multiplet is not gauge invariant, but rather in a specific gauge; this gauge is called the Wess-Zumino gauge. It is possible to eliminate the gauge transformation in the commutator of two supersymmetries by introducing extra fields which are needed for a gauge invariant description. We leave it as an exercise to derive the full gauge invariant set of fields. These fields can be neatly assembled in a real superfield, which under a gauge transformation undergoes the shift

$$V \to V + i(\Xi - \Xi^*) \ , \tag{A.80}$$

where $\Xi(x, \theta)$ is a chiral superfield. This nicely connects with the remarks of the previous section. The Wess-Zumino gauge is that for which the extraneous components of the real superfield are set to zero ($A = \psi = C = 0$). We leave it to the reader to verify the algebra on the gaugino field λ.

Generalization to the non-Abelian case is totally straightforward. The only difference is that the gaugino and auxiliary fields $\lambda^A(x)$ and $D^A(x)$ now transform covariantly as members of the *adjoint* representation of the internal symmetry group. Thus the ordinary derivative acting on $\lambda^A(x)$ has to be replaced by the covariant derivative

$$(\mathcal{D}_\mu \lambda)^A = \partial_\mu \lambda^A + ig(T^C)^A_B A^C_\mu \lambda^B \ , \tag{A.81}$$

where the representation matrices are expressed in terms of the structure functions of the algebra through

$$(T^C)^A{}_B = -if_B{}^{CA} \ . \tag{A.82}$$

The $N = 1$ supersymmetric non-Abelian Yang-Mills Lagrangian is then given by

$$-\frac{1}{4}G^A_{\mu\nu}G^{A\mu\nu} + \lambda^{\dagger A}\sigma^\mu(\mathcal{D}_\mu\lambda)^A + \frac{1}{2}D^AD^A \ . \tag{A.83}$$

In the Wess-Zumino gauge, there is an alternate way to represent the three fields of the vector supermultiplet, by introducing a chiral superfield which transforms as a Weyl spinor under the Lorentz group. It is given by

$$\mathcal{W}^A(x,\theta) = \lambda^A(x) + \frac{1}{2}\left[D^A(x) + \frac{i}{2}\sigma^{\mu\nu}G^A_{\mu\nu}(x)\right]\theta + \frac{1}{4}\widehat{\theta}\theta\overline{\sigma}^\mu\partial_\mu\widehat{\lambda}^{\dagger A}(x) \ , \tag{A.84}$$

suppressing the spinor index. Under a gauge transformation, this superfield transforms covariantly, as a member of the adjoint representation. One can also easily show that, under a supersymmetry transformation, $\mathcal{W}^A(x,\theta)$ does indeed transform as a chiral superfield, that is,

$$\mathcal{W}^A(x^\mu,\theta) \rightarrow \mathcal{W}^A(x^\mu + \alpha^\dagger\sigma^\mu\theta, \theta + \alpha) \ . \tag{A.85}$$

This reformulation gives us an easy way to build invariants out of products of this superfield. As for the Wess-Zumino multiplet, invariants are the F-term of the products of this superfield. This time, we must take care that Lorentz and gauge invariance be satisfied. In particular, the Yang-Mills Lagrangian is simply

$$\mathcal{L}_{SYM} = \int d^2\theta \ \widehat{\mathcal{W}}^A\mathcal{W}^A \ + \ \text{c.c.} \ . \tag{A.86}$$

The other invariant

$$\mathcal{L}_{SST} = i\int d^2\theta \ \widehat{\mathcal{W}}^A\mathcal{W}^A \ + \ \text{c.c.} \ , \tag{A.87}$$

is the usual Yang-Mills surface term

$$\mathcal{L}_{SST} = G^A_{\mu\nu}\widetilde{G}^{A\mu\nu} - i\partial_\mu(\lambda^\dagger\sigma^\mu\lambda) \ . \tag{A.88}$$

There are no other supersymmetric invariants made out of this spinor superfield that lead to renormalizable interactions. However we can easily manufacture invariant combinations of higher dimensions. For instance for $SU(N)$ with $N > 2$, we can form the gauge adjoint "anomaly" composite

$$A^B = d^{BCD}\widehat{\mathcal{W}}^C\mathcal{W}^D \ , \tag{A.89}$$

leading to the invariant

$$\int d^2\theta \ A^BA^B \ . \tag{A.90}$$

Similar constructions can be made with composites which transform as a self-dual antisymmetric second rank Lorentz tensor, and member of the adjoint representation of the gauge group, such as

$$f^{ABC}\widehat{\mathcal{W}}^B\sigma^i\mathcal{W}^C \ . \tag{A.91}$$

Some of these constructions appear in the context of non-perturbative supersymmetric models.

Finally, it is straighforward to implement R-symmetry on the gauge supermultiplet. All we need require is that

$$\mathcal{W} \to e^{i\beta}\mathcal{W} \ . \tag{A.92}$$

This means that the gaugino carries one unit of R-symmetry, while the D and gauge fields have no R-number.

PROBLEMS

A. Prove the identity $(A.74)$.

B. In the Wess-Zumino gauge, verify the commutator of the supersymmetry algebra on the gaugino field.

C. Show that the spinor superfield $\mathcal{W}$ has the correct transformation law under supersymmetry.

D. Show that the components of a real superfield in the Wess-Zumino gauge transform according to Eq. (A.68).

A.4 INTERACTION OF CHIRAL AND VECTOR SUPERMULTIPLETS

The renormalizable interactions of gauge fields with spin zero and one-half matter fields are generalized in supersymmetry to the study of the interaction of gauge supermultiplets with chiral matter supermultiplets.

Let us start with the coupling of a Wess-Zumino supermultiplet to an Abelian gauge superfield. Consider first the free action for one chiral superfield; it is clearly invariant under the global phase transformations

$$\Phi(x,\theta) \to e^{i\eta}\Phi(x,\theta) \ , \tag{A.93}$$

as long as the η is a global parameter, independent of the coordinates.

To duplicate the Yang-Mills construction, we want to modify this action to make it invariant under the most general *local* phase transformation on the chiral superfield

$$\Phi(x,\theta) \to e^{i\eta\Xi(x,\theta)}\Phi(x,\theta) \ , \tag{A.94}$$

where $\Xi(x,\theta)$ is a chiral superfield. The kinetic term loses its invariance, since

$$\Phi^*(y,\theta)\Phi(y,\theta) \to e^{i\eta(\Xi(y,\theta)-\Xi^*(y,\theta))}\Phi^*(y,\theta)\Phi(y,\theta) \ , \tag{A.95}$$

where y_μ has been previously defined. To restore invariance under the local symmetry, we generalize the kinetic term by adding the gauge supermultiplet. We have seen that it is described by a real superfield, with the suggestive gauge transformation

$$V \to V - i(\Xi - \Xi^*) \ . \tag{A.96}$$

The change of the argument translates in a redefinition of $\Lambda(x)$ and $D(x)$ in the real superfield, and does not affect the counting of the number of degrees of freedom.The Action is simply

$$\int d^4x \int d^2\theta d^2\bar{\theta} \sum_a \Phi^*(y,\theta)e^{\eta V(y,\theta,\theta^*)}\Phi(y,\theta) \ . \tag{A.97}$$

In the Wess-Zumino gauge, this expression can be shown to reduce to

$$\begin{aligned}\mathcal{L} = &-\frac{1}{4}F_{\mu\nu}F^{\mu\nu} + \lambda^\dagger\sigma^\mu\partial_\mu\lambda + \frac{1}{2}D^2 \\ &+ (\mathcal{D}_\mu\varphi)^*(\mathcal{D}^\mu\varphi)^* + \psi^\dagger\sigma^\mu\mathcal{D}_\mu\psi + F^*F \\ &+ gD\varphi^*\varphi - \sqrt{2}g\widehat{\lambda}\psi\varphi^* + \sqrt{2}g\lambda^\dagger\widehat{\psi}^\dagger\varphi \ ,\end{aligned} \tag{A.98}$$

with the usual gauge covariant derivatives

$$\mathcal{D}_\mu\varphi = (\partial_\mu + igA_\mu)\varphi \ ; \qquad \mathcal{D}_\mu\psi = (\partial_\mu + igA_\mu)\psi \ . \tag{A.99}$$

The last line of this Lagrangian yields new interactions, over and above those present in the usual construction of gauge invariant theories, where derivatives are simply replaced by covariant derivatives. The reason is that the usual interaction terms created in this way, all proportional to the charge, are not supersymmetric invariants; the extra terms restore invariance under supersymmetry. However it is a bit tricky to check the invariance because we are in the Wess-Zumino gauge. This entails changes in the transformation properties of the fields of order g.

Consider the variation of the interaction of the fermion current with the gauge potential

$$\delta\left(ig\psi^\dagger\sigma^\mu\psi A_\mu\right) = ig\psi^\dagger\sigma^\mu\alpha F A_\mu + \frac{1}{\sqrt{2}}g\psi^\dagger\sigma^\mu\psi\lambda^\dagger\sigma_\mu\alpha + \text{c.c.}\ .$$

To offset the last term we need the variation

$$-\sqrt{2}g\widehat{\lambda}\psi\delta\varphi^* = -\frac{1}{\sqrt{2}}g\alpha^\dagger\sigma_\mu\lambda\psi^\dagger\sigma^\mu\psi\ .$$

By the same token, the variation

$$-\sqrt{2}g\delta\widehat{\lambda}\psi\varphi^* = -gD\widehat{\alpha}\psi\varphi^* + \cdots\ ,$$

is compensated by

$$gD\varphi^*\delta\varphi = gD\varphi^*\widehat{\alpha}\psi\ .$$

This procedure goes on *ad nauseam.* The alert student may have noticed the presence of a term proportional to F. The only way to compensate for it is to add a term in the variation of F itself. The extra variation

$$\delta_{WZ}F^* = -ig\psi^\dagger\sigma^\mu\alpha A_\mu\ ,$$

does the job. Its effect is to replace the derivative by the covariant derivative in the transformation law, which we do for all of them. Even then we are not finished: we still have one stray term proportional to F. Indeed we have

$$-\sqrt{2}g\widehat{\lambda}\delta\psi\varphi^* = -\sqrt{2}gF\widehat{\lambda}\alpha\varphi^* + \cdots\ ,$$

which can only cancelled by adding a term in the variation of F, yielding the final modification

$$\delta_{WZ}F^* = -ig\psi^\dagger\sigma^\mu\alpha A_\mu - \sqrt{2}g\alpha^\dagger\widehat{\lambda}^\dagger\varphi\ . \qquad (A.100)$$

You have my word that it is the last change, but to the non-believer, I leave the full verification of the modified supersymmetric algebra in the Wess-Zumino gauge as an exercise.

This simple Lagrangian of course does not lead to a satisfactory quantum theory because of the ABJ anomaly associated with the $U(1)$, but this can be easily remedied by adding another chiral superfield with opposite charge. In this case, the extra terms beyond the covariant derivatives read

$$gD(\varphi_1^\dagger\varphi_1 - \varphi_2^\dagger\varphi_2) - \left(\sqrt{2}g\widehat{\lambda}(\psi_1\varphi_1^* - \psi_2\varphi_2^*) + \text{ c.c.}\right) . \qquad (A.101)$$

From the equations of motion, the value of the auxiliary field is

$$D = -g(\varphi_1^\dagger\varphi_1 - \varphi_2^\dagger\varphi_2) ,$$

yielding the extra contribution to the potential

$$V = \frac{g^2}{2}(\varphi_1^\dagger\varphi_1 - \varphi_2^\dagger\varphi_2)^2 . \qquad (A.102)$$

Generalization to the non-Abelian case is straightforward. We merely quote the results for a chiral matter superfield transforming as a representation $\mathbf{r}$ of the gauge group. The derivatives on the matter fields ψ_a and φ_a are replaced by the covariant derivatives

$$\mathcal{D}_\mu = \partial_\mu + ig\mathbf{T}^B A_\mu^B ,$$

where $\mathbf{T}^B$ represent the gauge algebra in the representation of the chiral superfield. The auxiliary fields $D^A(x)$ now couple through the term

$$gD^A\varphi^{\dagger a}(T^A)_a^{\ b}\varphi_b , \qquad (A.103)$$

and the gauginos by the terms

$$-\sqrt{2}g\varphi^{\dagger a}(T^A)_a^{\ b}\psi_b^t\widehat{\lambda}^A + \sqrt{2}g\lambda^{A\dagger}\widehat{\psi}^{\dagger\, a}(T^A)_a^{\ b}\varphi_b , \qquad (A.104)$$

where we have displayed the internal group indices (but not the spinor indices).

Lastly, we note that the gauge coupling preserves R-symmetry, irrespective of the R-value of the chiral superfield. Thus the only place R-invariance can be broken is the superpotential.

PROBLEMS

A. Evaluate the Abelian action $(A.97)$ in the Wess-Zumino gauge.

B. Show that the modification $(A.100)$ to the transformation of F in the Wess-Zumino gauge is sufficient to close the supersymmetry algebra.

A.5 SUPERSYMMETRY BREAKING

Exact supersymmetry implies equal masses for bosons and fermions, a feature not found in Nature. Thus any phenomenological application of supersymmetry requires an understanding of its breaking. We do not consider the so-called *hard breaking*, induced by higher-dimension (≥ 4) operators, which leaves no trace of the symmetry in the quantum field theory. Rather we discuss the more subtle breaking mechanisms which do not alter the ultraviolet properties of the theory.

Simplest is *soft breaking*, with the symmetry broken by infrared effects. This is accomplished through scale-dependent terms of dimension-two and three (relevant) operators. Intuitively, these do not affect the theory in the limit where all masses are taken to zero, relative to the scale of interest. These terms describe masses of the spin zero superpartners of the massless chiral fermions for the Wess-Zumino multiplet, and masses of the spin 1/2 gauginos for the gauge supermultiplets. The dimension-three terms describe interactions between the spin-zero partners of the chiral multiplets.

There are mass terms which do not break any symmetry other than supersymmetry by creating a mass gap between the particles within a supermultiplet (adding mass terms for the chiral fermions would break chiral symmetry, and often gauge symmetries). In the case of gauge supermultiplets, the gauge boson masses are protected by gauge symmetries, while the Majorana the gaugino mass term

$$M_i \widehat{\lambda}_i \lambda_i \ , \qquad (A.105)$$

leave the gauge group invariant but break the continuous R-symmetry down to R-parity, its discrete Z_2 subgroup.

Soft breaking is not fundamental, rather a manisfestation of symmetry breaking in the effective Lagrangian language. We already encountered an example with the soft breaking of chiral symmetry in the effective chiral Lagrangian that describes the strong interactions. We understand these terms to come from the quark masses in the QCD Lagrangian, which in turn are generated by the spontaneous breaking of the electroweak symmetry. In the case of supersymmetry, the actual mechanism by which supersymmetry is broken is not known, although one can devise models where spontaneous symmetry breaking occurs naturally.

Spontaneous Breaking

More fundamental is spontaneous breaking of supersymmetry. A symmetry is spontaneously broken if the field configuration which yields minimum energy no longer sustains the transformation under that symmetry. Let us remind ourselves how it works for an internal symmetry. The simplest order parameter is a complex field $\varphi(x)$ with dynamics invariant under the following transformation

$$\delta\varphi(x) = e^{i\beta}\varphi(x) \ . \tag{A.106}$$

Suppose that in the lowest energy configuration, this field has a constant value $< \varphi(x) >_0$. Expanding $\varphi(x)$ away from this vacuum configuration, and setting

$$\varphi(x) = e^{i\eta(x)}(v + \rho(x)) \ , \tag{A.107}$$

we find that under the transformation, the angle $\eta(x)$ undergoes a simple shift

$$\eta(x) \rightarrow \eta(x) + \delta \ , \tag{A.108}$$

meaning that the dynamics is invariant under that shift. Geometrically, this variable is the angle which parametrizes the closed line of minima. The dynamical variable associated with this angle is identified with the massless Nambu-Goldstone boson, $\zeta(x)$, divided by the vacuum value. It couples to the rest of the physical system universally

$$\mathcal{L}_{NG} = \frac{1}{v}\zeta(x)\partial_\mu J^\mu \ , \tag{A.109}$$

where $J_\mu(x)$ is the Noether current of the broken symmetry. Clearly, a constant shift in ζ generates a surface term and leaves the Action invariant.

Let us apply this reacquired wisdom to the supersymmetric case, starting with the chiral superfield. In analogy, we expect to see a massless fermion, since the supersymmetry parameter is fermionic, that shifts by a constant under supersymmetry. In a constant field configuration, the supersymmetry algebra reads

$$\delta\varphi_0 = \widehat{\alpha}\psi_0 \ , \qquad \delta\psi_0 = \alpha F_0 \ , \qquad \delta F_0 = 0 \ . \tag{A.110}$$

Any non-zero value of ψ_0 breaks both supersymmetry and Lorentz invariance. Since we are interested in Lorentz-invariant vacua, we set $\psi_0 = 0$, obtaining the only Lorentz-invariant possibility

$$\delta\varphi_0 = 0, \quad \delta\psi_0 = \alpha F_0, \quad \delta F_0 = 0 \; ; \tag{A.111}$$

with $\varphi_0 \neq 0$ and $F_0 \neq 0$. Therefore, the only way to break the supersymmetry is through the configuration

$$F_0 \neq 0 \quad \leftrightarrow \quad \text{broken supersymmetry} \; .$$

Since F is a function of the scalar fields, it means that some $\varphi_0 \neq 0$. It must be noted that when $F_0 = 0$, and $\varphi_0 \neq 0$, any internal symmetry carried by φ_0 is broken. This fits nicely with our earlier remarks because a non-zero value for F gives the potential a positive minimum.

When $F_0 \neq 0$, the chiral fermion shifts under supersymmetry: it is the Nambu-Goldstone fermion associated with the breakdown of supersymmetry, as expected, since the broken symmetry is fermionic. It often goes under the name Goldstino.

A similar analysis carries to the vector multiplet. There, the only vacuum configuration which does not break Lorentz invariance, is that where A_μ and λ vanish in the vacuum, for which we have

$$\delta A^\mu_0 = 0 \; , \; \delta\lambda_0 = \frac{1}{\sqrt{2}}\alpha D_0 \; , \; \delta D_0 = 0 \; , \tag{A.112}$$

and the only way to break supersymmetry is to give D_0 a vacuum value, and in this case, the gaugino λ is the Goldstino.

Thus, with both chiral and vector superfields, spontaneous breakdown of supersymmetry comes about when the dynamics is such that either F or D is non-zero in the vacuum. Another way of arriving at the same conclusion is to note that the potential from these theories is given by

$$V = F_i^* F_i + \frac{1}{2} D^2 \; , \tag{A.113}$$

when F_i and D take on their values obtained from the equations of motion. Since V is the sum of positive definite quantities it never becomes negative and if supersymmetry is spontaneously broken, its value at minimum is non-zero.

It is possible to formulate a general argument based on the fundamental anticommutation relations. In theories with exact supersymmetry, the vacuum state is annihilated by the generators of supersymmetry. However, the square of the same supersymmetry generators is nothing but the energy: the energy of the supersymmetric ground state is necessarily zero. Since it is also the state of lowest energy, it follows that the potential is necessarily positive definite. This is what we have just seen above.

Now suppose that supersymmetry is spontaneously broken. This requires that the action of supersymmetry on the vacuum not be zero, and therefore that the vacuum energy be positive. Comparing with the form of the potential, this can happen only if F and/or D is non-zero.

Finally, it is interesting to examine the transformation properties of composite chiral superfields which might arise in field theories as a result of strong coupling in effective infrared theories. Since products of chiral superfields are also superfields, we might consider the two simplest composites, $\Phi_{\text{matter}} = \Phi\Phi$, made out of matter chiral multiplets, and $\Phi_{\text{gauge}} = \widehat{\mathcal{W}}^A\mathcal{W}^A$, made out of gauge multiplets. Straightforward multiplication yields

$$\Phi_{\text{matter}} = \varphi^2 + 2\widehat{\theta}\psi\varphi + \frac{1}{2}\widehat{\theta}\theta(2F\varphi - \widehat{\psi}\psi) \ , \qquad (A.114)$$

so that its F-term might acquire a non-zero vacuum value if the fermion bilinear condense. Our general analysis suggests this would break supersymmetry (is this true?). In a similar way we find for the gauge field singlet composite, suppressing gauge indices,

$$\begin{aligned}\Phi_{\text{gauge}} =&\widehat{\lambda}\lambda + \widehat{\theta}(\lambda D - \frac{i}{2}\sigma^{\mu\nu}G_{\mu\nu}\lambda)\\ &+ \frac{1}{2}\widehat{\theta}\theta(D^2 - 2\partial_\mu\lambda^\dagger\sigma^\mu\lambda - \frac{1}{2}G_{\mu\nu}G^{\mu\nu} + iG_{\mu\nu}\widetilde{G}^{\mu\nu}) \ .\end{aligned} \qquad (A.115)$$

The gaugino condensate can get a vacuum value without breaking supersymmetry. However, $< G_{\mu\nu}G^{\mu\nu} >_0$ can break supersymmetry since it contributes to the F term, while the gaugino bilinear does not seem capable of breaking supersymmetry.

These conclusions must be examined with caution because in field theory, the algebra of products of local fields may not be the same as expected from the classical transformation laws. Indeed, Konishi (*Phys. Let.***135B**, 439(1984)) has found an anomaly in the supersymmetry transformations of the *gauge singlet* composite fermion made out of two chiral superfields. In the presence of gauge interactions, he finds that under a supersymmetry,

$$\delta(\varphi_i\psi_i) = \alpha(F_i\varphi_i - \frac{1}{2}\widehat{\psi}_i\psi_i + C\frac{g^2}{32\pi^2}\widehat{\lambda}_a\lambda_a) \ , \qquad (A.116)$$

where C is the Casimir operator, and where i, a are the group indices for the matter and gauge multiplets, respectively. The appearance of two different fermion bilinears on the right hand side is intriguing. If the gauginos condense in the vacuum, the last term causes a shift in the matter composite fermion field, which must then be identified with the Goldstino, implying that supersymmetry has been broken dynamically.

A.6 MODELS OF SPONTANEOUS SUPERSYMMETRY BREAKING

We now present simple models where the dynamics are such that we have spontaneous supersymmetry breaking at tree level, either through the F or D terms. These are arranged in such a way that the minimum value of the potential is greater than zero.

F-breaking

The generic mechanism, invented by O'Raifeartaigh, requires at least three chiral superfields, Φ_j, $j = A, B, C$, interacting through the superpotential

$$W = m\Phi_A\Phi_B + \lambda(\Phi_A^2 - M^2)\Phi_C \ . \tag{A.117}$$

This theory is invariant under one global phase symmetry

$$R' = R - \frac{2}{3}X \ , \tag{A.118}$$

where the X is an Abelian symmetry with values $x_j = (1, -2, -2)$ for the three superfields, and the R-symmetry, which does not commute with supersymmetry has the same value $(2/3)$ for all three superfields

$$\begin{aligned} X: \ \Phi_j(x, \theta) &\to e^{ix_j\alpha}\Phi_j(x, \theta) \ , \\ R: \ \Phi_j(x, \theta) &\to e^{i\frac{2\beta}{3}}\Phi_j(x, e^{i\beta}\theta) \ , \end{aligned} \tag{A.119}$$

Under the combined phase symmetry the components of the superfields have the following values, indicated in parentheses,

$$\begin{aligned} &\varphi_A(0) \ , \ \psi_A(-1) \ , \ F_A(-2) \ , \\ &\varphi_B(2) \ , \ \psi_B(1) \ , \ F_B(0) \ , \\ &\varphi_C(2) \ , \ \psi_C(1) \ , \ F_C(0) \ . \end{aligned} \tag{A.120}$$

The F equations of motion yield

$$\begin{aligned} &F_A = -m^*\varphi_B^* - 2\lambda^*\varphi_A^*\varphi_C^* \ , \\ &F_B = -m^*\varphi_A^* \ , \qquad F_C = \lambda^*(M^{2*} - \varphi_A^{*2}) \ . \end{aligned} \tag{A.121}$$

Clearly there is no field configuration for which all three F's vanish: if $F_C = 0$, then $F_B \neq 0$, and $F_B = 0$ leads to $F_C \neq 0$. The potential is the sum of the absolute square of the F-terms

$$V = |m\varphi_B + 2\lambda\varphi_A\varphi_C|^2 + |m\varphi_A|^2 + |\lambda(\varphi_A^2 - M^2)|^2 \ , \qquad (A.122)$$

from which we deduce only two independent extremum conditions

$$\begin{aligned} m\varphi_B + 2\lambda\varphi_A\varphi_C &= 0 \ , \\ |m|^2\varphi_A + 2\varphi_A^*|\lambda|^2(\varphi_A^2 - M^2) &= 0 \ , \end{aligned} \qquad (A.123)$$

and their complex conjugates. From these two independent conditions, only two of the scalar fields can be determined, and the third one is left with an undetermined vacuum value. This is a general characteristic of F-type breaking: at tree level one field combination is left undetermined, so that in field space the potential has a continuous minimum along that field direction, called a *flat direction.* The vacuum manifold at tree level is a barren two dimensional plane spanned by the values of the undetermined complex field. This degeneracy can be lifted once quantum corrections are included.

For simplicity, assume the parameters m, and λ to be real (see problem). The last equation, rewritten in the form,

$$\varphi_A(|m|^2 + 2|\lambda|^2|\varphi_A|^2) - 2\lambda^2M^2\varphi_A^* = 0 \ ,$$

has two different solutions, depending on the parameters, they are

$$\begin{aligned} \text{solution I :} \qquad & M^2 - \frac{m^2}{2\lambda^2} < 0 \ ; \\ & \varphi_A = 0 \ ; \quad \varphi_B = 0 \ ; \quad \varphi_C \ \text{undetermined} \ ; \\ & F_A = F_B = 0 \ ; \quad F_C = -\lambda M^2 \ ; \\ & V_0 = \lambda^2 M^4 \ . \end{aligned}$$

$$\text{solution II :} \qquad M^2 - \frac{m^2}{2\lambda^2} > 0 \; ;$$

$$\varphi_A = \sqrt{M^2 - \frac{m^2}{2\lambda^2}} \; ; \quad \varphi_B = -\frac{2\lambda}{m}\varphi_C\sqrt{M^2 - \frac{m^2}{2\lambda^2}} \; ;$$

$$F_A = 0 \; ; \quad F_B = -m\sqrt{M^2 - \frac{m^2}{2\lambda^2}} \; ; \quad F_C = \frac{m^2}{2\lambda} \; ;$$

$$V_0 = m^2(M^2 - \frac{m^2}{4\lambda^2}) \; ,$$

where V_0 is the value of the potential at minimum. In both cases, we see that supersymmetry is broken since one of the F fields does not vanish. In both cases, the R' symmetry is spontaneously broken, except at one point ($\varphi_C = 0$). Thus we expect to have two massless particles, one Goldstino and one Nambu-Goldstone boson in both cases.

• Solution I allows us to immediately identify ψ_C with the massless Goldstino, the only fermion field which gets shifted by a supersymmetry transformation, since only $F_C \neq 0$. This is consistent with the superpotential where the only fermion mass term is $-m\widehat{\psi_A\psi_B}$, which describes a Dirac fermion of mass m.

The scalar masses are given by the second derivative of the potential evaluated at the minimum. The relevant terms are

$$V = m^2|\varphi_B|^2 + m^2|\varphi_A|^2 - 2\lambda M^2(\varphi_A^2 + \varphi_A^{*2}) + \cdots \; . \qquad (A.124)$$

If we let

$$\varphi_A = \frac{1}{\sqrt{2}}(a + ib) \; ,$$

we obtain the mass terms

$$m^2|\varphi_B|^2 + \frac{1}{2}(m^2 - 2\lambda^2 M^2)a^2 + \frac{1}{2}(m^2 + 2\lambda^2 M^2)b^2 \; ,$$

which allows us to extract the following mass squared for the scalars

$$0, \;\; 0, \;\; m^2, \;\; m^2, \;\; m^2 - 2\lambda^2 M^2, \;\; m^2 + 2\lambda^2 M^2 \; . \qquad (A.125)$$

The effect of F-supersymmetry breaking has been to split the mass squared by equal and opposite amounts for the scalars. The scalars for which this happens belong to the superfield that couples to the superfield which gets a non-zero F-term, in this case Φ_C. This is easy to see because of the term of the form $\phi\phi F$ in the potential: when $F_0 \neq 0$, it generates a quadratic term for ϕ. Furthermore, because of the opposite split, the mass sum rule is still satisfied, namely

$$\{m^2 + m^2 + (m^2 - 2\lambda M^2) + (m^2 + 2\lambda M^2)\} - 2 \cdot 2\{m^2\} = 0 \ , \quad (A.126)$$

remembering that one Dirac = two Weyl.

• Solution II is slightly more complicated, because there are two non-zero F-terms. From the values of F_B and F_C in the vacuum, it proves convenient to introduce the angle η by

$$\tan\eta = \frac{m}{2\lambda\sqrt{M^2 - \frac{m^2}{2\lambda^2}}} \ , \quad (A.127)$$

such that in the vacuum the combination $\sin\eta F_B + \cos\eta F_C$ vanishes. From the transformation laws, we deduce that the fermion

$$\psi = \sin\eta\psi_B + \cos\eta\psi_C \quad (A.128)$$

does not shift under supersymmetry, while the orthogonal combination

$$\psi_{NG} = \cos\eta\psi_B - \sin\eta\psi_C \quad (A.129)$$

does shift; it must be identified with the massless Goldstino, which we can verify by direct computation of the tree-level fermion masses. These are given by

$$\begin{aligned} &- m\widehat{\psi}_A\psi_B - 2\lambda\sqrt{M^2 - \frac{m^2}{2\lambda^2}}\widehat{\psi}_A\psi_C, \\ &= - 2\lambda\sqrt{M^2 - \frac{m^2}{4\lambda^2}}\widehat{\psi}_A(\sin\eta\psi_B + \cos\eta\psi_C) \ , \end{aligned} \quad (A.130)$$

so that the missing orthogonal combination ψ_{NG}, is indeed the massless Goldstino as expected, while ψ becomes the Dirac partner of ψ_A; the value of the Dirac mass can also be written as

$$m_D = \frac{2\lambda}{m}(\cos\eta F_B - \sin\eta F_C) \ . \quad (A.131)$$

It should be obvious that much work can be avoided by introducing the scalar mass combinations

$$\varphi = \sin\eta\varphi_B + \cos\eta\varphi_C\ , \qquad \chi = \cos\eta\varphi_B - \sin\eta\varphi_C\ , \tag{A.132}$$

in terms of which the potential reads

$$\begin{aligned} V = &|(m\ \cos\eta - 2\lambda\varphi_A\sin\eta)\chi + (m\ \sin\eta + 2\lambda\varphi_A\cos\eta)\varphi|^2 \\ &+ m^2|\varphi_A|^2 + \lambda^2|\varphi_A^2 - M^2|^2\ . \end{aligned} \tag{A.133}$$

Expand φ_A as $\langle\varphi_A\rangle_0 + \hat{\varphi}_A$, and obtain the quadratic terms

$$4\lambda^2(M^2 - \frac{m^2}{4\lambda^2})|\varphi|^2 + (4\lambda^2M^2 - m^2)|\hat{\varphi}_A|^2 - \frac{m^2}{2}(\hat{\varphi}_A^2 + \hat{\varphi}_A^{*2})\ ,$$

leading to the following squared masses,

$$0,\ \ 0,\ \ 4\lambda^2M^2 - m^2,\ \ 4\lambda^2M^2 - m^2,\ \ 4\lambda^2M^2 - \frac{3}{2}m^2,\ \ 4\lambda^2M^2 - \frac{1}{2}m^2\ , \tag{A.134}$$

and the magic sum rule is satisfied. By now this should not appear too magical since the relevant F-term couples to $\varphi_A^2 + \varphi_A^{*2}$. The magnitude of the shift among the scalar fields again has to do with the coupling and value of the non-vanishing F-term

$$F' = \cos\eta F_B - \sin\eta F_C = -m\sqrt{M^2 - \frac{m^2}{4\lambda^2}}\ . \tag{A.135}$$

Incidentally, the presence of the massless χ fields should not be left unexplained (see problem). The main result of this example is that F-breaking leads to the tree-level fatidic sum rule

$$\mathcal{S}tr M^2 = \sum_{J=0,1/2}(-1)^{2J}(2J+1)m_J^2 = 0\ , \tag{A.136}$$

which restricts the uses of this mechanism for phenomenology. However it can be changed by quantum corrections and by non-renormalizable couplings which appear naturally when generalizing supersymmetry to include gravity (supergravity). Note that both solutions have two massless scalars. One must be the Nambu-Goldstone associated with R'-breaking. Finally we note the ubiquitous presence of R-symmetry in this example.

D breaking

This particular mechanism was invented by Fayet and Iliopoulos. It necessarily involves an Abelian gauge supermultiplet. The simplest example that is anomaly free is a U(1) gauge theory in interaction with two chiral superfields Ψ_1 and Ψ_2 of opposite charge. The Lagrange density is

$$\begin{aligned}\mathcal{L} = &-\frac{1}{4}F_{\mu\nu}F^{\mu\nu} + \lambda^\dagger\sigma^\mu\partial_\mu\lambda + \frac{1}{2}D^2 \\ &+ \sum_{i=1}^{2}((\mathcal{D}_\mu\varphi_i)^*(\mathcal{D}^\mu\varphi_i) + \psi_i^\dagger\sigma^\mu\mathcal{D}_\mu\psi_i + F_i^*F_i) \\ &+ gD(\varphi_1^*\varphi_1 - \varphi_2^*\varphi_2) - (\sqrt{2}g\widehat{\lambda}(\psi_1\varphi_1^* - \psi_2\varphi_2^*) + \text{c.c.}) \\ &+ m(\varphi_1 F_2 + \varphi_2 F_1 - \widehat{\psi}_1\psi_2 + \text{c.c.}) - \epsilon\mu^2 D\ .\end{aligned} \qquad (A.137)$$

Note that we have added a D-term, the integral of which is a supersymmetric invariant; $\epsilon = \pm 1$ and μ^2 is a positive mass squared. Clearly this can only be done for a U(1) theory for which D is gauge invariant. In the above,

$$\mathcal{D}_\mu\varphi_i = (\partial_\mu + q_i g A_\mu)\varphi_i\ ; \qquad \mathcal{D}_\mu\psi_i = (\partial_\mu + iq_i g A_\mu)\psi_i\ ;, \qquad (A.138)$$

with $q_1 = -q_2 = 1$. This has R-symmetry, under

$$\Phi_j(x,\theta) \to e^{i2\beta}\Phi_j(x, e^{i\beta}\theta)\ , \quad j = 1,2\ . \qquad (A.139)$$

The equations of motion for the F_i and D terms yield

$$\begin{aligned} D &= \epsilon\mu^2 - g(\varphi_1^*\varphi_1 - \varphi_2^*\varphi_2)\ , \\ F_1 &= -m\varphi_2^*\ , \quad F_2 = -m\varphi_1^*\ . \end{aligned} \qquad (A.140)$$

There is no field configuration for which all three vanish, and supersymmetry is necessarily broken. The potential is given by

$$V = \frac{1}{2}\{\epsilon\mu^2 - g(\varphi_1^*\varphi_1 - \varphi_2^*\varphi_2)\}^2 + m^2(\varphi_1^*\varphi_1 + \varphi_2^*\varphi_2)\ . \qquad (A.141)$$

Its minimization yields

$$\varphi_1^*\{m^2 - g(\epsilon\mu^2 - g\varphi_1^*\varphi_1 + g\varphi_2^*\varphi_2)\} = 0 \ , \tag{A.142}$$

$$\varphi_2^*\{m^2 + g(\epsilon\mu^2 - g\varphi_1^*\varphi_1 + g\varphi_2^*\varphi_2)\} = 0 \ . \tag{A.143}$$

Evidently we cannot have both φ_1 and φ_2 different from zero, unless $m^2 = 0$. There are two types of solutions, one for which $\varphi_1 = \varphi_2 = 0$, and the other with $\varphi_2 = 0$, $\varphi_1 \neq 0$ (the other case, $\varphi_1 = 0$, $\varphi_2 \neq 0$ is obtained from the latter by changing the sign of g). Let us examine both cases:

a-) When $\varphi_1 = \varphi_2 = 0$ the vacuum values are

$$D = \epsilon\mu^2 \ ; \quad F_1 = F_2 = 0 \ , \quad V_0 = \frac{\mu^4}{2} \ . \tag{A.144}$$

The supersymmetry transformation laws tell us that the gaugino λ is the massles Goldstino. Only supersymmetry is broken. The gauge U(1) and the R symmetry are untouched.

The only fermion mass comes from the original Dirac mass term, $-m\widehat{\psi}_1\psi_2$. The masses of the scalars are extracted from the quadratic terms in the potential,

$$V = \frac{1}{2}\mu^4 - \epsilon\mu^2 g(\varphi_1^*\varphi_1 - \varphi_2^*\varphi_2) + m^2(\varphi_1^*\varphi_1 + \varphi_2^*\varphi_2) + \cdots \ , \tag{A.145}$$

yielding two real scalars with masses $m^2 + \epsilon\mu^2 g$, and two with masses $m^2 - \epsilon g\mu^2$. The supertrace now reads

$$\mathcal{S}tr\ M^2 = 2(m^2 + \epsilon\mu^2 g) + 2(m^2 - \epsilon\mu^2 g) - 2\cdot 2m^2 = 0 \ . \tag{A.146}$$

Again it vanishes, but this time for a different reason. We see that the mass^2 of the first two scalars is shifted by gD_0, the second by $-gD_0$, which correspond to the value of the charge times D_0. They cancel because of our anomaly cancellation mechanism which has the sum of the charges vanishing. So we really have

$$\mathcal{S}tr\ M^2 = 2g(q_1 + q_2)D_0 \ . \tag{A.147}$$

We could have devised a different model, say with one multiplet of charge 2 and eight multiplets of charge -1. The $U(1)$ triangle anomaly diagram vanishes since

$$\sum q_i^3 = (2)^3 + 8(-1) = 0 \ , \tag{A.148}$$

however the D term is given by

$$D = \epsilon\mu^2 + 2g\varphi^*\varphi - g\sum_{i=1}^{8}\varphi_i^*\varphi_i \ , \tag{A.149}$$

so that

$$\mathcal{S}tr\ M^2 = [8(-1) + (2)]gD_0 = -6gD_0 \ . \tag{A.150}$$

We leave it to the reader to devise such a model, ugly as it may be. However this shows that D-breaking may alter the mass sum rule.

b-) The second case is described by $\varphi_2 = 0, \quad \varphi_1 = v$, which we take to be real without loss of generality. This solution breaks both gauge symmetry and the R-symmetry, but leaves a linear combination invariant. Hence we expect only one Nambu-Goldstone boson, to be eaten by the gauge boson. Minimization of the potential yields

$$v^2 = \frac{1}{g^2}(g\epsilon\mu^2 - m^2) \ , \tag{A.151}$$

which taking g positive can happen only for $\epsilon = 1$ and then only if $g\mu^2 > m^2$. In the vacuum, we have

$$D = D_0 = \frac{m^2}{g} \ ; \qquad F_1 = 0 \ ; \ \ F_2 = -\frac{m}{g}\sqrt{g\mu^2 - m^2} \ , \tag{A.152}$$

In this case, both supersymmetry and the gauge symmetry are spontaneously broken, and the Goldstino is a linear combination of the gaugino λ and the chiral ψ_2. We can see this directly from the fermion mass matrix in the Lagrangian

$$(m\widehat{\psi}_2 + \sqrt{2}gv\widehat{\lambda})\psi_1 + \text{c.c.} \ . \tag{A.153}$$

This describes a Dirac fermion with mass

$$m_D = \sqrt{2g^2v^2 + m^2} \ . \tag{A.154}$$

The orthogonal combination does not have a mass term; it must be the massless Goldstino

$$\lambda_G = \frac{1}{\sqrt{2g^2v^2 + m^2}}(\sqrt{2}gv\psi_2 - m\lambda) \ . \tag{A.155}$$

One can check explicitly that it shifts by a constant under a supersymmetry transformation

$$\delta\lambda_G = -\frac{m}{\sqrt{2}g}\alpha\sqrt{2g^2v^2+m^2} \ , \tag{A.156}$$

while the orthogonal combination experiences no shift

$$\delta\psi = \frac{\alpha}{\sqrt{2g^2v^2+m^2}}[-m^2v + gv\frac{m^2}{g})] = 0 \ , \tag{A.157}$$

as we expected. From the spontaneous breakdown of the $U(1)$, the gauge boson gets a mass

$$m_A^2 = 2g^2v^2 \ . \tag{A.158}$$

The scalar masses are obtained by expanding the potential from its vacuum value. Letting $\varphi_1 = v + \hat{\varphi}_1, \varphi_2 = \hat{\varphi}_2$, we obtain

$$\begin{aligned} V &= V_0 + \frac{1}{2}g^2v^2(\hat{\varphi}_1 + \hat{\varphi}_1^*)^2 + (m^2 - gD_0)|\varphi_1|^2 + (m^2 + gD_0)|\hat{\varphi}_2|^2 \\ &= V_0 + \frac{1}{2}g^2v^2(\hat{\varphi}_1 + \hat{\varphi}_1^*)^2 + 2m^2|\hat{\varphi}_2|^2, \end{aligned} \tag{A.159}$$

so that we have the following scalar masses squared

$$2m^2, \ \ 2m^2, \ \ 0, \ \ 2g^2v^2 \ .$$

The zero mass scalar is of course the Nambu-Goldstone boson coming from the breaking of the gauge $U(1)$; it appears here because we are not in the unitary gauge. Now our mass sum rule, adding the gauge multiplet yields

$$\sum_{J=0,1/2,1} (-1)^{2J}(2J+1)m_J^2 = 3m_A^2 - 2\cdot 2m_D^2 + (2m^2 + 2m^2 + 2g^2v^2) = 0 \ ! \tag{A.160}$$

Again, we ask the reader to convince himself or herself that the zero on the right hand side occurs really because of the particular anomaly cancellation we have chosen. The three scalar mass squared are

$$= m \pm gD_0 \ , \qquad m^2 + gD_0 \ , \qquad m^2 + gD_0 + 2(g\mu^2 - m^2) \ , \tag{A.161}$$

so that

$$\mathcal{S}tr\ M^2 = g(q_1 + q_2)D_0\ .$$

A.7 DYNAMICAL SUPERSYMMETRY BREAKING

Over the last few years, there has been much progress in understanding supersymmetric theories which have both weak and strong coupling sectors. In these theories, supersymmetry might be dynamically broken as a result of the strong coupling. Its study therefore requires the use of non-perturbative methods, which are beyond the scope of this introductory Appendix. We will nevertheless give a glimpse of the possibilities by discussing a simple example.

In *global* supersymmetry, dynamical breaking can be studied in some generality by evaluating the order parameter of supersymmetry, the so-called Witten index

$$\Delta = Tr(-1)^F\ , \tag{A.162}$$

which counts the number of fermionic less the number of bosonic states in the vacuum, $n_F - n_B$. The case $\Delta = 0$ is ambiguous since it may mean one of two possibilities: either $n_B = N_f \neq 0$, implying the existence of states of zero energy, and unbroken supersymmetry, or both n_B and N_f are zero, in which case it is broken. However, when $\Delta \neq 0$, the unambiguous conclusion is that there exist states of zero energy, and supersymmetry is unbroken. The beauty of this index is that it is invariant under adiabatic changes of the theory. It can be computed in the perturbative regime, and then carried over to non perturbative situations. Its drawback is that it applies only to global supersymmetry, and when large field configurations become important. If Δ is computed different from zero, there is no breaking.

Asymptotically free theories come with a scale Λ, obtained by dimensional transmutation. Below that scale, the theory is strongly coupled, above it is perturbative. In order to determine its infrared behavior it is important to be able to construct the the effective low energy Lagrangian. We consider this problem, starting with the super Yang-Mills theory without matter. The Lagrangian density is

$$\mathcal{L}_{SYM} = -\frac{1}{4}G^A_{\mu\nu}G^{\mu\nu A} + \lambda^{\dagger A}\sigma^\mu(\mathcal{D}_\mu\lambda)^A + \frac{1}{2}D^A D^A\ , \tag{A.163}$$

where

$$(\mathcal{D}_\mu\lambda)^A = \partial_\mu\lambda^A + ig(T^C)^A_{\ B}A^C_\mu\lambda^B \ . \tag{A.164}$$

It is supersymmetric, and one can build a conserved supersymmetry current, S_μ, by Noether methods. In addition, the N=1 super-Yang-Mills Lagrangian density is invariant under the chiral R-symmetry which acts only on the gaugino

$$\lambda \to e^{i\beta}\lambda \ .$$

The ABJ anomaly contributes to the divergence of the R-current through the usual triangle diagram as

$$\partial^\mu J^R_\mu = 3N_c\frac{g^2}{32\pi^2}\,\mathrm{Tr}\,(G_{\mu\nu}\widetilde{G}^{\mu\nu}) \ , \tag{A.165}$$

where N_c is the number of colors (here for $SU(N_c)$). In this theory without matter, the β-function is itself proportional to the number of colors

$$\beta(g) = -3N_c\frac{g^3}{16\pi^2} \ . \tag{A.166}$$

Hence we can rewrite the anomaly as

$$\partial^\mu J^R_\mu = -\frac{\beta(g)}{g}\,\mathrm{Tr}\,(G_{\mu\nu}\widetilde{G}^{\mu\nu}) \ . \tag{A.167}$$

This anomalous symmetry is much like the PQ symmetry, except that here it is not broken spontaneously. Since the β-function has the same sign as in QCD, it leads to a strong force at large distances, which, by analogy with QCD, we expect to form fermion condensates. The prime candidate is the Lorentz-invariant condensate of two gauginos

$$\varphi_c = \widehat{\lambda}^A\lambda^A \ . \tag{A.168}$$

Does this break supersymmetry? The answer lies in the transformation of φ_c under a supersymmetric transformation. We have

$$\begin{aligned}\delta\varphi_c &= 2\widehat{\lambda}^A\delta\lambda^A \ , \\ &= \sqrt{2}(D^A\widehat{\lambda}^A\alpha + \frac{i}{2}\widehat{\lambda}^A\sigma_{\mu\nu}G^{A\mu\nu}\alpha) \ , \\ &= \sqrt{2}\widehat{\alpha}(\lambda^A D^A - \frac{i}{2}\sigma^{\mu\nu}G^A_{\mu\nu}\lambda^A) \ ,\end{aligned} \tag{A.169}$$

using the identity,

$$\sigma^2\sigma^{\mu\nu T}\sigma^2 = -\sigma^{\mu\nu}\ .$$

By comparing with the chiral multiplet transformation law, we define the composite fermion field

$$\psi_c = D^A\lambda^A - \frac{i}{2}\sigma^{\mu\nu}G^A_{\mu\nu}\lambda^A\ ; \qquad (A.170)$$

its variation yields, after some tedious but straightforward algebra, the F component of the composite multiplet

$$F_c = -2\partial_\mu\lambda^{A\dagger}\sigma^\mu\lambda^A + D^AD^A - \frac{1}{2}G^A_{\mu\nu}G^{A\mu\nu} + iG^A_{\mu\nu}\widetilde{G}^{A\mu\nu}\ . \qquad (A.171)$$

It is of course, up to a surface term, the Lagrange density of the super Yang-Mills theory, which is not too surprising since its integral is supersymmetric invariant. It now becomes clear that the condensate $<\widehat{\lambda}^A\lambda^A>_0$ cannot break supersymmetry since it does not tranform as an F-term. By the same token, we conclude that the "glue" condensate $< G^A_{\mu\nu}G^{A\mu\nu}>_0$ can break supersymmetry since it contributes to an F-term. Seen from another point of view, it raises the energy density of the vacuum, thus breaking supersymmetry.

This composite supermultiplet plays another role, that of the anomaly, *i.e.* it appears on the right hand side of the equations giving the divergences of the dilatation current, the axial current and the γ-trace of the supersymmetric current (the divergence of the supersymmetric current is not anomalous). In equations

$$\begin{aligned}
\partial^\mu D_\mu &= \frac{\beta(g)}{g}\frac{1}{2}\,\mathrm{Tr}\,(G^A_{\mu\nu}G^{A\mu\nu})\ ,\\
\partial^\mu J^R_\mu &= -\frac{\beta(g)}{g}\frac{1}{2}\,\mathrm{Tr}\,(G^A_{\mu\nu}\widetilde{G}^{A\mu\nu})\ , \qquad (A.172)\\
\gamma^\mu S_\mu &= \frac{\beta(g)}{2g}2G^A_{\mu\nu}\sigma^{\mu\nu}\lambda^A\ .
\end{aligned}$$

Through supersymmetry the chiral anomaly is linked to the scale anomaly. In regular QCD, the chiral anomaly is of order $\frac{1}{N_c}$, since it is generated by fermion loops, but in the supercase, the gaugino contribution is of order $\frac{1}{N_c}\times N_c \sim 1$, and it does not go away in the large N_c limit. Hence there is no parameter which can be switched off to get rid of the anomaly. It would seem that in Super-QCD, there is no sense in which we should consider a Lagrangian invariant under both scale and chiral transformations, implemented with an "anomaly" term. Still, this approach was pursued by G. Veneziano and S. Yankielowicz (*Phys. Lett.* **B113**, 231 (1982)), in order to find out if gaugino condensates can form under the influence of the strong force.

They proceed to write the effective low energy Lagrangian for pure super Yang-Mills in terms of the gauge singlet composite chiral superfield Φ_c, given by

$$\begin{aligned}\Phi_c =& \widehat{W}^A W^A \ , \\ =& \widehat{\lambda}\lambda + \widehat{\theta}(\lambda D - \frac{i}{2}\sigma^{\mu\nu}G_{\mu\nu}\lambda) \\ &+ \frac{1}{2}\widehat{\theta}\theta(D^2 - 2\partial_\mu\lambda^\dagger\sigma^\mu\lambda - \frac{1}{2}G_{\mu\nu}G^{\mu\nu} + iG_{\mu\nu}\widetilde{G}^{\mu\nu}) \ .\end{aligned} \tag{A.173}$$

All "color" indices have been suppressed. This composite superfield has dimension three. Since we are going to use scale invariance as a guide to building the effective Lagrangian the kinetic term meeds to have the correct dimension and be supersymmetric. The candidate kinetic term is the D-term of a real superfield, with dimension d such that $4 = \frac{1}{2} \times 4 + d$, since each θ has dimension one-half. Hence we need $d = 2$. Out of the chiral superfield Φ_c we can make a real superfield of dimension six, $|\Phi_c(y_\mu, \theta)|^2$. We conclude that the only candidate for a kinetic term is

$$\mathcal{L}_{\text{kin}} = \int d^2\theta d^2\overline{\theta}(\Phi_c^*(y_\mu, \theta)\Phi_c(y_\mu, \theta))^{1/3} \ . \tag{A.174}$$

Under R-symmetry, the composite superfield transforms a

$$\Phi_c(y_\mu, \theta) \to e^{2i\beta}\Phi_c(y_\mu, e^{i\beta}\theta) \ , \tag{A.175}$$

and the kinetic term is manifestly invariant. Now we have to add a function of Φ_c which actually mocks up the scale and chiral (R) anomalies. We follow the procedure we have used previously in the QCD case. The relevant fact in this construction is that the right-hand side of the anomaly equations are given by the F-term of the composite superfield, Φ_c. We seek to construct a term of the form

$$\mathcal{L}_A = \int d^2\theta W(\Phi_c) + \text{c.c.} \ . \tag{A.176}$$

Under R-symmetry,

$$\begin{aligned}\mathcal{L}_A &\to \int d^2\theta W(e^{2i\beta}\Phi_c(x_\mu, e^{i\beta}\theta)) + \text{c.c.} \ , \\ &= \int d^2\theta' e^{-2i\beta} W(e^{2i\beta}\Phi_c(x_\mu, \theta')) + \text{c.c.} \ ,\end{aligned} \tag{A.177}$$

so that (dropping the primes)

$$\delta\mathcal{L}_A = 2i\beta \int d^2\theta(-W + \Phi_c \frac{\partial W}{\partial \Phi_c}) + \text{c.c.} \ . \tag{A.178}$$

The anomaly can be written in the form

$$-\frac{\delta\mathcal{L}_A}{\delta\alpha} = -\frac{\beta(g)}{2g} G_{\mu\nu}\widetilde{G}^{\mu\nu} \ . \tag{A.179}$$

Since

$$i \int d^2\theta \Phi_c + \text{c.c.} = -2G_{\mu\nu}\widetilde{G}^{\mu\nu} \ , \tag{A.180}$$

it suggests that

$$-W + \Phi_c \frac{\partial W}{\partial \Phi_c} = \xi \Phi_c \ , \tag{A.181}$$

where ξ is an unknown constant. Differentiating, we get

$$\Phi_c \frac{\partial^2 W}{\partial \Phi_c^2} = \xi \ ,$$

which is easily integrated to yield

$$W(\Phi_c) = \xi \Phi_c (\ln \frac{\Phi_c}{\mu^3} - 1) \ ,$$

where μ is an integration constant. The full effective low energy Lagrangian which reproduces the anomaly is then

$$\int d^2\theta d^2\overline{\theta} (\Phi_c^* \Phi_c)^{1/3} + \xi \int d^2\theta \Phi_c (\ln \frac{\Phi_c}{\mu^3} - 1) + \text{c.c.} \ . \tag{A.182}$$

It can be checked that we also get the correct anomalous equations for the dilatation and superconformal currents. We can examine the potential of this model to find its ground state. First we extract the auxiliary fields.

$$\xi \int d^2\theta \Phi_c (\ln \frac{\Phi_c}{\mu^3} - 1) = \xi (F_c \ln \frac{\varphi_c}{\mu^3} + \frac{1}{\varphi_c} \widehat{\psi}_c \psi_c) \ . \tag{A.183}$$

On the other hand, the kinetic term yields

$$\frac{1}{3}(\varphi_c^*\varphi_c)^{1/3}\frac{1}{(\varphi_c^*\varphi_c)}[F_c^*F_c + \text{ normal kinetic terms }] \ , \qquad (A.184)$$

from which we deduce the equation of motion

$$F_c = 3\xi(\varphi_c^*\varphi_c)^{2/3}\ln\frac{\varphi_c^*}{\mu^3} \ , \qquad (A.185)$$

leading to the potential

$$V = 9|\xi|^2(\varphi_c^*\varphi_c)^{2/3}\ln\frac{\varphi_c}{\mu^3}\ln\frac{\varphi_c^*}{\mu^3} \ . \qquad (A.186)$$

One has to be careful because the kinetic term for φ_c is not canonical

$$\frac{1}{3}(\varphi_c^*\varphi_c)^{-2/3}\partial_\mu\varphi_c^*\partial^\mu\varphi_c \ , \qquad (A.187)$$

but we regain the canonical form in terms of

$$\hat{\varphi}_c = \sqrt{3}\varphi_c^{1/3} \ , \qquad \partial_\mu\hat{\varphi}_c = \frac{1}{\sqrt{3}}\varphi_c^{-2/3}\partial_\mu\varphi_c \ , \qquad (A.188)$$

to arrive at the Lagrangian

$$\mathcal{L} = \partial_\mu\hat{\varphi}_c^*\partial^\mu\hat{\varphi}_c - |\xi|^2(\hat{\varphi}_c^*\hat{\varphi}_c)^2\ln\frac{\hat{\varphi}_c}{\mu}\ln\frac{\hat{\varphi}_c^*}{\mu} \ . \qquad (A.189)$$

The potential is at its minimum when

$$|\hat{\varphi}_c^2|\ln\frac{\hat{\varphi}_c}{\mu} = 0 \ ,$$

which has two possible solutions, $\hat{\varphi}_c = 0$ or $\hat{\varphi}_c = \mu$. The former does not make any sense because of the fermion kinetic terms. There is only one viable solution, $\hat{\varphi}_c = \mu$, showing that supersymmetry is not broken, the conclusion we wanted to reach. The non zero vacuum value of the composite superfield suggests that the gaugino condensate indeed forms, but without breaking supersymmetry. Expanding the potential away from minimum, letting

$$\hat{\varphi}_c = \mu + \varphi_c' \ ,$$

we obtain

$$V = -|\xi|^2|\mu + \varphi'|^2 \ln(1 + \frac{\varphi'}{\mu}) \ln(1 + \frac{\varphi'^*}{\mu}) ,$$
$$= -|\xi|^2\mu^2|\varphi'|^2 - \cdots , \tag{A.190}$$

which gives the mass

$$m^2 = |\xi|^2\mu^2 . \tag{A.191}$$

This mass is like that of η' in QCD, which arises because of the anomaly. Because of supersymmetry, we also have two real scalar degrees of freedom and one massive Weyl fermion.

To conclude, in pure Super-Yang-Mills without matter, a gaugino condensate may form, but without breaking supersymmetry, in accordance with the transformation properties of the condensate. It is reassuring to see how it happens, albeit through a dangerous procedure.

The lesson is that to break supersymmetry dynamically, one must devise more complicated theories. In particular, the addition of chiral matter in Super-Yang-Mills theories generates many ways to break Susy dynamically. The recent dramatic increase (see K. Intrilligator and N. Seiberg, *Lectures on Supersymmetric Gauge Theories and Electric-Magnetic Duality* in *Nucl. Phys. Proc. Suppl.* **45BC**, 1(1996)) in our understanding of non-perturbative methods in supersymmetric quantum field theories has produced many such examples.

PROBLEMS

A. Show that the effective Lagrangian does reproduce the conformal anomaly and that in the trace of the supersymmetric current.

B. Derive the form of the fermion kinetic term, and show that the solution $\hat{\varphi}_c = 0$ is untenable.

BIBLIOGRAPHY

I have assembled below a list of books and review articles relevant to topics in the chapters indicated. The student is urged to consult the latest review articles, not only for their contents and their extensive references, but also to get a taste of current research directions.

Chapter 1

Books on Quantum Field Theory:

- L. Brown, *Quantum Field Theory*, (Cambridge University Press, New York, 1992).

- C. Itzykson and J.-B. Zuber, *Quantum Field Theory*, (McGraw-Hill, New-York, 1980)

- F. Mandl and G. Shaw, *Quantum Field Theory*, (Wiley, New York, 1984).

- M. E. Peskin and D. V. Shroeder, *An Introduction to Quantum Field Theory*, (Addison-Wesley, Reading, MA, 1995).

- A. M. Polyakov, *Gauge Fields and Strings*, (Harwood, 1987).

- P. Ramond, *Field Theory: A Modern Primer*, second edition, (Addison-Wesley, 1989).

- S. Weinberg, *The Quantum Theory of Fields*, Vol I (Cambridge University Press, Cambridge, 1995)

- S. Weinberg, *The Quantum Theory of Fields*, Vol II (Cambridge University Press, Cambridge, 1996)

Books on Group Theory:

- R. Gilmore, *Lie Groups, Lie Algebras, and Some of their Applications*, (Wiley, New-York, 1974)

- J. E. Humphreys, *Introduction to Lie Algebras and Representation Theory*, (Springer-Verlag, New-York, 1972)

- S. Sternberg, *Group Theory and Physics*, (Cambridge University Press, Cambridge, 1994)

- W. G. McKay and J. Patera, *Tables of Dimensions, Indices, and Branching Rules for Representations of Simple Lie Algebras*, (Marcel Dekker, New-York, 1981)

Review:

- R. Slansky, *Group Theory for Unified Model Building, Phys. Rept.* **79**, 1(1981).

Chapters 2 and 3

Books on the Standard Model:

- I. J. R. Atchison & A.J.G. Hey, *Gauge Theories in Particle Physics*, (Adam Hilger, Bristol, 1989).
- V. D. Barger and R. J. N. Phillips, *Collider Physics*, (Addison-Wesley, Redwood City, 1987).
- R. Cahn, G. Goldhaber, *The Experimental Foundation of Particle Physics*, (Oxford University Press, Cambridge, 1983).
- T. P. Cheng and L.-F. Li, *Gauge Theory of Elementary Particle Physics*, (Clarendon Press, Oxford, 1984).
- J. F. Donoghue and B. R. Holstein, *Dynamics of the Standard Model*, (Cambridge University Press, Cambridge, UK, 1992).
- H. Georgi, *Weak Interactions and Modern Particle Theory*, (Benjamin/Cummings, 1984).
- J. F. Gunion, H. E. Haber, G. L. Kane, and S. Dawson, *The Higgs Hunter Guide*, (Addison-Wesley, Redwood City, CA, 1990).
- J. C. Taylor, *Gauge Theories of Weak Interactions*, (Cambridge University Press, 1976).
- S. B. Trieman, R. Jackiw, B. Zumino, and E. Witten, *Current Algebra and Anomalies*, (Princeton University Press, 1985).
- For historical perspective, see V. Fitch and J. Rosner in *Twentieth Century Physics*, L. M. Brown, A. Pais and B. Pippard eds. (IOP/AIP, Philadelphia, 1994).
- For a flavor of pre-1070 weak interactions, see R. E. Marshak, Riazuddin, and C. P. Ryan, *Theory of Weak Interactions in Particle Physics*, (Wiley, New York, 1969).

Review:

- M. K. Gaillard, P. D. Grannis, and F. J. Sciulli, *The Standard Model of Particle Physics, Rev. Mod. Phys.* **71**, S96(1999)

Reprint Collection:

- *Unified Theories of Elementary Particles* **212**, M. Kobayashi and M. Yoshimura, editors, (Physical Society of Japan).

Chapter 4

Books:

- E. D. Commins, and P. H. Bucksbaum, *Weak Interactions of Leptons and Quarks*, (Cambridge University Press, Cambridge, 1983).

- F. Halzen and A. D. Martin, *Quarks and Leptons: An Introductory Course in Particle Physics*, (Wiley, New-York, 1984).

- D. H. Perkins, *Introduction to High Energy Physics*, (Addison-Wesley, Menlo Park, CA, 1987, third edition).

Reviews:

- L.-L Chau, *Quark Mixing in Weak Interactions*, *Phys. Rept.* **95**, 1(1983).

- J. Mnich, *Experimental Tests of the Standard Model in* $e^+ e^- \to f\bar{f}$ *at the Z Resonance*, *Phys. Rept.* **271**,181(1996).

- S. Schael, *B physics at the Z-resonance*, *Phys. Rept.* **313**, 293(1999).

Chapter 5

Books:

- R. D. Field, *Applications of Perturbative QCD*, (Benjamin/Cummings, Menlo Park, 1989).

- C. Quigg, *Gauge Theory of the Strong, Weak, and Electromagnetic Interactions*, (Addison-Wesley, Redwood City, 1983).

- G. Sterman, *Introduction to Quantum Field Theory*, (Cambridge University Press, Cambridge, 1993).

Reviews:

- H. Leutwyler, *Annals Phys.* **235**, 165(1994).

- T. Hebbeker, *Tests of Quantum Chromodynamics in Hadronic Decays of Z bosons Produced in e+e- Annihilation*, *Phys. Rept.* **217**, 69(1992).

- M. Neubert, *Heavy-quark Symmetry*, *Phys. Rept.* **245**, 259(1994).

Chapter 6

Reviews:

- M. K. Gaillard, B. W. Lee, J. L. Rosner, *The Search for Charm, Rev.Mod.Phys.* **47**, 277(1975).
- J. L. Rosner, in *Testing the Standard Model*, M. Cvetič and P. Langacker, eds. (World Scientific, Singapore, 1991).
- B. A. Kniehl, *Higgs Phenomenology at One Loop in the Standard Model, Phys. Rept.* **240**, 211(1994).
- B. Winstein, and L. Wolfenstein, *The Search for Direct CP Violation Rev. Mod. Phys.* **65**, 1113(1993).
- A. J. Buras and R. Fleischer, in *Heavy Flavours II* (World Scientific, Singapore, 1997).

Chapter 7

Reviews:

- M. Martinez, R. Miquel, L. Rolandi, and R. Tenchini, *Precision Tests of the Electroweak Interaction at the Z Pole, Rev. Mod. Phys.* **71**, 575(1999).
- M. Sher, *Electroweak Potential and Vacuum Stability, Phys. Rept.* **179**, 273(1989).
- M. Quiros, in *Perspectives in Supersymmetry*, G.L. Kane ed. (World Scientific, Singapore 1998).

Chapter 8

Books:

- Felix Boehm and Petr Vogel, *Physics of Massive Neutrinos*, (Cambridge University Press, 1987).
- J. N. Bahcall, *Neutrino Astrophysics* (Cambridge University Press, 1989).

Reviews:

- S.M. Bilenkii, S.T. Petcov, *Rev. Mod. Phys.* **59**, 671(1987); Erratum-ibid. **61**, 169(1989).
- J. D. Vergados, *The Neutrino Mass and Family, Lepton and Baryon Number Non-Conservation in Gauge Theories, Phys. Rept.* **133**, 1(1986).

- S. Turck-Chièze, W. Däppen, E. Fossat, J. Provost, E. Schatzman, and D. Vignaud, *The Solar Interior*, *Phys. Rept.* **230**, 57(1993).

- T. K. Gaisser, F. Halzen, T. Stanev, *Particle Astrophysics with High Energy Neutrinos*, *Phys. Rept.* **258**, 173(1995).

- N. Hata, and P. Langacker, *Phys. Rev.* **D56**, 6107(1997).

- J. N. Bahcall, M. H. Pinsonneault, and G. J. Wasserburg, *Rev. Mod. Phys.* **67**, 781(1995).

- K. Zuber, *On the Physics of Massive Neutrinos*, *Phys. Rept.* **305**(1998).

- L. Wolfenstein *Neutrino Physics*, *Rev. Mod. Phys.* **71**, S140(1999).

- P. Fisher, B. Kayser, and K. S. McFarland, *Neutrino Mass and Oscillation*, hep-ph/9906244.

Chapter 9

Reviews:

- H.-Y. Cheng, *The Strong CP Problem Revisited*, *Phys. Rept.* **158**, 1(1988).

- J. E. Kim, *Light Pseudoscalars, Particle Physics and Cosmology*, *Phys. Rept.* **150**, 1(1987).

- P. Sikivie in *Cosmology and Particle Physics* Edited by E. Alvarez, R. Dominguez Tenreiro, J.M. Ibanez Cabanell, M. Quiros. (World Scientific, Singapore, 1987).

Chapter 10 and Appendix

Books:

- R. Arnowitt, A. Chamseddine, and P. Nath, $N = 1$ *Supergravity*, (World Scientific, Singapore, 1984).

- D. Bailin and A. Love, *Supersymmetric Gauge Theory and String Theory*, (Institute of Physics Publishing, Bristol, England, 1994).

- P. G. O. Freund, *Introduction to Supersymmetry*, (Cambridge University Press, Cambridge England, 1986).

- S. J. Gates, M. T. Grisaru, M. Rocek, and W. Siegel, *Superspace or One Thousand and One Lessons in Supersymmetry*, (Benjamin/Cummings, 1983).

- *Perspectives in Supersymmetry*, G.L. Kane, (ed.) (World Scientific, Singapore 1998)

- R.N. Mohapatra, *Unification and Supersymmetry: The Frontiers of Quark-Lepton Physics*, (Springer-Verlag, New-York, 1992).

- G. G. Ross, *Grand Unified Theories*, (Addison-Wesley, Redwood City CA, 1985).

- P. P. Srivastava, *Supersymmetry and Superfields*, (Adam-Hilger, Bristol England, 1986).

- J. Wess and J. Bagger, *Introduction to Supersymmetry*, (Princeton University Press, Princeton, NJ, 1992).

- P.C. West, *Introduction to Supersymmetry and Supergravity*, (World Scientific, Singapore, 1990).

Reviews:

- P. Fayet, S. Ferrara, *Supersymmetry*, *Phys. Rept.* **32**, 249-334(1977).

- H.P. Nilles, *Supersymmetry, Supergravity, and Particle Physics*, *Phys. Rept.* **110**, 1-152(1984).

- H. E. Haber and G.L. Kane, *The Search for Supersymmetry: Probing Physics Beyond the Standard Model*, *Phys. Rept.* **117**, 75-263(1985).

- M. F. Sohnius, *Introducing Supersymmetry* *Phys. Rept.* **128**, 39(1985).

- M. Chen, C. Dionisi, M. Martinez, and X. Tata, *Signals from Non-strongly Interacting Supersymmetric Particles at LEP*, *Phys. Rept.* **159**, 201(1988).

- R. Barbieri, *Riv. Nuovo Cimento* **11**, 1(1989).

- F. Cooper, A. Khare, U. Sukhatme, *Supersymmetry and Quantum Mechanics*, *Phys. Rept.* **251**, 267(1995).

- S. Martin, *A Supersymmetry Primer*, hep-ph/9709356, and in *Perspectives in Supersymmetry*, G.L. Kane ed., (World Scientific, Singapore 1998).

Reprint Collections:

- *Supersymmetry*, S. Ferrara, ed. (World Scientific, Singapore, 1987).

- *Superstrings*, J. Schwarz, ed. (World Scientific, Singapore, 1985).

Index